REAL ANALYSIS

– PART I –

Prof. Dr. Roselin Antony
Professor (Mathematics)
Department of Mathematics and Statistics
Faculty of Sciences
Sam Higginbottom University of
Agriculture, Technology and Sciences
Prayagraj (Allahabad), Uttar Pradesh, India

SCIENTIFIC
PUBLISHERS

Published by

SCIENTIFIC PUBLISHERS (INDIA)
5 A, New Pali Road, P.O. Box 91
Jodhpur 342 001 (India)
E-mail: info@scientificpub.com
Website: www.scientificpub.com

ISBN : 978-93-89412-89-5 [H/B] Vol. 1

E-ISBN : 978-93-89412-91-8 VOL 1

ISBN: 978-93-89412-95-6 HB SET

ISBN: 978-93-894112-92-5 HB (Vol 2)

Visit the Scientific Publishers (India) website at
http://www.scientificpub.com
Printed in India

FOREWORD

Mathematics is the language of Science. There is a lot of joy in understanding mathematics, for instance, how the proof of Fermat's Last Theorem or the secrets of pi, e, epsilon.... I am sure that this book would be quite interesting, in all the topics being covered. This book is good for math teachers and students who undertake mathematics courses. This book contains only the “best” proofs from the field of Real Analysis in mathematics.

Countless math books are published each year, however only a tiny percentage of these titles are destined to become the kind of classics that are loved the world over by students and mathematicians. This book is devoted to explain and understand the topics of Analysis of Mathematics in a best way and to solve mathematical problems easily. In brief, its intention is that you learn how to do mathematics and how to use mathematics.

This is part one of a two-volume book series on **Real Analysis**. The author beautifully introduces the number systems of Mathematics and then moves on to the sequences, series, functions and their limits and continuity. The emphasis is on rigour and foundations of analysis. Beginning with the construction of the number systems and set theory, the book discusses the basics of analysis (limits, series, continuity, differentiation, Riemann integration), through to power series, Lebesgue integral and then finally the several variable calculus.

I am excited for the opportunities this book provides. Every teacher and student, or researcher of mathematics, who reads this book, has the opportunity to dive into rich and complex examples of teaching and develop new ideas that will be powerful and helpful for the students. I hope that this book will become a primer for teachers, teacher educators, and professional developers, helping teachers across the country and worldwide to learn, and teach Real Analysis in an easy and better way.

I encourage and appreciate the author for her good work and tireless effort to make this book a grand success.

Most Rev. (Prof. Dr.) RAJENDRA B. LAL

Vice-Chancellor

Sam Higginbottom University of Agriculture, Technology and Sciences (SHUATS)

Prayagraj (Allahabad), UP, India

Date: 20th February, 2022

PREFACE

During 1870, a revolutionary change in the conceptual framework of Analysis began to take shape, one that ultimately led to the vast transformation and generalization of the understanding of such basic objects as functions and such notions as continuity, differentiability and integrability.

In the beginning, the functions in Analysis were given by formulas, or other analytic expressions such that the functions were continuous or nearly so, that necessarily these functions had derivatives for most points. Moreover, these functions were integrable by some known methods of integration. All of these ideas gave way for various examples and problems that arose in the subject, which could not be ignored and required new concepts to be understood. Parallel to these developments came new insights that were both geometric and more abstract. A clearer understanding of the nature of curves and their rectifiability, also the theory of sets, plane and the measure that could be assigned to each is highly needed.

Considering the role of this book as an important text of the subject, the proofs are considerably given in detailed manner. This may make some of the proofs longer than expected, since the essentials of the subject are a knowledge of the basic results and an ability to apply them.

Chapter 1 is an introductory part, which involves different number systems and the related properties. Chapter 2 deals with the basic sequences and series and theorems related to them whereas Chapter 4 deals with the sequences and series of functions. Chapter 3 describes the limit and continuity of functions. Chapter 5 and Chapter 8 deal with the differentiation of functions of one variable and several variables. Chapter 7 describes different kinds of special functions and their properties. Chapter 6 and Chapter 9 deal with the integration of functions of one variable and several variables.

I am pleased to present this book *Real Analysis, Part* I to Undergraduate, Postgraduate and Doctoral students of science, engineering and business discipline in University and Colleges. The topics have been dealt in a detailed manner, supported by illustrations for better comprehension. In many aspects, the topics and structure of this book will be highly meritorious and unique. This book will contribute significantly to academic teaching, scientific research, practical works and for competitive exams preparation.

This Real Analysis book is intended to provide students, professors, school teachers, technicians, scientists and engineers with a readily available reference to the essential mathematics formulae, tables and general information needed during their studies and work situation. In short, I demand honest effort, and I also hope that your effort would be rewarded.

I have to start by thanking my awesome husband Prof. Dr. Antony Joseph Raj, Dean of College of Forestry in Sam Higginbottom University of Agriculture, Technology and Sciences (SHUATS), Prayagraj, India. He was so important to this book getting done from reading early drafts to giving me advice on the cover and encouragement, so I could work continuously on this book. Thank you so much, Dear! I also greatly appreciate and thank my daughters Rithika Marlene Antony and Runisha Smiline Antony for their patience and tolerance and always making me cheer up, to make this book a successful one. At this moment, I would also like to remember my parents, whose prayers are always with me and who would have been happy to see this book a successful one. Critical evaluation and suggestions for the improvement of the book are welcome and will be gratefully acknowledged.

December, 2021 **Prof. Dr. Roselin Antony**

For my Husband

Prof. Dr. Antony Joseph Raj

and

My Daughters

Rithika Marlene Antony

&

Runisha Smiline Antony

Real Analysis (Part - I)

Contents

Unit 1 INTRODUCTION OF NUMBERS

1.1 Introduction

In olden days, mathematics was considered as the science which is concerned with numbers and figures. Nowadays, the modern algebra deals with abstract structures instead of numbers, and some branches of geometry study objects in the plane or in space. Other branches of mathematics, like analysis, discrete mathematics, and probability theory, also study objects that are different from numbers and figures. In general, we can say that mathematics is abstract, but not always. We can conclude that, mathematics should be defined not by the objects we study, but how we study them. Hence, mathematics is the science that establishes absolute truth called *theorems* and the logical arguments are called *proofs*.

Real analysis is based on the study of the ideas in calculus, which mainly deals with the real numbers, because the real numbers have the properties that support the derivatives and integrals to work with. The study of derivatives and integrals depends on the fundamental properties of the real numbers. Inside the set of real numbers $\mathbb{R}$, there are three familiar sets of numbers: the natural numbers (1, 2, 3, ...) denoted by $\mathbb{N}$, the integers (....., -2, -1, 0, 1, 2, ...) denoted by $\mathbb{Z}$, and the rational numbers, that is, the fractions, $\left\{\frac{a}{b}; a, b \in \mathbb{Z}, b \neq 0\right\}$, denoted by $\mathbb{Q}$. These sets are subsets of one another given by the order $\mathbb{N} \subseteq \mathbb{Z} \subseteq \mathbb{Q} \subseteq \mathbb{R}$, where each set is a proper subset of the next.

The most fundamental objects in the study of mathematics are sets and functions. We begin with the concepts in logic, basics of set theory and then we briefly discus familiar number sets: the natural numbers, the integers, the rational numbers and the real numbers.

1.2 Basic concepts in logic

In Mathematics, logics are used with *statements*. Statements are declarative sentences that can be either *true* or *false*. We can link statements together using *logical operators* to obtain new statements. The logical operators are *conjunction* (and), *disjunction* (or), *negation* (not), *implication* (if, then), and *equivalence* (if and only if).

1.2.1. Logical Operators

Conjunction joins two statements together to explain that both are true. That is, the conjunction of statements A and B is the statement "A and B," denoted by $A \wedge B$. The statement $A \wedge B$ is true if both A and B are true. It is false otherwise.

Disjunction explains that at least one of two statements is true. That is, the disjunction of statements A and B is the statement "A or B," denoted by $A \vee B$. The statement $A \vee B$ is true if at least one of A and B is true, while if both A and B are false, then so is $A \vee B$. It is important to note that the word "or" can be used with several different meanings:

(i) At least one of two statements is true (inclusive or).

(ii) Exactly one of two statements is true (complementary or).

(iii) At most one of two statements is true (exclusive or).

In mathematics, the logical operator *or* is meant as an *inclusive or*, unless it is mentioned.

Negation is the negative of the given statement. The negation of statement A is "A is false," or in other words, "A is not true." We can express it as "not A," and we denote it by $\overline{A}$. The statement A is true exactly when A is false. We have to distinguish between the concepts of a statement that negates A and a statement that contradicts A. We say that a statement B contradicts A if both A and B cannot be true simultaneously. The negation of a statement A is also a contradiction of A, but the converse is not necessarily true: a contradiction is not necessarily a negation. If A denote the statement "the real number x is positive," and B denote "the real number x is negative," then B contradicts A, but B is not the negation of A, since the correct negation is "the real number x is negative or zero."

Implications are used when one statement follows from another. We can express this by the statement "if A is true, then so is B," or "if A, then B" for short. We denote this by $A \Rightarrow B$. We can also say that "A implies B." The statement $A \Rightarrow B$ means that, in every case in which A is true, B is true. We can also express $A \Rightarrow B$ as "B follows from A," "A is sufficient for B," and "B is necessary for A."

Equivalence explains the condition of two statements always being true at the same time. The statement "A is equivalent to B," denoted by $A \Leftrightarrow B$, is true if both A and B are true or both A and B are false. If one of A and B is true but the other false, then $A \Leftrightarrow B$ is false. If $A \Leftrightarrow B$, we can say "A if and only if B," or "A is necessary and sufficient for B."

1.2.2. Quantifiers

The statements that contain variables are called *predicates*. The truth or falsity of the statements depends on the values of the variables. For example, "x is a perfect square" is a predicate that is true if $x = 4$, but false if $x = 5$. Let $A(x)$ be a predicate, where x is the variable. Then there are two ways to form a "normal" statement (fixed to be either true or false) from it. The first of these is "$A(x)$ is true for all x," which we denote by $(\forall x)(Ax)$. The symbol $\forall$ is called the *universal quantifier*. It represents an upside-down A as in "All." The second is "there exists an x such that $A(x)$ is true," which we denote by $(\exists x)(Ax)$. Here $\exists$ is called the *existential quantifier*. It represents the E in "Exists" written backward.

1.3 Set theory

- If x belongs to a set A, we write $x \in A$ and say that "x is an element of A." Otherwise, we write $x \notin A$.
- If A and B are any two sets, then $A \subseteq B$ ("A is a subset of B" or "A is contained in B") means that each element of A is also an element of B. We can also write $B \supseteq A$ ("B contains A") instead of $A \subseteq B$.
- We say two sets A and B are *equal*, and we write as $A = B$, if $A \subseteq B$ and $B \subseteq A$.
- Any statement S has a *negation* $\sim S$ ("not S") defined by $\sim S$ is true if S is false and $\sim S$ is false if S is true.

- The following are some of the properties of negation:
  (i) $\sim\sim A = \sim(\sim A) = A$
  (ii) $\sim(A \text{ and } B) = (\sim A) \text{ or } (\sim B)$
  (iii) $\sim(A \text{ or } B) = (\sim A) \text{ and } (\sim B)$
  (iv) $\sim(\forall x \in X : P(x)) = (\exists x \in X :\sim P(x))$
  (v) $\sim(\exists x \in X : P(x)) = (\forall x \in X :\sim P(x))$.
- Let A and B be two statements. A *implies* B will be denoted by $A \Rightarrow B$. If A implies B, means that if we wish to prove B, it suffices to prove A (A is a sufficient condition for B).
- The equivalence relation as $A \Leftrightarrow B$ ("A and B are equivalent" or "A if and only if B," often written as A iff B) of the statements A and B is defined by

$$(A \Leftrightarrow B) = (A \Rightarrow B) \text{ and } (B \Rightarrow A).$$

  A is a necessary and sufficient condition for B, or vice versa.
- The *contrapositive* of the statement $A \Rightarrow B$ is $\sim B \Rightarrow \sim A$. In general, any statement is considered to be equivalent to its contrapositive. It is often easier to prove a statement's contrapositive instead of directly proving the statement itself.
- Given two sets A and B, we define $A \cup B$ ("the union of A with B") as the set $A \cup B = \{x : x \in A \text{ or } x \in B \text{ or both}\}$. In case of unions, if we say that $x \in A$ or $x \in B$, it also includes that x is in both A and B. We define $A \cap B$ ("the intersection of A with B") as the set $A \cap B = \{x : x \in A \text{ and } x \in B\}$.
- We say two sets A and B are disjoint, if $A \cap B = \phi$.
- Let A and B be subsets of X. Then $A \setminus B = \{x \in X : x \in A \text{ and } x \notin B\}$ is the *relative complement* of B in A. When X is any given set, we also write $A^c = X \setminus A$ and we say A^c as the complement of A.
- Another set operation is the symmetric difference, which is given as follows;

$$A \Delta B = (A \cup B) \setminus (A \cap B)$$

- De Morgan's Laws state that how complements interact with unions and intersections.

Let $A, B \subseteq X$. Then, $(A \cup B)^c = A^c \cap B^c$ and $(A \cap B)^c = A^c \cup B^c$.

- Some other identities are;

  $A \setminus A = \varphi$,

  $A \setminus \varphi = A$,

  $A \cup A = A$,

  $A \cap A = A$,

  $A \cup B = B \cup A$, $A \cap B = B \cap A$, (Commutativity)

  $A \cup (B \cup C) = (A \cup B) \cup C = A \cup B \cup C$,

  $A \cap (B \cap C) = (A \cap B) \cap C = A \cap B \cap C$, (Associativity)

  $A \cup (B \cap C) = (A \cup B) \cap (A \cup C)$, $A \cap (B \cup C) = (A \cap B) \cup (A \cap C)$, (Distributivity)

  $A \subseteq B \Leftrightarrow B^c \subseteq A^c$,

  $A \setminus B = A \cap B^c$,

  $A \subset A$,

  $A \subset B, B \subset C \Rightarrow A \subset C$.

  $\varphi \subset A$.

- If *X* is a set, then *P*(*X*) is the power set of *X*. The elements of *P*(*X*) are the subsets of *X*.

  Sometimes the power set is written as 2^X.

- If A and B are any two sets, then

  $P(A) \cup P(B) \subseteq P(A \cup B)$ and

  $P(A) \cap P(B) = P(A \cap B)$.

- Let $f : X \to Y$ be a function, then $im(f) = \{y \in Y; \exists x \in X : y = f(x)\}$ is called the *image of f*. We say *f* is *surjective* (or onto) if *im*(*f*) = *Y*, *injective* (or one-to-one) if *f*(*x*) = *f*(*y*) implies *x* = *y* for all *x, y* ∈*X*, and *f* is *bijective* if *f* is both injective and surjective.

- If *A* and *B* are sets, the *Cartesian product* $A \times B$ of *A* and *B* is the set of all ordered pairs (*a, b*) with $a \in A$ and $b \in B$.

  If *A* and *B* are any two sets, then $A \times B = B \times A$.

  If *A*, *B* and *C* are any sets, then

  i) $A \times B = \varphi \Leftrightarrow A = \varphi$ or $B = \varphi$.

ii) $(A \cup B) \times C = (A \times C) \cup (B \times C)$.

iii) $(A \cap B) \times C = (A \times C) \cap (B \times C)$.

- Let X be a set and $A = \{A_i : i \in I\}$ be a family of sets and I is an index set. Then, the *union and Intersection of this family* are given by

  $$\bigcup_{i \in I} A_i = \{x \in X ; \exists i \in I : x \in A_i\}$$

  and

  $$\bigcap_{i \in I} A_i = \{x \in X ; \forall i : x \in A_i\}.$$

- Let $f : X \to Y$ be a function, and $A \subset X$ and $B \subset Y$ are proper subsets. *Image of A under f*, denoted by *f*(*A*), defined as $f(A) = \{f(x) \in Y : x \in A\}$.
- The *pre-image of B under f* or *Inverse image of B*, denoted by $f^{-1}(B)$, is defined as $f^{-1}(B) = \{x \in X : f(x) \in B\}$. It is possible to form $f^{-1}(B)$ for a set $B \subset Y$ even though *f* is not one-to-one or onto.
- Two sets *A* and *B* have the same *cardinality* if there is a bijection from *A* to *B*. In this case we write *A* ~*B*. We say *A* is *countable* if N ~*A*. An infinite set that is not countable is called an uncountable set.

1.4 Techniques for proofs

The different techniques for proving any mathematical statement are *proof by contradiction* and *proof by induction*.

1.4.1. Principle of contradiction

To prove $A \Rightarrow B$ *by contradiction*, one supposes *B* is false (that ~*B* is true). Then, also assuming that *A* is true, one reaches a conclusion *C* which is already known to be false. This contradiction shows that if *A* is true ~*B* cannot be true, and hence *B* is true if *A* is true. We can give some examples for proving any statement by contradiction.

Examples 1.4.1.1

1. $\sqrt{2}$ is irrational. That is, we have to prove that it cannot be written as a quotient of two integers. In order to prove the above statement, we assume that $\sqrt{2}$ is rational is false. Finally, at the end of the proof we conclude that $\sqrt{2}$ is irrational."

2. There are infinitely many prime numbers. The proof of this statement can be obtained by assuming the contradiction.

1.4.2. Principle of induction

Another important technique for proving any given statement is known as *mathematical induction.* Induction arguments prove infinitely many statements at once. In induction, we prove the statements A_1, A_2, A_3, . . .in two steps. In the first step, we show that A_1 is true, and then in the second step, we will prove that A_{n+1} follows from A_n for every n. This second step, namely $A_n \Rightarrow A_{n+1}$, is called the *inductive step*, and A_n is called the *induction hypothesis*. These two steps truly prove all of the statements A_n. Indeed, the first statement, A_1, is proved directly. Since we have shown that A_n implies A_{n+1} for each n, it follows that in particular, A_1 implies A_2 and so on.

Examples 1.4.2.1

1. For every positive integer n, $2^n > n$.

   In order to prove the above statement, we first prove that it is true for $n = 1$. Then, in the inductive step, we show that if the statement holds for n, then it holds for n+1.
2. If a_1, . . . ,a_n are arbitrary nonnegative numbers, then $\sqrt[n]{a_1.....a_n} \le \dfrac{a_1 + ... + a_n}{n}$.

   Equality holds if and only if $a_1 = \cdot\cdot\cdot = a_n$.
3. If a_1, . . . ,a_n are arbitrary positive numbers, then $n\left(\dfrac{1}{a_1} + ... + \dfrac{1}{a_n}\right)^{-1} \le \sqrt[n]{a_1.....a_n}$.

   Equality holds if and only if $a_1 = \cdot\cdot\cdot = a_n$.

   Like example 1, the above examples 2 and 3 can also be proved by induction.

1.5 Number System

1.5.1 The set N of Natural Numbers

The set of all positive integers $\{1,2,3,.......\}$ is denoted by $\mathbb{N}$. Each positive integer n has $n + 1$ as its successor. The following are some properties of $\mathbb{N}$;

i) 1 belongs to $\mathbb{N}$.

ii) If n belongs to $\mathbb{N}$, then its successor $n + 1$ belongs to $\mathbb{N}$.

iii) 1 is not the successor of any element in $\mathbb{N}$.

iv) If n and m in $\mathbb{N}$ have the same successor, then $n = m$.

v) A subset of $\mathbb{N}$ which contains 1, and which contains $n + 1$ whenever it contains n, must equal $\mathbb{N}$.

All the above properties are known as the *Peano Axioms* or *Peano Postulates*. Property (v) is the basis of mathematical induction. Hence, we simply define the natural numbers to be the set $\mathbb{N}$ given in the Peano Postulates.

The following theorem gives some of the very familiar properties of addition and multiplication of natural numbers. The main technique of proof for these properties follows from the *Peano* Postulates. The different parts of the theorem have been arranged so that to prove each, it is permissible to use everything stated previously, but not subsequently.

Theorem 1.5.1.1

Let $a, b, c \in \mathbb{N}$.

i) If $a+c=b+c$, then $a=b$ (Cancellation Law for Addition).

ii) $(a+b)+c=a+(b+c)$ (Associative Law for Addition).

iii) $1+a=s(a)=a+1$.

iv) $a+b=b+a$ (Commutative Law for Addition).

v) $a+b\neq 1$.

vi) $a+b\neq a$.

vii) $a\cdot 1=a=1.a$ (Identity Law for Multiplication).

viii) $(a+b)c=ac+bc$ (Distributive Law).

ix) $ab=ba$ (Commutative Law for Multiplication).

x) $c(a+b)=ca+cb$ (Distributive Law).

xi) $(ab)c=a(bc)$ (Associative Law for Multiplication).

xii) If $ac=bc$ then $a=b$ (Cancellation Law for Multiplication).

xiii) $ab=1$ if and only if $a=1=b$.

Addition and multiplication are the two most important binary operations on the natural numbers. The most important relations on the natural numbers are less than and less than or equal to.

Definition 1.5.1.2

The relation < (less than) on $\mathbb{N}$ is defined by $x < y$ if and only if there is some $p \in \mathbb{N}$ such that $a + p = b$, for all $a, b \in \mathbb{N}$. The relation $\leq$ on $\mathbb{N}$ is defined by $a \leq b$ if and only if $a < b$ or $a = b$, for all $a, b \in \mathbb{N}$. We will write $a > b$ to mean the same thing as $b < a$, and similarly for $a \geq b$. As expected, if $a, b \in \mathbb{N}$ and $a < b$, then the element $p \in \mathbb{N}$ such that $a + p = b$ is unique. The following theorem gives some of the basic properties of $<$ and $\leq$.

Theorem 1.5.1.3

Let $a, b, c, d \in \mathbb{N}$.

1. $a < a+1$.
2. $1 \leq a$.
3. If $a < b$ and $b < c$, then $a < c$; if $a \leq b$ and $b < c$, then $a < c$; if $a < b$ and $b \leq c$, then $a < c$; if $a \leq b$ and $b \leq c$, then $a \leq c$.
4. $a < b$ if and only if $a + c < b + c$.
5. $a < b$ if and only if $ac < bc$.
6. One of $a > b$ or $a = b$ or $a > b$ is true (Trichotomy Law).
7. If $a \leq b$ and $b \leq a$, then $a = b$.
8. $b < a < b+1$ is not true.
9. $a \leq b$ if and only if $a < b + 1$.
10. $a < b$ if and only if $a + 1 \leq$ b.

Theorem 1.5.1.4 (Well-Ordering Principle).

Let $A \subseteq \mathbb{N}$ be nonempty. Then there is an $n \in A$ such that $n \leq m$ for every $m \in A$. The prime numbers are in divisible quantities.

Lemma 1.5.1.5 (Euclid's Lemma).

Let $a, b, p \in \mathbb{N}$ with p a prime number. If p divides the product ab, then p divides a or p divides b.

Definition 1.5.1.6

Let X and Y be any two sets. A *function* $f : X \to Y$ is a rule that assigns to each element of X a unique element of Y. We call X the *domain* of f and Y as the *codomain*. If $x \in X$, we use the symbol $f(x)$ to represent the point in Y associated with x. Sometimes, we also use the symbol $x \mapsto f(x)$ which means that x maps to $f(x)$. If $y \in Y$ is associated with the point $x \in X$, then we write $y = f(x)$. We define the *range* of a function f to be the set $f(X) = \{y \in Y : y = f(x) \text{ for some } x \in X\}$ and thus $f(X) \subseteq Y$.

Definition 1.5.1.7

Let X and Y be any two sets and $f : X \to Y$. We say that f maps X *onto* Y if for each $y \in Y$, there is a point $x \in X$ such that $f(x) = y$. We say that f is *one-to-one* if each y in $f(X)$ is mapped to by a unique element of X. This second property is given as the conditional statement

$$\text{if } f(x) = f(y), \text{ then } x = y.$$

Onto is otherwise termed as *surjective* and *one-to-one* as *injective*. A function that is both one-to-one and onto is called *bijective* or a bijection.

Definition 1.5.1.8

Given sets X, Y, Z and functions $f : X \to Y$ and $g : Y \to Z$, we define a new function $h : X \to Z$ called the *composition* of f and g by

$$h(x) = g(f(x)).$$

It is also written as $(g \circ f)(x) = g(f(x))$.

Definition 1.5.1.9

When a function $f : X \to Y$ is one-to-one and onto Y, it is possible to define a new function called the *inverse* of f, denoted by $f^{-1} : Y \to X$, defined by

$$f^{-1}(y) = x \text{ if and only if } f(x) = y.$$

Since f is onto, we know that every $y \in Y$ is mapped to by some x in X and the fact that f is one-to-one implies that f^{-1} is well defined. Further more, f^{-1} satisfies the equations

$$f^{-1}(f(x) = x, \quad \text{for all } x \in X$$

and $$f^{-1}(f(y)) = y, \quad \text{for all } y \in Y.$$

Whenever such a function exists, then it is unique.

Definition 1.5.1.10

Let $f : X \to Y$ be a function and let's consider *f* as a function on the subsets of *X*, rather than as a function on the points of *X*. Let $A \subseteq X$ and $B \subseteq Y$. Then the range to the subsets of the domain is defined by $f(A) = \{y \in Y : f(x) = y \text{ for some } x \in A\}$. When a function is viewed in this manner rather than as a function acting on some particular domain, we can define f^{-1}, whether f is one-to-one or not as, $f^{-1}(B) = \{x \in X : f(x) = y, \text{ for some } y \in B\}$. The set $f^{-1}(B)$ is called the *inverse image of B* or the *pre-image* or *pull back of B*.

Theorem 1.5.1.11

Let $f : X \to Y$ where *X* and *Y* are sets. If $\{A_n : n \in \mathbb{N}\}$ is a collection of subsets of *Y*, then

$$f^{-1}\left(\bigcup_{n=1}^{\infty} A_n\right) = \bigcup_{n=1}^{\infty} f^{-1}(A_n) \text{ and } f^{-1}\left(\bigcap_{n=1}^{\infty} A_n\right) = \bigcap_{n=1}^{\infty} f^{-1}(A_n).$$

Definition 1.5.1.12

Let *A* and *B* be any two sets. We say that *A* and *B* have the same *cardinality* if there is a bijection $f : A \to B$. If such a bijection exists, then we write this as card(*A*) =card(*B*).

Definition 1.5.1.13

A set *A* is a finite set with cardinality *n* if card(*A*) =card(*B*) where $B = \{1, 2, 3, \ldots\ldots, n\}$. The empty set has cardinality 0. A set A is *countable* if card (*A*) = card (N)or if *A* is finite, and *uncountable* otherwise. That is, a set *A* is countable if there is a one-to-one function $f : A \to B$ where $B \subseteq \mathbb{N}$.

Countable sets are also called *enumerable* and empty sets are finite sets. From this definition, we understand that the natural numbers form the basis for the smallest type of infinite sets.

1.5.2 The set Q of Rational Numbers

The set $Z = \{......., -2, -1, 0, 1, 2, ...\}$ is called the set of all *integers*. The rational numbers Q solve the problem of multiplicative inverses and provide us with the concept of division, that is,

$$Q = \left\{ \frac{m}{n}; m \in Z \text{ and } n \in N \right\}$$

The space Q is a highly satisfactory algebraic system in which the basic operations addition, multiplication, subtraction and division can be fully studied. No system is perfect, however, and Q is inadequate in some ways. One drawback to Q above is that there are repeated elements, e.g. 1/2 and 2/4. To solve this issue, we will often have a condition that p and q have no common factors in order to get a true representation for the rational numbers. There are two operations that are derived from the addition and multiplication of the natural numbers, and there are additive and multiplicative identities and inverses. Such a set structure is called a *field.* The rational number system also has an order $<$ which satisfies;

i) if $x, y \in Q$, then one of the following holds; $x < y$, $y < x$ or $x = y$.

ii) if $x, y \in Q$, with $x < y$ and $y < z$, then $x < z$.

The basic algebraic operations in Q are addition and multiplication. When a pair a, b of rational numbers are given, then the sum $a + b$ and the product ab also represent rational numbers. The following are the properties, which are satisfied by the rational numbers;

1. $a + (b + c) = (a + b) + c$, for all a, b, c. (Associative law for Addition)
2. $a + b = b + a$, for all a, b. (Commutative Law for Addition)
3. $a + 0 = a$, for all a.
4. For each a, there is an element $-a$ such that $a + (-a) = 0$.
5. $a(bc) = (ab)c$, for all a, b, c. (Associative law for Multiplication)
6. $ab = ba$, for all a, b. (Commutative Law for Multiplication)
7. $a \cdot 1 = a$, for all a.

8. For each $a \neq 0$, there is an element a^{-1} such that $aa^{-1} = 1$.
9. $a(b+c) = ab + ac$, for all a, b, c .(Distributive law)

Archimedean Property: If $x \in \mathbb{Q}$, then there is an integer n with $x < n$.

1.5.3 The set R of Real Numbers

The rational numbers and the irrational numbers together constitute the set $\mathbb{R}$, called the set of *real numbers*. The *irrational numbers* are numbers whose infinite decimal expansions are infinite and nonrepeating. The geometric representation of the real numbers as points on the number line naturally implies that there is an *order* among the real numbers. In particular, those real numbers that are greater than 0, that is, which correspond to points to the right of 0, are called *positive*. We can imagine the real numbers in various ways. The basic properties that we accept in real numbers without proof are called the axioms of real numbers. We state the axioms in some groups. The first group consists of axioms regarding addition and multiplication. The second group is made up of properties about ordering like less than and greater than.

Axioms of Field 1.5.3.1

The first group of axioms is the field axioms. We assume that there are two operations defined on the real numbers, called *addition* and *multiplication*. By this, we mean that for any two (not necessarily distinct) real numbers $a, b \in \mathbb{R}$, there correspond a number denoted by $a+b$ (the sum of a and b) as well as a number denoted by $a \cdot b$ (the product of a and b). We also suppose that two distinct numbers 0 and 1 are specified. The first group of axioms deals with these concepts.

Let $a, b, c \in \mathbb{R}$.

1. $a+b=b+a$. (Commutative law of addition)
2. $(a+b)+c = a+(b+c)$. (Associative law of addition)
3. $a+0=a$.
4. For each $a \in \mathbb{R}$, there exists $a, b \in \mathbb{R}$ such that $a+b=0$.
5. $a \cdot b = b \cdot a$. (Commutative law of multiplication)
6. $(a \cdot b) \cdot c = a \cdot (b \cdot c)$. (Associative law of multiplication)
7. $a \cdot 1 = a$.

8. For each $a \in \mathbb{R}$, $a \neq 0$, there exists $a, b \in \mathbb{R}$ such that $a \cdot b = 1$.
9. $a \cdot (b + c) = a \cdot b + a \cdot c$. (Distributive law)

If two operations that satisfy all the nine axioms above are defined on a set, then we say that the set with the two operations is a *field*. (This includes the specification of the distinct elements 0 and 1.) The first group of axioms thus tells us that the real numbers form a field. This is why we call axioms 1–9 the *field axioms*.

Axioms of Order 1.5.3.2

The second group of axioms is the Order axioms. We suppose that there is a so-called *order relation*, denoted by $<$ (less than), defined on the real numbers. By this, we mean that for any two real numbers *a* and *b*, $a < b$ is either true or false. The order axioms fix some properties of this order relation.

1. For any two real numbers *a* and *b*, exactly one of $a < b,\ a = b,\ b < a$ is true. (Trichotomy)
2. For each $a, b, c \in \mathbb{R}$, if $a < b$ and $b < c$, then $a < c$. (Transitivity)
3. For each $a, b, c \in R$, if $a < b$, then $a + c < b + c$. (Addition)
4. For each $a, b, c \in R$, if $a < b$, and $c > 0$, then $a \cdot c < b \cdot c$. (Multiplication)
5. $0 \neq 1$. (Non – degeneracy)

Let $(X, +, \cdot)$ be a field. We say *X* is an *ordered field*, if there is a relation $<$ on *X* which satisfies all the above axioms.

1.5.4 Properties of Real Numbers

Definition 1.5.4.1

We call the number *a positive*, if $a > 0$; *negative or non positive*, if $a < 0$; and *nonnegative*, if $a \geq 0$.

Definition 1.5.4.2

The *absolute value* of a real number *a*, denoted by $|a|$, is defined as;

$$|a| = \begin{cases} a, & \text{if } a \geq 0 \\ -a, & \text{if } a < 0 \end{cases}$$

The properties of absolute value of real numbers are given as follows;

Let $a, b \in \mathbb{R}$.

1. $|a| \geq 0$ and $|a| = 0$ if and only if $a = 0$.
2. $|a| = |-a|$.
3. $|a \cdot b| = |a| \cdot |b|$.
4. If $b \neq 0$, then $\left|\frac{1}{b}\right| = \frac{1}{|b|}$ and $\left|\frac{a}{b}\right| = \frac{|a|}{|b|}$.
5. *Triangle inequality*: $|a+b| \leq |a| + |b|$; $\quad ||a| - |b|| \leq |a-b|$.

Definition 1.5.4.3

Let $E \subseteq \mathbb{R}$ be a set.

The set E is *bounded above*, if there is a real number $M \in \mathbb{R}$, such that $x \leq M$ for all x in E.

The number M is called an *upper bound* of E.

The set E is *bounded below*, if there is a real number $m \in \mathbb{R}$ such that $x \geq m$ for all x in E.

The number m is called the *lower bound* of E.

A subset of $\mathbb{R}$ is said to be *bounded*, if it has an upper bound and a lower bound.

Definition 1.5.4.4

Let $M \in \mathbb{R}$. Then, the number M is a *least upper bound* of E, if M is an upper bound of E, and if $M \leq U$ for all upper bounds U of E. The least upper bound is also called the *supremum*.

Let $m \in \mathbb{R}$. Then, the number m is a *greatest lower bound* of E, if m is a lower bound of E, and if $m \geq L$ for all lower bounds L of E. The greatest lower bound is also called the *infimum*.

Theorem 1.5.4.5

Any nonempty subset E of $\mathbb{R}$ with a lower bound has a greatest lower bound.

Any nonempty subset E of $\mathbb{R}$ with an upper bound has a least upper bound.

Definition 1.5.4.6

An *open interval* in $\mathbb{R}$ is any set of the form

$$(a,b)=\{x\in \mathbb{R}; a<x<b\}.$$

A *closed interval* in $\mathbb{R}$ is any set of the form

$$[a,b]=\{x\in \mathbb{R}; a\le x\le b\}.$$

A *half – open intervals* in $\mathbb{R}$ is any set of the form

$$[a,b)=\{x\in \mathbb{R}; a\le x<b\} \text{ or } (a,b)]=\{x\in \mathbb{R}; a<x\le b\}.$$

Definition 1.5.4.7

Open sets in $\mathbb{R}$ are defined to be arbitrary unions of open bounded intervals, and a *closed set* is any set whose complement in $\mathbb{R}$ is open.

Theorem 1.5.4.8

i) The intersection of two open intervals is either empty or is another open interval.

ii) The union of two open intervals is either another open interval or is just two disjoint open intervals.

Definition 1.5.4.9

By a *sequence*, we mean, a function defined on *J*, the set of positive integers. If $f(n)=x_n$, then the sequence is defined as $\{x_n\}$ or $x_1, x_2,$............The elements of a sequence need not be distinct. If $x_n \in A$, for all $n\in J$, then we say $\{x_n\}$ is a sequence in *A*.

Note 1.5.4.10

Let *A* be a countable set. Therefore, $J \sim A$ implies there is a bijection $f : J \to A$. Hence, *A* is the range of the sequence defined by *f*. That is, $A=\{f(1), f(2),.......\}$. Therefore, the elements of any countable set A can be arranged in the sequence of $\{x_n\}$ of distinct points.

Theorem 1.5.4.11

Every infinite subset of a countable set *A* is countable.

Proof

Given that A is countable.

Hence, its elements can be arranged in the sequence of distinct terms. That is,

$$A = \{x_1, x_2, ...\}.$$

Let B be an infinite subset of A.

Let n_1 be the smallest positive integer such that $x_{n_1} \in B$. Since B is infinite, there exist a smallest positive integer $n_2 > n_1$ such that $x_{n_2} \in B$.

Having chosen $n_1, n_2, ..., n_k$, there exist a least positive integer $n_{k+1} > n_k$ such that $\{x_{n_{k+1}}\} \in B$.

Let $B = \{x_{n_k}\}_{k=1}^{\infty}$.

We have to prove that, B is countable.

Define $f : J \to B$ as $f(k) = x_{n_k}$.

Hence, f is a bijection.

Therefore, $J \sim B$ and hence B is countable.

Hence the theorem.

Note 1.5.4.11

Every subset of a countable set is atmost countable. Also, we can say that, a countable set cannot contain an uncountable set.

Theorem 1.5.4.12

Let $\{E_n\}, n = 1,2,3,.....$ be a sequence of countable sets. Then, $E = \bigcup_{n=1}^{\infty} E_n$ is countable. We can also say that, "Countable union of countable sets is countable."

Proof

Since E_n is countable, its elements can be arranged in a sequence $\{x_{n_k}\}_{k=1}^{\infty}$ of distinct points.

Now consider the infinite array,

$$\begin{matrix} x_{11} & x_{12} & x_{13} & x_{14} & \cdots & \cdots & \cdots \\ x_{21} & x_{22} & x_{23} & x_{24} & \cdots & \cdots & \cdots \\ x_{31} & x_{32} & x_{33} & x_{34} & \cdots & \cdots & \cdots \\ x_{41} & x_{42} & x_{43} & x_{44} & \cdots & \cdots & \cdots \\ \cdots & \cdots & \cdots & \cdots & \cdots & \cdots & \cdots \\ \cdots & \cdots & \cdots & \cdots & \cdots & \cdots & \cdots \\ \cdots & \cdots & \cdots & \cdots & \cdots & \cdots & \cdots \end{matrix}$$

in which E_n form the nth row. This infinite array contains all the elements of E. The elements of this infinite array can be arranged in the sequence $x_{11}, x_{21}, x_{12}, x_{31}, x_{22}, x_{13}, x_{41}, x_{32}, x_{23}, x_{14},$ ....(1.5.4.12.1).

If any two of the sets E_n have the elements in common these will appear more than once in equation (1.5.4.12.1). Hence there is a subset T of J such that $E \sim T$. Since $T \subset J$ and J is countable, T is atmost countable.

Therefore, E is atmost countable, that is, E is finite or countable.

Since $E_1 \subset E$ and E_1 is countable, E_1 is infinite. Therefore, E is infinite and hence countable.

Hence the theorem.

Theorem 1.5.4.13

Let A be a countable set. For each $n \in J$, let $B_n = \{(a_1, a_2, \ldots\ldots, a_n) / a_1, a_2, \ldots\ldots, a_n \in A\}$. Then, B_n is countable.

Proof

Given that A is countable. Also, $B_n = A^n$.

We have to prove that, B_n is countable.

We can prove this by induction method.

When $n = 1$, $B_n = B_1 = A' = A$ which is countable.

Hence the result is true for $n = 1$.

When $n > 1$, assume B_{n-1} is countable.

We have to prove that, $B_n \sim B_{n-1} \times A$.

Let $f : B_n \to B_{n-1} \times A$ be defined as

$f\left((a_1,a_2,\ldots\ldots,a_n)\right)=\left((a_1,a_2,\ldots\ldots,a_{n-1}),a_n\right)$ where $a_1,a_2,\ldots\ldots,a_n\in A$.

We have to prove that, *f* is one – one.

Now, $f\left((a_1,a_2,\ldots\ldots,a_n)\right)=f\left((b_1,b_2,\ldots\ldots,b_n)\right)$.

That is, $\left((a_1,a_2,\ldots\ldots,a_{n-1}),\ a_n\right)=\left((b_1,b_2,\ldots\ldots,b_{n-1}),\ b_n\right)$.

That implies, $(a_1,a_2,\ldots\ldots,a_{n-1})=(b_1,b_2,\ldots\ldots,b_{n-1})$ and $a_n=b_n$.

That is, $a_1=b_1,\ a_2=b_2,\ldots\ldots,\ a_{n-1}=b_{n-1},\ a_n=b_n$.

Hence, $(a_1,a_2,\ldots\ldots,a_n)=(b_1,b_2,\ldots\ldots,b_n)$.

Therefore, *f* is one – one.

We have to prove that, *f* is onto.

Let $(b,a)\in B_{n-1}\times A$.

Then, $b\in B_{n-1}$ and $a\in A$.

Therefore, $b=(a_1,a_2,\ldots\ldots,a_{n-1})$, where $a_1,a_2,\ldots\ldots,a_{n-1}\in A$ and $a\in A$.

Consider, $(a_1,a_2,\ldots\ldots,a_{n-1},a)\in B_n$.

Now, $f(a_1,a_2,\ldots\ldots,a_{n-1},a)=\left((a_1,a_2,\ldots\ldots,a_{n-1}),a\right)=(b,a)$.

Now, for all $(b,a)\in B_{n-1}\times A$, there exist $(a_1,a_2,\ldots\ldots,a_{n-1},a)$ such that $f(a_1,a_2,\ldots.,a_{n-1},a)=(b,a)$.

Therefore, *f* is onto.

Hence, $B_n\sim B_{n-1}\times A$ .... (1.5.4.13.1)

Also, $B_{n-1}\times A=\bigcup\limits_{b\in B_{n-1}}\{b\}\times A$ .... (1.5.4.13.2)

We have to prove that, $\{b\}\times A\sim A$, for each $b\in B_{n-1}$.

Let $g:\{b\}\times A\to A$ be defined as $g\left((b,a)\right)=a$.

First, we have to prove that, *g* is one – one.

Now, $g(b,a_1)=g(b,a_2)\Rightarrow a_1=a_2\Rightarrow(b,a_1)=(b,a_2)$.

Therefore, *g* is one – one.

Next, we have to prove that, *g* is onto.

For all $a\in A$ there exist $(b,a)\in\{b\}\times A$ such that $g\left((b,a)\right)=a$.

Therefore, *g* is onto.

Hence, $\{b\}\times A\sim A$.

Since A is countable, $\{b\} \times A$ is countable for each $b \in B_{n-1}$.

Since countable union of countable sets is countable, $\bigcup_{b \in B_{n-1}} \{b\} \times A$ is countable.

From equation (1.5.4.13.2), $B_{n-1} \times A$ is countable.

Therefore, by equation (1.5.4.13.1), B_n is countable.

Hence, by induction, B_n is countable for all integers.

Hence the theorem.

Theorem 1.5.4.14

The set of rational numbers is countable.

Proof

Let $\mathbb{Q}$ be the set of rationals.

That is, $\mathbb{Q} = \{(a/b) / a \in \mathbb{Z}, b \in J\}$.

For each $b \in J$, let $A_i = \{(a/b) / a \in \mathbb{Z}\}$.

We have to prove that, $A_i \sim \mathbb{Z}$.

Define a map $f : A_i \to \mathbb{Z}$ as $f\left(\frac{a}{b}\right) = a$.

First, we have to prove that, f is one – one.

Now, $f\left(\frac{a_1}{b}\right) = f\left(\frac{a_2}{b}\right) \Rightarrow a_1 = a_2 \Rightarrow \frac{a_1}{b} = \frac{a_2}{b}$.

Hence, f is one – one.

Next, we have to prove that, f is onto.

For all $a \in \mathbb{Z}$, there exist $\frac{a}{b} \in A_i$ such that $f\left(\frac{a}{b}\right) = a$.

Hence, f is onto.

Therefore, $A_i \sim \mathbb{Z}$.

Since $\mathbb{Z}$ is countable, A_i is countable.

As $\mathbb{Q} = \bigcup_{b \in J} A_i$ and countable union of countable sets is countable, $\mathbb{Q}$ is countable.

Hence the theorem.

Theorem 1.5.4.15

Let A be the set of all sequences, where the elements are distinct 0 and 1. Then, A is uncountable.

Proof

The distinct elements of A are

$$s_1 = \{1,0,0,......\}$$
$$s_2 = \{0,1,0,......\}$$
$$s_3 = \{0,0,1,......\}$$
$$...... \quad ...... \quad ......$$

Therefore, A is an infinite set.

Let B be a countable subset of A. Hence, elements of B can be arranged in a sequence $\{s_n\}$ of distinct terms.

Now, let us construct a sequence s whose nth term is

$$\begin{cases} 1, & \text{if the n}^{\text{th}} \text{ term of sequence } s_n \text{ is } 0 \\ 0, & \text{if the n}^{\text{th}} \text{ term of sequence } s_n \text{ is } 1 \end{cases}$$

Therefore, $s \in A$ and $s \neq s_n$, for $n = 1,2,3,.....$

Hence, $s \in A$, but $s \notin B$.

Therefore, B is a proper subset of A. .... (1.5.4.15.1)

We have to prove that, A is uncountable.

Suppose A is countable. Then, by equation (1.5.4.15.1), A is a proper subset of A, which is a contradiction.

Therefore, A is uncountable.

Hence the theorem.

1.6 Metric Space

Definition 1.6.1

A set X is called a *metric space* if any two elements of X, there is an associated real number $d(x, y)$ called the distance from x to y, such that

i) $d(x, y) > 0$, if $x \neq y$ and $d(x, y) = 0$, if $x = 0$,

ii) $d(x, y) = d(y, x)$,

iii) $d(x, z) \leq d(x, y) + d(y, z)$, for all $z \in X$.

Any function satisfying these three properties is called a *metric or a distance function.*

Examples 1.6.2

1. For $x, y \in \mathbb{R}$ define $d(x, y) = |(x, y)|$. Then, d is a metric in $\mathbb{R}$.
2. For $x = (x_1, x_2, \ldots\ldots, x_k), y = (y_1, y_2, \ldots\ldots, y_k) \in \mathbb{R}^k$, define

   $d(x, y) = |(x - y)| = \sqrt{\sum_{i=1}^{k} (x_i - y_i)^2}$. Then, d is a metric on $\mathbb{R}^k$.
3. Let X be any set. For $x, y \in X$, define $d(x, y) = \begin{cases} 1, & \text{if } x \neq y \\ 0, & \text{if } x = y \end{cases}$

Then, d is a metric on X.

Note 1.6.3

1. By the segment (a, b), we mean the set of all real numbers x, such that $a < x < b$.
2. By the interval $[a, b]$, we mean the set of all real numbers x, such that $a \leq x \leq b$.
3. By the half open intervals $(a, b]$ or $[a, b)$, we mean the set of all real numbers x such that $a < x \leq b$ or $a \leq x < b$.
4. Let $a_i < b_i$, for $i = 1, 2, 3, \ldots\ldots, k$. Cell I is defined as the set of all $x = (x_1, x_2, \ldots\ldots, x_k) \in \mathbb{R}^k$, such that $a_i \leq x_i \leq b_i$, for all $i = 1, 2, \ldots\ldots, k$.

   Therefore, $I = [a_1, b_1] \times [a_2, b_2] \times \ldots\ldots \times [a_k, b_k]$.
   Cell is an interval
   Cell is a rectangle
   Cell is a rectangular parallelo piped.
5. For $x \in \mathbb{R}^k$ and $r > 0$, then open ball with center x and radius r is defined as the set of all $y \in \mathbb{R}^k$ with $|y - x| < r$.
6. For $x \in R^k$ and $r > 0$, then closed ball with center x and radius r is defined as the set of all $y \in \mathbb{R}^k$ with $|y - x| \leq r$.
7. $E \subset \mathbb{R}^k$ is said to be *convex*, if $\lambda x + (1 - \lambda) y \in E$ for $x, y \in E$ and $0 < \lambda < 1$.

Theorem 1.6.4

Every open ball in $\mathbb{R}^k$ is convex.

Proof

Let E be an open ball in $\mathbb{R}^k$, with center x and radius r.

Let $x_1, x_2 \in E$ and $0 < \lambda < 1$.

Then, $|x_1 - x| < r$ and $|x_2 - x| < r$.

We have to prove that, $\lambda x_1 + (1-\lambda)x_2 \in E$.

That is to prove that, $|\lambda x_1 + (1-\lambda)x_2 - x| < r$.

Now,

$$|\lambda x_1 + (1-\lambda)x_2 - x| = |\lambda x_1 + (1-\lambda)x_2 - (\lambda x + (1-\lambda)x| = |\lambda(x_1 - x) + (1-\lambda)(x_2 - x)|.$$

That is, $|\lambda x_1 + (1-\lambda)x_2 - x| \leq |\lambda(x_1 - x)| + |(1-\lambda)(x_2 - x)|$.

Since, $|\lambda| = \lambda$ and $|1-\lambda| = 1-\lambda$, we have $|\lambda x_1 + (1-\lambda)x_2 - x| < r$.

That implies, $\lambda x_1 + (1-\lambda)x_2 \in E$.

Therefore, E is convex.

Hence the theorem.

Theorem 1.6.5

Every closed ball in $\mathbb{R}^k$ is convex.

Proof

Let E be a closed ball in $\mathbb{R}^k$, with center x and radius r.

Let $x_1, x_2 \in E$ and $0 < \lambda < 1$.

Then, $|x_1 - x| \leq r$ and $|x_2 - x| \leq r$.

We have to prove that, $\lambda x_1 + (1-\lambda)x_2 \in E$.

That is to prove that, $|\lambda x_1 + (1-\lambda)x_2| \leq r$.

Now, $|\lambda x_1 + (1-\lambda)x_2 - x|$

$$= |\lambda x_1 + (1-\lambda)x_2 - (\lambda x + (1-\lambda)x| \leq \lambda|x_1 - x| + (1-\lambda)|x_2 - x| = r.$$

That implies E is convex.

Hence the theorem.

Theorem 1.6.6

Every k – cell is convex.

Proof

Let $I=[a_1,b_1]\times[a_2,b_2]\times......\times[a_k,b_k]$ be any cell.

Let $x, y\in I$ and $0<\lambda<1$.

Hence, $x=(x_1,x_2,.....,x_k), y=(y_1,y_2,.....,y_k)$ and $a_i\le x_i,\ y_i\le b_i$, for all $i=1,2,.....,k$.

We have to prove that, $\lambda x+(1-\lambda)y\in I$.

That is to prove that, $a_i\le \lambda x_i+(1-\lambda)y_i\le b_i$, for all $i=1,2,.....,k$.

Now, $\lambda x+(1-\lambda)y=\lambda(x_1,x_2,.....,x_k)+(1-\lambda)(y_1,y_2,.....,y_k)$.

$$=(\lambda x_1,\lambda x_2,.....,\lambda x_k)+\big((1-\lambda)y_1,\ (1-\lambda)y_2,.....,(1-\lambda)y_k\big).$$

$$=\big(\lambda x_1+(1-\lambda)y_1,\lambda x_2+(1-\lambda)y_2,.....,\lambda x_k+(1-\lambda)y_k\big).$$

$$\text{.................. (1.6.6.1)}$$

We have, $a_i\le x_i\le b_i$, for all $i=1,2,.....,k$, which implies $\lambda a_i\le\lambda x_i\le\lambda b_i$.

$$\text{................. (1.6.6.2)}$$

Also, we have, $a_i\le y_i\le b_i$, for all $i=1,2,.....,k$, which implies

$(1-\lambda)a_i\le(1-\lambda)y_i\le(1-\lambda)b_i$. .... (1.6.6.3)

Adding equations (1.6.6.2) and (1.6.6.3), we get

$$a_i\le\lambda x_i+(1-\lambda)y_i\le b_i,\quad \text{for all } i=1,2,....,k.$$

Hence, from equation (1.6.6.1), we have, $\lambda x+(1-\lambda)y\in I$.

Therefore, I is convex.

Hence the theorem.

Definition 1.6.7

Let X be a metric space. A *neighborhood* of a point p is a set $N_r(p)$ consisting of all $q\in X$ such that $d(p,q)<r$, where r is the radius of $N_r(p)$.

Definition 1.6.8

Let X be a metric space and $E\subset X$. Then, E is called an *open set*, if every point of E is an interior point of E.

Theorem 1.6.9

Every neighborhood is an open set.

Proof

Let X be a metric space and $N_r(p)$ be a neighborhood of $p \in X$.

We have to prove that, $N_r(p)$ is open.

Let $q \in N_r(p)$ and hence, $d(p,q) < r \Rightarrow r - d(p,q) > 0$.

Let $\delta = r - d(p,q) > 0$.

Hence, we have to prove that, $N_\delta(q) \subset N_r(p)$.

Let $x \in N_\delta(q)$.

That implies, $d(x,q) < \delta = r - d(p,q) \Rightarrow d(p,q) + d(x,q) < r$. .. (1.6.9.1)

Now, $d(x,p) \leq d(x,q) + d(q,p)$.

Hence, by equation (1.6.9.1), we have $d(x,p) < r$.

That is, $x \in N_r(p)$.

Hence, q is an interior point of $N_r(p)$.

Since q is arbitrary, every point of $N_r(p)$ is its interior point.

Therefore, $N_r(p)$ is open.

Hence the theorem.

Definition 1.6.10

Let X be a metric space. p is called a *limit point* of $E \subset X$ if every neighborhood of p contains a point of E other than p.

That is, $N_r(p) \cap (E - \{p\}) \neq 0$, for all $r > 0$.

Definition 1.6.11

Let X be a metric space and $E \subset X$. Then, E is *closed* if it contains all its limit points.

Theorem 1.6.12

If p is a limit point of E, then every neighborhood of p contains infinitely many points of E.

Proof

Given that p is a limit point of E.

We have to prove that, every neighborhood of p contains infinitely many points of E.

Suppose not,

Then, there exists $r>0$ such that $N_r(p)\cap E$ contains only finite points.

Hence, $N_r(p)\cap(E-\{p\})$ is finite.

Let $N_r(p)\cap(E-\{p\})=\{p_1,p_2,\ldots\ldots,p_n\}$. .... (1.6.12.1)

Therefore, for all $i=1,2,\ldots\ldots,n,\quad p_i\in N_r(p)\cap(E-\{p\})$.

That implies, $d(p,p_i)<r$ and $p_i\neq p$, also $d(p_i,p)>0$, for all $i=1,2,..,n$.

Let $\delta=\min\limits_{1\le i\le n}.d(p,p_i)$.

Hence, $0<\delta<r$.

We have to prove that, $N_\delta(p)\cap(E-\{p\})=\phi$.

Let $x\in N_\delta(p)$.

Therefore, $d(x,p)<\delta\le d(p,p_i)$, for all $i=1,2,\ldots\ldots,n$.

Hence, $x\neq p_i$, for all $i=1,2,\ldots\ldots,n$. .... (1.6.12.2)

Also, $d(x,p)<\delta<r$.

Therefore, $x\in N_r(p)$. .... (1.6.12.3)

From equations (1.6.12.1), (1.6.12.2) and (1.6.12.3), we get

$$x\notin E-\{p\}\Rightarrow N_\delta\cap(E-\{p\})=\phi.$$

Therefore, p is not a limit point of E, which is a contradiction.

Hence, every neighborhood of p contains infinitely many points of E.

Hence the theorem.

Note 1.6.13

1. If $A\cap B=\varphi$, then $A\subset B^c$.
2. A finite set has no limit point.

Proof

Let E be any finite subset of a metric space X.

Let $p\in X$.

Since E is a finite set, no neighborhood of p can contain infinitely many points of E.

Hence, p is not a limit point of E.
Therefore, E has no limit point.
Hence proved.

Definition 1.6.14

Let X be a metric space and $E \subset X$. If $p \in E$ and if p is not a limit point of E, then p is called an *isolated point* of E.

Definition 1.6.15

Let X be a metric space and $E \subset X$. Then, $p \in E$ is called an *interior point* of E, if there exists $r > 0$ such that $N_r(p) \subset E$.

Definition 1.6.16

Let X be a metric space and $E \subset X$. Then, E is *perfect* if E is closed and if every point of E is a limit point of E.

Definition 1.6.17

Let X be a metric space and $E \subset X$. Then, E is *bounded*, if there is a real number k and a point $q \in X$ such that $d(p,q) < k$, for all $p \in E$.

Definition 1.6.18

Let X be a metric space and $E \subset X$. Then, E is *dense* in X, if every point of X is either a point of E or a limit point of E.

Theorem 1.6.19

A set E is open if and only if its complement is closed.

Proof

Direct part,

Given that E is open.

We have to prove that, E^c is closed.

That is to prove that, E^c contains all its limit points.

Let $x \in X$ be a limit point of E^c.

Therefore, we have to prove that, $x \in E^c$.

Suppose $x \notin E^c$.

Since x is a limit point of E^c, there exists $r > 0$, such that $N_r(x) \cap (E^c - \{x\}) \neq \phi$.

That is, $N_r(x) \cap E^c \neq \varphi \Rightarrow N_r(x) \not\subset (E^c)^c \Rightarrow N_r(x) \not\subset E$.

Therefore, x is not an interior point of E, which is a contradiction, since, "E is open implies every point of E is an interior point of E".

Hence, $x \in E^c$.

Therefore, E^c is closed.

Converse part,

Given that, E^c is closed.

We have to prove that, E is open.

Let $x \in E$. That implies, $x \notin E^c$.

Since E^c is closed, x is not a limit point of E^c.

Therefore, there exists $r > 0$ such that $N_r(x) \cap (E^c - \{x\}) = \phi$.

Since $x \notin E^c$, $N_r(x) \cap E^c = \phi$.

Therefore, $N_r(x) \subset (E^c)^c \Rightarrow N_r(x) \subset E$.

That implies, E is open.

Hence the theorem.

Note 1.6.20

A set F is closed if and only if its complement is open.

Proof

By theorem 1.6.19, F^c is open if and only if $(F^c)^c$ is closed.

F^c is open if and only if F is closed.

That is, F is closed if and only if F^c is open.

Hence proved.

Theorem 1.6.21

i) For any arbitrary collection $\{G_\alpha\}$ of open sets $\bigcup_\alpha G_\alpha$ is open. That is, arbitrary union of open sets is open.

ii) For any arbitrary collection $\{F_\alpha\}$ of closed sets $\bigcap_\alpha F_\alpha$ is closed.

iii) For any finite collection $G_1, G_2, \ldots.., G_n$ of open sets, $\bigcap_{i=1}^{n} G_i$ is open.

iv) For any finite collection $F_1, F_2, \ldots\ldots, F_n$ of closed sets, $\bigcup_{i=1}^{n} F_i$ is closed.

Proof

i) Given that $\{G_\alpha\}$ is a collection of open sets.

We have to prove that, $\bigcup_{\alpha} G_\alpha$ is open.

Let $x \in \bigcup_{\alpha} G_\alpha$.

That implies, $x \in G_\alpha$, for some α.

Since G_α is open, there exists $r > 0$ such that $N_r(x) \subset G_\alpha \subset \bigcup_{\alpha} G_\alpha$.

Therefore, $\bigcup_{\alpha} G_\alpha$ is open.

Hence part (i) is proved.

ii) Given that, $\{F_\alpha\}$ is a collection of closed sets.

Hence, $\{F_\alpha^c\}$ is a collection of open sets.

Therefore, by part (i), $\bigcup_{\alpha} F_\alpha^c$ is an open set.

Hence, by Demorgan's law, $\left(\bigcap_{\alpha} F_\alpha\right)^c$ is an open set.

That implies, $\bigcap_{\alpha} F_\alpha$ is a closed set.

Hence part (ii) is proved.

iii) Given that $G_1, G_2, \ldots\ldots, G_n$ is a finite collection of open sets.

We have to prove that, $\bigcap_{i=1}^{n} G_i$ is an open set.

Hence, $x \in G_i$, for $i = 1, 2, \ldots\ldots, n$.

Since each G_i is open, there exists $r_i > 0$, such that $N_{r_i}(x) \subset G_i$, for all $i = 1, 2, \ldots., n$. .... (1.6.21.1)

Let $r = \min(r_1, r_2, \ldots\ldots, r_n)$.

Then, $r > 0$ and $N_r(x) \subset N_{r_i}(x)$, for all $i = 1, 2, \ldots\ldots, n$.

Hence by equation (1.6.21.1), $N_r(x) \subset G_i$, for all $i = 1,2,......,n$.

Therefore, $N_r(x) \subset \bigcap_{i=1}^{n} G_i$.

Hence, $\bigcap_{i=1}^{n} G_i$ is an open set.

Hence part (iii) is proved.

iv) Given that, $F_1, F_2,......, F_n$ are closed.

Hence, $F_1^c, F_2^c,......, F_n^c$ are open sets.

Therefore, by Part (iii), $F_1^c \cap F_2^c \cap ......\cap F_n^c$ is an open set.

That is, $\bigcap_{i=1}^{n} F_i^c$ is open.

By Demorgan's law, $\left(\bigcup_{i=1}^{n} F_i\right)^c$ is an open set.

Therefore, $\bigcup_{i=1}^{n} F_i$ is a closed set.

Hence part (iv) is proved.

Hence the theorem.

Note 1.6.22

1. Arbitrary intersection of open sets need not be an open set.

   Consider $G_n = \left(-\frac{1}{n}, \frac{1}{n}\right), \quad n = 1, 2, \text{.......}$

   Then, G_n is a collection of open sets and $\bigcap_{n=1}^{\infty} G_n = \{0\}$ is not an open set.

2. Arbitrary union of closed sets need not be a closed set.

   Consider R with usual metric.

   Let $F_n = \left\{\frac{1}{n}, 1\right\}$.

   Here, F_n is a collection of closed sets in $\mathbb{R}$.

   $\bigcup_{n=1}^{\infty} F_n = (0,1]$ need not be a closed set.

Definition 1.6.23

Let $E \subseteq X$. Then, the set of all limit point of *E* is denoted by E' . The *closure* of *E* is defined as $\overline{E} = E \cup E'$.

Theorem 1.6.24

i) $\overline{E}$ is closed

ii) $E = \overline{E}$ if and only if E is closed.

iii) If *F* is any closed subset of *X* such that $F \supset E$, then $F \supset \overline{E}$.

Proof

i) We have to prove that, $\overline{E}$ is closed.

That is, to prove that, $\overline{E}^c$ is open.

Let $x \in \overline{E}^c$.

Hence, $x \notin \overline{E} = E \cup E' \Rightarrow x \notin E$ and $x \notin E'$.

Now, $x \notin E \Rightarrow E - \{x\} = E$ . ... (1.6.24.1)

x is not a limit point of *E*.

That implies, there exists $r > 0$ such that $N_r(x) \cap (E - \{x\}) = \phi$.

Hence, by equation (1.6.24.1), we have, $N_r(x) \cap E = \phi$ . .... (1.6.24.2)

We have to prove that, $N_r(x) \cap E' = \phi$.

Let $y \in N_r(x)$.

Since $N_r(x)$ is open, there exists $\delta > 0$ such that $N_\delta(x) \subset N_r(x)$.

Hence, from equation (1.6.24.2), $N_\delta(y) \cap E = \phi$.

That is, $N_\delta(y) \cap (E - \{y\}) = \phi$.

Therefore, *y* is not a limit point of *E*.

That implies, $y \notin E'$.

Hence, $N_r(x) \cap E' = \phi$.

Now, $N_r(x) \cap \overline{E} = N_r(x) \cap [E \cup E'] = (N_r(x) \cap E) \cup (N_r(x) \cap E') = \phi$.

Hence, $N_r(x) \subset (\overline{E})^c$. That implies, $(\overline{E})^c$ is open.

Therefore, $\overline{E}$ is closed.

Hence, part (i) is proved.

ii) Direct part,

Given that, $E=\overline{E}$. .... (1.6.24.3)

We have to prove that, E is closed.

By part (i), $\overline{E}$ is closed and hence, by equation (1.6.24.3), E is closed.

Converse part,

Given that, E is closed.

We have to prove that, $E=\overline{E}$.

Since E is closed, E contains all its limit points. Hence, $E \supset E'$.

That implies, $E \cup E' = E$, hence $\overline{E}=E$.

Hence, part (ii) is proved.

iii) Given that, $E \subset F$.

Therefore, $E' \subset F'$.

Since F is closed, F contains all its limit points, $E' \subset F' \subset F$.

Hence, $E \cup E' \subset F$. That is, $\overline{E} \subset F$.

Therefore, E is the smallest closed set contained in F.

Hence part (iii) is proved.

Hence the theorem.

Theorem 1.6.25

Let E be a non – empty set of real numbers which are bounded above. Let $y=\sup E$. Then, $y\in \overline{E}$. Hence, $y\in E$, if E is closed.

Proof

If $y\in E$, then obviously $y\in \overline{E}$.

If $y\notin E$, then we have to prove that, $y\in E'$.

Let $r>0$ be given. Then, $y-r<y$.

Hence, $y-r$ is not an upper bound of E.

Therefore, there exists $x\in E$ such that $y-r<x<y<y+r$.

Hence, $x\in N_r(y)\cap(E-\{y\})$. That is, $N_r(y)\cap(E-\{y\})\neq \varphi$.

Therefore, y is a limit point of E.

Hence, $y\in E'$ and thus $y\in \overline{E}$.

Since E is closed, $\overline{E}=E$.

Thus, $y \in \overline{E} \Rightarrow y \in E$.

Hence the theorem.

Note 1.6.26

Let E be a non-empty set of real numbers which are bounded below.

Let $y = \inf E$.

Then, $y \in \overline{E}$. Hence, $y \in E$, if E is closed.

Theorem 1.6.27

Suppose $Y \subset X$. A subset E of Y is open relative to Y, if and only if $E = Y \cap G$ for some open subset G of X.

Proof

Direct part,

Given that, $E \subset Y \subset X$ is open in Y.

For $y \in Y$ and $y > 0$, $N_r(y) = \{x \in X / d(x, y) < r\}$ and

$V_r(y) = \{y \in Y / d(x, y) < r\}$.

Therefore, $V_r(y) = N_r(y) \cap Y$. .... (1.6.27.1)

Since E is open in Y, for each $y \in E$, there exists $r_y > 0$ such that

$V_{r_y}(y) \subset E$.

Hence, $E = \bigcup_{y \in E} V_{r_y}$.

Therefore, by equation (1.6.27.1), we have

$$E = \bigcup_{y \in E} N_{r_y}(y) \cap Y = Y \cap \left[\bigcup_{y \in E} N_{r_y}(y) \right].$$

Let $G = \bigcup_{y \in E} N_{r_y}(Y)$.

Since $N_{r_y}(y)$ is open in X and arbitrary union of open sets is open,

$\bigcup_{y \in E} N_{r_y}(y)$ is open in X.

Hence, G is open in X.

Therefore, $E = Y \cap G$, where G is open in X.

Converse part,

Given that, $E = Y \cap G$, for some open subsets G of X.

We have to prove that, E is open in Y.

Let $y \in E$. That implies, $y \in Y \cap G$.

Hence, $y \in Y$ and $y \in G$.

Since G is open in X, there exists $r > 0$ such that $N_r(y) \subset G$.

Therefore, $Y \cap N_r(y) \subset Y \cap G$.

That is, $V_r(y) \subset E$.

Therefore, E is open in Y.

Hence the theorem.

Note 1.6.28

Suppose $Y \subset X$. A subset E of Y is closed relative to Y, if and only if $E = Y \cap F$, for some closed subset F of X.

Theorem 1.6.29

If E is open in Y, then E need not be open in X.

Let us consider an example:

Take $X = \mathbb{R}$, $Y = [0,1]$ and $E = [0,1/2)$.

Hence, $E \subset Y \subset X$.

We have to prove that, 0 is the interior point.

Now, $V_{1/4}(0) = \left\{ x \in Y / |x - 0| < \frac{1}{4} \right\} = [0,1/4) \subset E$.

Therefore, 0 is the interior point of E in Y. Hence, E is open in Y. But, E is not open in X.

Definition 1.6.30

By an open cover of a set E, in a metric space X, we mean, a collection $\{G_\alpha\}$ of open subsets of X, such that $E \subset \bigcup_{\alpha} G_\alpha$.

Definition 1.6.31

A subset K of a metric space X is said to be compact, if every open cover of K contains a finite sub-cover.

Examples 1.6.32

1. (0,1) is not compact.

   Let, $\mathcal{A}=\left\{\left(\frac{1}{n},1\right)/n=2,3,.....\right\}$. Thus, $\mathcal{A}$ is an open cover for (0, 1), but, no finite sub collection of $\mathcal{A}$ will cover (0, 1).

   Hence, (0, 1) is not compact.

2. $\mathbb{R}$ is not compact.

   Let, $\mathcal{A}=\{(-n,n)/n=1,2,3,.....\}$. Thus, $\mathcal{A}$ is an open cover for $\mathbb{R}$, but, no finite sub collection of $\mathcal{A}$ will cover $\mathbb{R}$.

   Hence, $\mathbb{R}$ is not compact.

3. Every finite subset of a metric space is compact.

Theorem 1.6.33

Suppose $K\subset Y\subset X$. Then, *K* is compact relative to *X*, if and only if *K* is compact relative to *Y*.

Proof

Direct part,

Given that $K\subset Y\subset X$ and K is compact relative to X.

We have to prove that, K is compact relative to Y.

Let $\{A_\alpha\}$ be a collection of open sets in Y, such that $K\subset\bigcup_\alpha A_\alpha$.

.... (1.6.33.1)

Since $\{A_\alpha\}$ is open in Y, $A_\alpha=Y\cap G_\alpha$, for some G_α in X.

Hence equation (1.6.33.1) becomes, $K\subset\bigcup_\alpha(Y\cap G_\alpha)=Y\cap\left(\bigcup_\alpha G_\alpha\right)$.

Therefore, $K\subset\bigcup_\alpha G_\alpha$.

Since G_α is a collection of open sets in X, $\{G_\alpha\}$ is a cover for K in X.

Since K is compact in X, there exists $\{G_{\alpha_i}\}$, for $i=1,2,3,.....,n$, such that

$K\subset\bigcup_{i=1}^{n}G_{\alpha_i}$.

Hence, $Y\cap K\subset Y\cap\left(\bigcup_{i=1}^{n}G_{\alpha_i}\right)$.

That is, $K \subset \bigcup_{i=1}^{n}(Y \cap G_{\alpha_i}) = \bigcup_{i=1}^{n} A_{\alpha_i}$.

Therefore, $\{A_{\alpha_i}\}$, $i = 1,2,.....,n$ is a finite sub-cover for K in Y.

Hence, K is compact in Y.

Converse part,

Given that, $K \subset Y \subset X$ and K is compact relative to Y.

We have to prove that, K is compact relative to X.

Let $\{G_\alpha\}$ be an open cover for K in X.

Hence, $K \subset \bigcup G_\alpha$.

That implies, $Y \cap K \subset Y \cap \left(\bigcup_\alpha G_\alpha\right)$. That is, $K \subset \cup(Y \cap G_\alpha)$.

Let $A_\alpha = Y \cap G_\alpha$.

Therefore, A_α is open in Y.

Hence, $K \subset \bigcup_\alpha A_\alpha$.

Therefore, $\{A_\alpha\}$ is an open cover for K in Y.

Since K is compact in Y, there exists a finite sub-cover

$\{A_{\alpha_i}\}$, for $i = 1,2,.....,n$, such that $K \subset \bigcup_{i=1}^{n} A_{\alpha_i}$.

Hence, $K \subset \bigcup_{i=1}^{n}(Y \cap G_{\alpha_i}) = Y \cap \left(\bigcup_{i=1}^{n} G_{\alpha_i}\right) \subset \bigcup_{i=1}^{n} G_{\alpha_i}$.

Therefore, every open cover for K in X has a finite sub-cover.

That is, K is compact in X.

Hence the theorem.

Theorem 1.6.34

Compact subsets of metric spaces are closed.

Proof

Let $K \subset X$.

Given that, K is compact.

We have to prove that, K is closed.

That is to prove that, K^c is open.

Let $x \in K^c$ and $y \in K$.

Hence, $x \neq y$. Thus, $d(x, y) > 0$.

Let $\delta = d(x, y) > 0$.

We have to prove that, $N_{\delta/2}(x) \cap N_{\delta/2}(y) = \phi$.

Suppose not,

Let $z \in N_{\delta/2}(x) \cap N_{\delta/2}(y)$.

That implies, $z \in N_{\delta/2}(x)$ and $z \in N_{\delta/2}(y)$.

Therefore, $d(x, z) < \dfrac{\delta}{2}$ and $d(y, z) < \dfrac{\delta}{2}$.

Hence, $d(x, y) \leq d(x, z) + d(y, z) < \delta$.

That is, $\delta < \delta$, which is a contradiction.

Hence, $N_{\delta/2}(x) \cap N_{\delta/2}(y) = \phi$.

Now, for each $y \in K$, there exists a neighborhood U_y of x and a neighborhood V_y of y, such that $U_y \cap V_y = \phi$.

Therefore, $\{V_y / y \in K\}$ is a collection of open subsets of X, such that $K \subset \bigcup_{y \in K} V_y$.

Since K is compact, there is a finite sub collection $\{V_{y_1}, V_{y_2}, \ldots\ldots, V_{y_n}\}$

such that $K \subset \bigcup_{i=1}^{n} V_{y_i}$. .... (1.6.34.1)

For every $i = 1, 2, \ldots.., n$, $U_{y_i} \cap V_{y_i} = \phi$.

Let $U = \bigcap_{i=1}^{n} U_{y_i}$.

Since $U_{y_1}, U_{y_2}, \ldots\ldots, U_{y_n}$ are neighborhood of x, U is a neighborhood of x, for all $i = 1, 2, \ldots\ldots, n$.

Now, since $U \subset U_{y_i}$, $U_{y_i} \cap V_{y_i} = \phi \Rightarrow U \cap V_{y_i} = \phi$.

That is, $U \cap \left(\bigcup_{i=1}^{n} V_{y_i} \right) = \phi$.

Therefore, by equation (1.6.34.1), we have, $U \cap K = \phi$.

That implies, $U \subset K^c$.

Hence, K^c is open.

That implies, K is closed.

Hence the theorem.

Theorem 1.6.35

Closed subsets of compact sets are compact.

Proof

Given that, $F \subset K \subset X$, where K is compact and F is closed.

We have to prove that, F is compact.

Let $\{A_\alpha\}$ be a collection of open sets in X, such that $F \subset \bigcup_{\alpha} A_\alpha$.

.... (1.6.35.1)

Since F is closed, F^c is open.

Therefore, $F^c \cup \left(\bigcup_{\alpha} A_\alpha \right)$ is a cover for K.

Hence, $K \subset F^c \cup \left(\bigcup_{\alpha} A_\alpha \right)$.

Since K is compact, the above open cover has a finite sub-cover, $\{F^c, A_{\alpha_1}, A_{\alpha_2}, ..., A_{\alpha_n}\}$.

Therefore, $K \subset F^c \cup \left(\bigcup_{i=1}^{n} A_{\alpha_i} \right)$.

By the given data, $F \subset K \subset F^c \cup \left(\bigcup_{i=1}^{n} A_{\alpha_i} \right)$ and $F \cap F^c = \phi$.

We know that, if $A \subset B \cup C$ and $A \cap B = \phi$, then $A \subset C$.

By using this result, we get, $F \subset \bigcup_{i=1}^{n} A_{\alpha_i}$.

Therefore, F is compact.

Hence the theorem.

Note 1.6.36

If F is closed and K is compact, then $F \cap K$ is compact.

Proof

Since compact subset of a metric space is closed, K is closed.

Since K and F are closed, $K \cap F$ is closed.

Now, $K \cap F \subset K$.

Hence, $K \cap F$ is a closed subset of a compact set.

Therefore, by the previous theorem 1.6.35, $K \cap F$ is compact.

Theorem 1.6.37

If $\{K_\alpha\}$ is a collection of compact subsets of a metric space X, such that the intersection of every finite sub-collection of $\{K_\alpha\}$ is non – empty, then $\bigcap_\alpha K_\alpha$ is non – empty.

Proof

Given that $\{K_\alpha\}$ is a collection of compact subsets of X, satisfying the finite intersection property.

We have to prove that, $\bigcap_\alpha K_\alpha$ is non – empty.

Fix a member K_1 of $\{K_\alpha\}$.

In order to prove the result, it is enough, if we prove that some element of K_1 belong to every K_α.

Suppose not,

For each $x \in K_1$, there exists some α such that $x \notin K_\alpha$.

Therefore, $x \in K_\alpha^c \Rightarrow K_1 \subset K_\alpha^c \Rightarrow K_1 \subset \bigcup_\alpha K_\alpha^c$. .... (1.6.37.1)

Since K_α is a compact subset of X, K_α is closed.

Hence, K_α^c is open.

Now, $\{K_\alpha^c\}$ is a collection of open subsets of X such that $K_1 \subset \bigcup_\alpha K_\alpha^c$.

Since K_1 is compact, there exists a finite sub-collection $\left\{K_{\alpha_1}^c, K_{\alpha_2}^c, ..., K_{\alpha_n}^c\right\}$ such that $K_1 \subset \bigcup_{i=1}^n K_{\alpha_i}^c$.

Therefore, $K_1 \cap \left(\bigcup_{i=1}^{n} K_{\alpha_i}^c \right)^c = \phi$.

Hence, by DeMorgan's law, we get $K_1 \cap \left(\bigcap_{i=1}^{n} K_{\alpha_i} \right) = \phi$, which is a contradiction to finite intersection property.

Therefore, $\bigcap_{\alpha} K_\alpha \neq \phi$.

Hence the theorem.

Note 1.6.38

Let $\{K_n\}$ be a sequence of non – empty compact sets, such that $K_n \supset K_{n+1}$, for $n = 1,2,3,.....$ Then, $\bigcap_{n=1}^{\infty} K_n \neq \phi$.

Proof

Given that, $\{K_n\}$ is a sequence of non – empty compact sets, such that $K_n \supset K_{n+1}$, for $n = 1,2,3,......$

We have to prove that, $\{K_n\}$ satisfies the finite intersection property.

Let $\left\{K_{n_1}, K_{n_2}, ..., K_{n_p}\right\}$ be a finite collection of K_n.

Since $K_n \supset K_{n+1}$, $\bigcap_{i=1}^{p} K_{n_i} = K_t$, where $t = \max\{n_1, n_2, ..., n_p\}$.

Since $K_n \neq \varphi$, for $n = 1, 2, 3, ...,$ $\bigcap_{i=1}^{p} K_{n_i} = K_t \neq \varphi$.

Therefore, by the previous theorem 1.6.37, $\bigcap_{n=1}^{\infty} K_n \neq \phi$.

Hence proved.

Theorem 1.6.39

If E is an infinite subset of a compact set K, then K has a limit point in K.

Proof

Given that, E is an infinite subset of a compact set K.

We have to prove that, E has a limit point in K.

That is, we have to prove that, some point of *K* is a limit point of *K*.

Suppose not,

That is, no point of *K* is a limit point of *K*.

Therefore, for each $x \in K$, there exists $r_x > 0$ such that $N_{r_x}(x) \cap (E - \{x\}) = \phi$.

Now, $N_{r_x}(x) \cap E \subseteq N_{r_x}(x) \cap \left[(E - \{x\}) \cup \{x\} \right]$

$$= \left[N_{r_x}(x) \cap (E - \{x\}) \right] \cup \left(N_{r_x}(x) \cap \{x\} \right).$$

That implies, $N_{r_x}(x) \cap E \subseteq \phi \cup \{x\} = \{x\}$. .... (1.6.39.1)

Now, $\{N_{r_x}(x) / x \in K\}$ is a collection of open subsets of *X*, such that $K \subset \bigcup_{x \in K} N_{r_x}(x)$.

Since *K* is compact, there exists a finite sub collection $\{N_{r_{x_1}}(x_1), N_{r_{x_2}}(x_2), \ldots\ldots, N_{r_{x_n}}(x_n)\}$ such that $K \subset \bigcup_{i=1}^{n} N_{r_{x_i}}(x_i)$. ...... (1.6.39.2)

Since $E \subset K_n$ and by equation (1.6.39.2), we have $E \subset K_n \bigcup_{i=1}^{n} N_{r_{x_i}}(x_i)$.

Now, $E = \left(\bigcup_{i=1}^{n} N_{r_{x_i}}(x_i) \right) \cap E = \bigcup_{i=1}^{n} \left(N_{r_{x_i}}(x_i) \cap E \right) \subseteq \{x_1, x_2, \ldots\ldots, x_n\}$.

Therefore, *E* has a finite number of elements, which is a contradiction to, *E* is an infinite set.

Therefore, *E* has a limit point in *K*.

Hence the theorem.

Theorem 1.6.40

If $\{I_n\}$ is a sequence of intervals in $\mathbb{R}^1$, such that $I_n \supset I_{n+1}$, for $n = 1, 2, 3, \ldots\ldots$, then $\bigcap_{n=1}^{\infty} I_n$ is non – empty.

Proof

Since I_n is an interval, $I_n = [a_n, b_n]$, for some $a_n, b_n \in \mathbb{R}$

Let $I_{n+1} = [a_{n+1}, b_{n+1}]$.

Since $I_n \supset I_{n+1}$, for $n = 1, 2, 3, ...$, $a_n \le a_{n+1} \le b_{n+1} \le b_n$.

Therefore, a_n is increasing and b_n is decreasing monotonically.

Let $E = \{a_n / n = 1, 2, ...\}$.

Now, $a_n \le b_n \le b_1$.

Hence, E has b_1 as its upper bound.

Since $\mathbb{R}$ has Least Upper Bound (l.u.b.) property, every bounded set of real numbers has a l.u.b. in $\mathbb{R}$.

Therefore, x = l.u.b. of E exists in $\mathbb{R}$.

Let $n, m \in J$.

Hence, $a_n \le a_{n+m} \le b_{n+m} \le b_m$, for each m.

Now, $x = \sup.E$, then $x \le b_n$.

Also, $a_m \le x$, for each m.

Therefore, $a_m \le x \le b_n$. Thus, $x \in I_m$, for all $m \in J$.

This implies, $x \in \bigcap_{n=1}^{\infty} I_n$.

That is, $\bigcap_{n=1}^{\infty} I_n \ne \varphi$.

Hence the theorem.

Theorem 1.6.41

Let k be a positive integer. If $\{I_n\}$ is a sequence of k – cells such that $I_n \supset I_{n+1}$, for $n = 1,2,3,.....$, then $\bigcap_{n=1}^{\infty} I_n$ is non – empty.

Proof

Let $I_n = I_{n_1} \times I_{n_2} \times ...... \times I_{n_k}$.

Let us take, $I_{n_j} = [a_{n_j}, b_{n_j}]$, for all $j = 1, 2, 3, ...,k$ and $n \in J$.

Since $I_n \supset I_{n+1}$, $I_{n_j} \supset I_{(n+1)j}$, for $n \in J$.

Therefore, for each j = 1, 2, ….., k, $\{I_{n_j}\}$ is a sequence of intervals in $\mathbb{R}$, such that $I_{n_j} \supset I_{(n+1)j}$, for $n \in J$.

Hence, by the previous theorem 1.6.40, $\bigcap_{n=1}^{\infty} I_{n_j} \neq \phi$.

Let $x_j \in \bigcap_{n=1}^{\infty} I_{n_j}$, for $j = 1, 2, \ldots\ldots, k$.

Let $x = (x_1, x_2, \ldots\ldots, x_k)$. Then, $x \in \mathbb{R}^k$.

Let $n \in J$, then, for all $j = 1, 2, \ldots . k$, $x_j \in I_{n_j}$.

Therefore, $x \in I_{n_1} \times I_{n_2} \times \ldots\ldots \times I_{n_k}$.

Thus, $x \in I_{n_\infty}$. That is, $x \in \bigcap_{n=1}^{\infty} I_n$.

Therefore, $\bigcap_{n=1}^{\infty} I_n \neq \varphi$.

Hence the theorem.

Theorem 1.6.42

Every k-cell is compact.

Proof

Let $I = [a_1, b_1] \times [a_2, b_2] \times \ldots\ldots \times [a_k, b_k]$ be any k – cell.

Let $\delta = \left(\sum_{j=1}^{k} (b_j - a_j)^2 \right)^{1/2}$.

Then, $|x - y| < \delta$, for all $x, y \in I$.

We have to prove that, *I* is compact.

Suppose not,

Hence, there is an open cover $\{G_\alpha\}$ of *I* such that no finite sub-collection of $\{G_\alpha\}$ covers *I*.

Now, for $j = 1, 2, \ldots\ldots, k$, put $c_j = \dfrac{b_j + a_j}{2}$.

The subintervals $[a_j, c_j]$ and $[c_j, b_j]$ will divide *I* into 2^k k – cells Q_i whose union is *I*.

Since I is not covered by finitely many $\{G_\alpha\}'s$, atleast one of these Q_i call it I_1 is not covered by finitely many $\{G_\alpha\}'s$.

Let $I_1 = [a_{11}, b_{11}] \times [a_{12}, b_{12}] \times .....\times [a_{1k}, b_{1k}]$.

Now, for all $x, y \in I_1$,

$$|x-y| \le \left[\sum_{j=1}^{k}(b_{1j}-a_{1j})^2\right]^{1/2} = \left[\sum_{j=1}^{k}\left(\frac{b_j-a_j}{2}\right)^2\right]^{1/2} = 2^{-1}\delta.$$

Put $d_j = \dfrac{b_{1j}+a_{1j}}{2}$. Therefore, the sub-intervals $[a_{1j}, d_j]$ and $[d_j, b_{1j}]$ will divide I_1 into 2^k k – cells P_i whose union is I_1.

Since I_1 is not covered by finitely many $\{G_\alpha\}'s$, atleast one of this P_i, call it I_2 is not covered by finitely many $\{G_\alpha\}'s$.

As before, for all $x, y \in I_2$, $|x-y| \le 2^{-2}\delta$.

Next, we divide I_2 and proceed.

Proceeding like this, we get a sequence $\{I_n\}$ of k – cells satisfying the following;

1. $I \supset I_1 \supset I_2 \supset ......$
2. For each $n \in J$, I_n is not covered by a finite collection of $\{G_\alpha\}'s$.
3. For all $x, y \in I_n$, $|x-y| \le 2^{-n}\delta$.

Hence, by the first condition, I_n is a sequence of k – cells such that $I_n \supset I_{n+1}$, for all $n = 1, 2, ...$

Therefore, by theorem 1.6.41, $\bigcap_{n=1}^{\infty} I_n \neq \phi$.

Let $x \in \bigcap_{n=1}^{\infty} I_n$. Then, $x \in I_1 \subset I \subset \bigcup_{\alpha} G_\alpha$.

Therefore, $x \in G_{\alpha_0}$, for some α_0.

Since G_{α_0} is open, there exists $r > 0$ such that $N_r(x) \subset G_{\alpha_0}$. ...... (1.6.42.1)

Choose a positive integer m such that $2^{-m}\delta < r$.

We have to prove that, $I_m \subset G_{\alpha_0}$.

Let $y \in I_m$.

Since $x \in \bigcap_{n=1}^{\infty} I_n$, $x \in I_m$.

Therefore, $x, y \in I_m$.

Hence, by condition (3), $|x - y| \leq 2^{-m}\delta < r$.

Thus, by equation (1.6.42.1), we have, $y \in N_r(x) \subset G_{\alpha_0}$.

That is, $y \in G_{\alpha_0}$.

Hence, we get, $I_m \subset G_{\alpha_0}$, which is a contradiction to condition (2).

Therefore, k – cell I is compact.

Hence the theorem.

Theorem 1.6.43

The following theorem is known as ***Heine Borel Theorem.***

If a set E in R has any one of the following three properties, then it has the other two.

i) E is closed and bounded

ii) E is compact

iii) Every infinite subset of E has a limit point in E.

Proof

First we will prove that, $(i) \Rightarrow (ii)$

Given that, E is closed and bounded.

We have to prove that, E is compact.

Since E is bounded, there exists $l > 0$ and $y \in \mathbb{R}^k$ such that $|x - y| \leq l$, for all $x \in E$.

That is, $|x| - |y| \leq |x - y| \leq l$, for all $x \in E$.

That implies, $|x| \leq |y| + l = M$ (say).

Then, $M > 0$. Hence, $|x| \leq M$, for all $x \in E$.

That is, $x \in [-M, M]$, for all $x \in E$.

Let $x = (x_1, x_2, ..., x_k) \in E$.

Now, $|x_j| \le |x| \le M \Rightarrow |x_j| \le M$.

Hence, $x_j \in [M, -M]$, for all $j = 1, 2, ..., k$.

Therefore, $x = (x_1, x_2, ..., x_k) \in [-M, M] \times [-M, M] \times ..... \times [-M, M] = I$ (say).

That is, $x \in I \Rightarrow E \subset I$.
Since every k-cell is compact, I is compact. Also, E is closed.
We know that, closed subset of a compact set is compact.
Hence E is compact.

ii) We have to prove that, $(ii) \Rightarrow (iii)$.
The proof is already given in a previous theorem.

iii) We have to prove that, $(iii) \Rightarrow (i)$.
Given that, every infinite subset of E has a limit point in E.
We have to prove that, E is closed and bounded.
First, let us prove E is bounded.
Suppose not,
Then, for each $n \in J$, there exists $x_n \in E$ such that $|x_n| > n$.

Let $S = \{x_n / n = 1, 2, ...\}$. Then, $S \subset E$ and S is infinite.
Hence, we have to prove that, S has no limit point in $\mathbb{R}^k$.
Suppose there exists an element $y \in \mathbb{R}^k$ such that y is a limit point of S.
Choose a positive integer m, such that $|y| < m - 1$. .... (1.6.43.1)

Since y is a limit point of S, $N_1(y)$ consists of infinite number of elements of S.
Hence, we can find a positive integer $p > m$ such that $x_p \in N_1(y)$.

Therefore, $|x_p - y| < 1$. .... (1.6.43.2)

Hence, we get $|x_p| - |y| \le |x_p - y| < 1$.

Therefore, $|y| > |x_p| - 1 > p - 1 > m - 1,$

Which is a contradiction to equation (1.6.43.1).

Thus, *S* has no limit point in $\mathbb{R}^k$. Since, $E \subset \mathbb{R}^k$, *S* has no limit point in *E*, which is a contradiction to the hypothesis.

Thus, *E* is bounded.

Next, we have to prove that, *E* is closed.

Suppose not,

Then there exists an element $x \in E^c$ such that *x* is not a limit point of *E*.

$N_r(x) \cap (E - \{x\}) \neq \varphi,$ for all $r > 0$.

Since $x \notin E$, $N_r(x) \cap E \neq \phi$.

Take $r = \dfrac{1}{n}$ and hence we have, $N_{(1/n)}(x) \cap E \neq \phi,$ for all $n \in J$.

Let $x_n \in N_{(1/n)}(x) \cap E$.

Therefore, $x_n \in E,\ |x_n - x| < \dfrac{1}{n}$ and $x_n \neq x$.

Let $S = \{x_n / n = 1, 2, ...\}$.

Hence $S \subset E$ and *S* is an infinite set.

We have to prove that, *x* is a limit point of *S*.

Let $r > 0$. Choose $m \in J$ such that $\dfrac{1}{m} < r$.

Now, for all $n \geq m$, $|x_n - x| < \dfrac{1}{n} \leq \dfrac{1}{m} < r$.

Therefore, $x_n \in N_r(x),$ for all $n \geq m$.

That is, $N_r(x)$ contains infinite number of elements of *S*.

Hence, *x* is a limit point of *S*.

Let $y \neq x \in \mathbb{R}^k$.

Now, we have to prove that, y is not a limit point of *S*.

Since $y \neq x$, choose $\delta = \dfrac{|y - x|}{2} > 0$ and a positive integer *m* such that $\dfrac{1}{m} < \delta$.

Now, for all $n \geq m$,

$$|x-y| = |(x_n - x) - (y - x)| \geq |(y-x)| - |(x_n - x)| > 2\delta - \frac{1}{n}.$$

That is, $|x-y| \geq 2\delta - \frac{1}{m} > 2\delta - \delta = \delta$.

Therefore, $x_n \notin N_\delta(y)$, for all $n \geq m$.

Hence, $N_\delta(y)$ does not contain infinite number of elements of *S*.

That implies, *y* is not a limit point of *S*. That is, *S* has no limit point in $\mathbb{R}^k$.

Since $E \subset \mathbb{R}^k$, *S* has no limit point in *E*, which is a contradiction to the hypothesis.

Therefore, *S* is closed.

Hence, *S* is both closed and bounded.

Hence the theorem.

Theorem 1.6.44

Weierstrass Theorem

Every bounded infinite subset of $\mathbb{R}^k$ has a limit point in $\mathbb{R}^k$.

Proof

Let *E* be a bounded infinite subset of $\mathbb{R}^k$.

Since *E* is bounded, by part $(i) \Rightarrow (ii)$ of Heine Borel theorem, $E \subset I$, where *I* is the k – cell.

Since k – cell is compact, *I* is compact. Since, every infinite subset of a compact set has a limit point, *E* has a limit point in $\mathbb{R}^k$.

Hence the theorem.

Definition 1.6.45

Two sets *A* and *B* of a metric space X are said to be *separated*, if $\overline{A} \cap B = \phi$ and $A \cap \overline{B} = \phi$.

Note 1.6.46

Separated sets are disjoint but disjoint sets need not be separated.

For example,

Let $A = (0,1)$ and $B = [1,2)$.

Now, $\bar{A} \cap B = [0,1] \cup [1,2) = \{1\} \neq \varphi$.

Also, $A \cap B = \{\varphi\}$.

Hence, A and B are disjoint but are not separated.

Definition 1.6.47

A subset E of X is said to be *connected* if E is not a union of two non-empty separated sets.

Theorem 1.6.48

A subset E of $\mathbb{R}$ is connected if and only if it has the following property; if $x, y \in E$ and $x < z < y$, then $z \in E$.

Proof

Direct part,

Given that $E \subset \mathbb{R}$ is connected.

We have to prove that, $x, y \in E$ and $x < z < y$, then $z \in E$.

Suppose, we assume that, there exists $x, y \in E$ and $x < z < y$ and $z \notin E$.

Let $A = (-\infty, z) \cap E$ and $B = (z, \infty) \cap E$.

Since $x < z$, $x \in (-\infty, z)$. Also $x \in E$. Hence, $x \in A$.

Since $z < y$, $y \in (z, \infty)$. Also, $y \in E$. Hence, $y \in B$.

Therefore, A and B are non-empty sets.

Now, $\bar{A} \cap B = \left(\overline{(-\infty, z) \cap E}\right) \cap \left((z, \infty) \cap E\right) \subset \overline{(-\infty, z)} \cap \bar{E} \cap \left((z, \infty) \cap E\right)$.

$$= (-\infty, z) \cap (z, \infty) \cap \bar{E} \cap E = \varphi.$$

Similarly, $A \cap \bar{B} = \phi$.

Therefore, A and B are separated sets.

Now, since, $E \subset \mathbb{R}$, $E = E \cap \mathbb{R} = E \cap \{(-\infty, z) \cup \{z\} \cup (z, \infty)\}$.

$$= \{E \cap (-\infty, z)\} \cup \{E \cap \{z\}\} \cup \{E \cap (z, \infty)\} = A \cup \phi \cup B.$$

Hence, $E = A \cup B$.

Therefore, E can be expressed as the union of two non-empty separated sets.

Hence, E is not connected, which is a contradiction to the hypothesis.

Converse part,

Given that, $x, y \in E,\ x < z < y \Rightarrow z \in E$.

We have to prove that, E is connected.

Suppose not,

Therefore, $E = A \cup B$, where A and B are non-empty separated sets.

Since A and B are non-empty, let $x \in A$ and $y \in B$.

Since, $A \cap B = \varphi,\ x \neq y$.

Let $x < y$.

Consider, $[x, y] \cap A$.

Since $x \in [x, y] \cap A$, $[x, y] \cap A$ is non-empty.

Also, $[x, y] \cap A \subset [x, y]$. Since $[x, y]$ is bounded, $[x, y] \cap A$ is non-empty bounded subset of $\mathbb{R}$.

Since $\mathbb{R}$ has least upper bound property,

$$z = \sup .[x, y] \cap A \text{ exists in } \mathbb{R}.$$

Hence, $z \in \overline{[x, y] \cap A} \in [x, y] \cap \overline{A}$.

Therefore, $z \in [x, y]$ and $z \in \overline{A}$.

That implies, $x \leq z \leq y$ and $z \in \overline{A}$. .... (1.6.48.1)

Since $\overline{A} \cap B = \varphi,\ z \notin B$.

Hence, $z \neq y$.

Therefore, from equation (1.6.48.1), we have, $x \leq z < y$. .... (1.6.48.2)

If $z \notin A$, then $z \neq x$.

Hence, by equation (1.6.48.2), we have $x < z < y$.

Now, $z \notin A, z \notin B$, hence $z \notin A \cup B$.

Therefore, $z \notin E$, which is a contradiction to the given data.

Hence, E is connected.

If $z \in A$ and since $A \cap \overline{B} = \varphi,\ z \notin \overline{B}$.

Therefore, there exists $\delta > 0$ such that $(z - \delta, z + \delta) \cap B = \varphi$. ... (1.6.48.3)

Choose $z_1 > z$ such that $z_1 < \min(z + \delta, y)$.

Now, $x < z - \delta < z < z_1 < z + \delta < y$.

Hence, $z_1 \in (z - \delta, z + \delta)$.

From equation (1.6.48.3), we have $z_1 \notin B$.

Now, $z_1 \in [x, y]$ and $z_1 > A$, then $z_1 \notin A$.

Since $z_1 \notin A$ and $z_1 \notin B$, we have $z_1 \notin A \cup B$. Therefore, $z_1 \notin E$.

That is, $x, y \in E$, $x < z_1 < y \Rightarrow z_1 \notin E$, which is a contradiction to the given data.

Therefore, E is connected.

Hence the theorem.

Unit	
2	**SEQUENCES AND SERIES**

Unit 2 SEQUENCES AND SERIES

2.1 Introduction

The word sequence refers to the succession of certain objects. In mathematics, these objects are considered to be real numbers. A sequence of real numbers seems to be an infinite succession as

$$x_1, x_2, x_3, x_4, ...$$

where x_n's are real numbers. When we come across a sequence, it is obvious to ask where the sequence leads. This idea is made clear by defining the term, convergence of sequences.

2.2 Sequences

Definition 2.2.1

A sequence is an ordered, countably infinite list of real numbers. Hence, the formal definition of a sequence is given as follows;

A *sequence* is a function whose domain is the set $\mathbb{N}$ of natural numbers. Hence, a sequence is a function whose domain is a set of the form $\{n \in \mathbb{Z} : n \geq m\}$, that is, $f : \mathbb{N} \to \mathbb{R}$ and it is denoted by a_n instead of $f(n)$. The value of a sequence $\{a_n\}$ at $n \in \mathbb{N}$ is given by a_n and this is called the nth term of that sequence.

Definition 2.2.2

If $\{a_n\}_{n\geq 1}$ is a sequence of real numbers and if $n_1 < n_2 < ..... < n_k < .....$ is an increasing sequence of positive integers, then the sequence $\{a_{n_k}\}_{k\geq 1}$ is called a *subsequence* of $\{a_n\}_{n\geq 1}$.

Definition 2.2.3

A sequence of real numbers $\{a_n\}_{n\geq 1}$ said to be *non-decreasing* or *increasing*, if $a_n \leq a_{n+1}$ or $a_n < a_{n+1}$, for all $n \geq 1$ respectively. The sequence $\{a_n\}_{n\geq 1}$ is said to be *non-increasing* or *decreasing*, if $a_n \geq a_{n+1}$ or $a_n > a_{n+1}$ respectively.

For $n \in \mathbb{N}$, let i) $a_n = 1$, ii) $a_n = (-1)^n$. The sequences defined by these a_n's are bounded.

If $a_n = n$, then the sequence defined by a_n is bounded above, while the sequence defined by $a_n = (-1)^n n$ is neither bounded above nor bounded below.

Examples 2.2.4

For $n \in \mathbb{N}$,

1. Let $a_n = \dfrac{1}{n}$, then $\{a_n\} = \left(1, \dfrac{1}{2}, \dfrac{1}{3}, \ldots\ldots, \dfrac{1}{n}, \ldots.\right)$.
2. Let $a_n = (-1)^n$, then $\{a_n\} = (-1, 1, -1, \ldots\ldots, -1, 1, \ldots\ldots)$.
3. Let $a_n = \dfrac{(-1)^{n+1}}{n}$, then $\{a_n\} = \left(1, -\dfrac{1}{2}, \dfrac{1}{3}, -\dfrac{1}{4}, \ldots\ldots\right)$.
4. Let $a_n = (n+1)^2$, then $\{a_n\} = (4, 9, 16, \ldots.)$.
5. Let $a_n = \dfrac{n+1}{n}$, then $\{a_n\} = \left(2, \dfrac{3}{2}, \dfrac{4}{3}, \dfrac{5}{4}, \ldots\ldots\right)$.

2.3 Convergence of sequences

Definition 2.3.1

Let $a \in \mathbb{R}$ and $\{a_n\}$ be a sequence in $\mathbb{R}$. Then, we say that $\{a_n\}$ *converges to a* or *a is the limit of* $\{a_n\}$ if for every $\varepsilon > 0$, there is an $N \in \mathbb{N}$ such that $|a_n - a| < \varepsilon$ for every $n \geq \mathbb{N}$ and it is symbolically represented as $\{a_n\} \to a$ or $\lim\limits_{n\to\infty} a_n = a$. If $\{a_n\}$ does not converge, it is said to diverge.

Theorem 2.3.2

Let $\{a_n\}$ be a sequence in a metric space X. Then

i) $\{a_n\}$ converges to a in X if and only if every neighbourhood of a contains all but finitely many terms of $\{a_n\}$.

ii) If $a, a' \in X$, and $\{a_n\}$ converges to a and also to a', then $a = a'$.

iii) If $\{a_n\}$ converges, then it is bounded.

iv) If $E \subset X$ and a is a limit point of E, then there is a sequence $\{a_n\}$ in E such that $\{a_n\} \to a$.

Proof

i) Given $\{a_n\} \to a$.

Let $N_r(a)$ be a neighbourhood of a.

Hence $r > 0$.

Since $\{a_n\} \to a$, there exists a positive integer N, such that $d(a_n, a) < r$, for all $n \geq N$.

That is, $a_n \in N_r(a)$, for all $n \geq N$.

Therefore, $N_r(a)$ contains all but finitely many terms of the sequence $\{a_n\}$.

Hence, every neighbourhood of a contains all but finitely many terms of the sequence $\{a_n\}$.

Conversely,

Given that, every neighbourhood of a contains all but finitely many terms of the sequence $\{a_n\}$, and we have to prove that $\{a_n\} \to a$.

Let $\varepsilon > 0$ be given. Then, $N_\varepsilon(a)$ is a neighbourhood of a_n.

From the given data, we have $a_n \in N_\varepsilon(a)$, for all $n \geq N$.

Hence, $d(a_n, a) < \varepsilon$, for all $n \geq N$.

Therefore, $\{a_n\} \to a$.

Hence part (i) is proved.

ii) To prove that, in a metric space, every convergent sequence has a unique limit.

Given that $\{a_n\} \to a$ and $\{a_n\} \to a'$.

Since $\{a_n\} \to a$, for given $\varepsilon > 0$, there exists a positive integer N_1 such that $d(a_n, a) < \varepsilon/2$, for all $n \geq N_1$.

Also, since $\{a_n\} \to a'$, for the same $\varepsilon > 0$, there exists a positive integer N_2 such that $d(a_n, a') < \varepsilon/2$, for all $n \geq N_2$.

Let $N = \max. \{N_1, N_2\}$, for all $n \geq N$.

Now, $d(a, a') \leq d(a, a_n) + d(a_n, a') < \varepsilon$.

Since ε is arbitrary, we have $a = a'$.

Hence part (ii) is proved.

iii) Given that, $\{a_n\} \to a$ and to prove that $\{a_n\}$ is bounded.

Since $\{a_n\} \to a$, there exists a positive integer N such that $d(a_n, a) < 1$, for all $n \geq N$.

Let $k = \max. \{1, d(a_1, a), d(a_2, a), \ldots\ldots, d(a_{N-1}, a)\}$.

That implies, $d(a_n, a) \leq k$, for all $n \in J$.

Therefore, $\{a_n\}$ is bounded.

Hence part (iii) is proved.

iv) Given that $E \subset X$, and a is a limit point of E.

Therefore, $N_r(a) \cap E \neq \phi$ for all $r > 0$.

Let $r = 1/n$.

Hence, we have $N_{1/n}(a) \cap E \neq \phi$ for all $n \in J$.

Let $a_n \in N_{1/n}(a) \cap E$, for all $n \in J$.

Hence, $\{a_n\}$ is a sequence in E and $d(a_n, a) < 1/n$.

Now, we have to prove that $\{a_n\} \to a$.

Let $\varepsilon > 0$ be given. We choose $N \in J$ such that $\frac{1}{\text{N}} < \varepsilon$.

Therefore, for all $n \geq N$, $d(a_n, a) < \frac{1}{n} \leq \frac{1}{N} < \varepsilon$.

That implies, $\{a_n\} \to a$.

Hence, part (iv) is proved.

Hence the theorem.

Theorem 2.3.3

If $\{a_n\},\{b_n\}$ are any two complex sequences, and $\lim_{n\to\infty} a_n = a$ and $\lim_{n\to\infty} b_n = b$, then

i) $\lim_{n\to\infty} a_n + b_n = a + b$.

ii) $\lim_{n\to\infty} ca_n = ca$ and $\lim_{n\to\infty} c + a_n = c + a$, where c is a constant.

iii) $\lim_{n\to\infty} a_n b_n = ab$.

iv) $\lim_{n\to\infty} \frac{1}{a_n} = \frac{1}{a}$, if $a_n \neq 0$, for $n = 1, 2, 3, \ldots\ldots$ and $a \neq 0$.

This theorem is known as *Limit Theorem for Sequences.*

Proof

i) Given that $\{a_n\} \to a$ and $\{b_n\} \to b$ and hence for given $\varepsilon > 0$, there exists positive integers N_1 and N_2, such that

$$|a_n - a| < \varepsilon/2, \text{ for all } n \geq N_1 \text{ and}$$

$$|b_n - b| < \varepsilon/2, \text{ for all } n \geq N_2 .$$

Let N = max. (N_1, N_2).

For all $n \geq N$,

$$|(a_n + b_n) - (a + b)| = |(a_n - a) + (b_n - b)| \leq |a_n - a| + |b_n - b| < \varepsilon .$$

Therefore, $(a_n + b_n) \to (a + b)$.

Hence, $\lim_{n\to\infty}(a_n + b_n) = a + b$.

Hence part (i) is proved.

ii) a) Suppose $c = 0$, then $ca_n = 0$, for all $n \in J$ and $ca = 0$.

$\{(ca_n)\} = \{0,0,0,\ldots..\} \to 0 = ca$.

Suppose $c \neq 0$, let $\varepsilon > 0$ be given.

Then, $\frac{\varepsilon}{|c|} > 0$.

Since $\{a_n\} \to a$, there exists a positive integer N such that $|a_n - a| < \frac{\varepsilon}{|c|}$, for all $n \ge N$.

Now, $|ca_n - ca| = |c(a_n - a)| = |c||a_n - a| < \varepsilon$, for all $n \ge N$.

Therefore, $\{ca_n\} \to ca$.

Hence, $\lim_{n\to\infty} ca_n = ca$.

b) Since $\{a_n\} \to a$, there exists a positive integer N such that $|a_n - a| < \varepsilon$, for all $n \ge N$.

Now, $|(c + a_n) - (c + a)| = |a_n - a| < \varepsilon$, for all $n \ge N$.

Therefore, $\{c + a_n\} \to c + a$.

Hence, $\lim_{n\to\infty} c + a_n = c + a$.

Hence part (ii) is proved.

iii) We know that, $a_n b_n - ab = (a_n - a)(b_n - b) + a(b_n - b) + b(a_n - a)$

.... (2.3.3.1)

Since $b_n \to b$, by (ii) ; (b), we have $b_n - b \to 0$.

Again, by (ii); (*a*), we have $a(b_n - b) \to 0$.

Similarly, $a_n \to a \Rightarrow a_n - a \to 0$ and hence $b(a_n - a) \to 0$.

Next, we have to prove that, $(a_n - a)(b_n - b) \to 0$.

Since $a_n \to a$ and $b_n \to b$, for given $\varepsilon > 0$, there exists positive integers N_1 and N_2 such that $|a_n - a| < \sqrt{\varepsilon}$, for all $n \ge N_1$ and $|b_n - b| < \sqrt{\varepsilon}$, for all $n \ge N_2$.

Let $N = \max.\{N_1, N_2\}$.

Therefore, $|(a_n - a)(b_n - b) - 0| = |a_n - a||b_n - b| < \varepsilon$.

Hence, $(a_n - a)(b_n - b) \to 0$, for all $n \ge N$.

Thus, (2.3.3.1) becomes, $a_n b_n - ab \to 0 \Rightarrow a_n b_n \to ab$.

Hence, $\lim_{n\to\infty} a_n b_n = ab$.

Hence part (iii) is proved.

iv) Now, $\left|\frac{1}{a_n}-\frac{1}{a}\right|=\left|\frac{a-a_n}{a_n a}\right|=\frac{|a-a_n|}{|a_n||a|}$. .... (2.3.3.2)

Since $a \neq 0$, $\frac{|a|}{2}>0$.

Since $\{a_n\} \to a$, there exists a positive integer N_1 such that $|a_n - a|<\frac{|a|}{2}$, for all $n \geq N_1$.

That is, $|a|-|a_n| \leq |a_n - a| < \frac{|a|}{2} \Rightarrow |a_n| > \frac{|a|}{2}$, for all $n \geq N_1$.

That implies, $\frac{1}{|a_n|}<\frac{2}{|a|}$, for all $n \geq N_1$.

Again, since $\{a_n\} \to a$, there exists a positive integer $N_2 \geq N_1$ such that $|a_n - a| < \frac{\varepsilon}{2}|a|^2$, for all $n \geq N_2$.

Therefore, (2.3.3.2) becomes, $\left|\frac{1}{a_n}-\frac{1}{a}\right|<\varepsilon$, for all $n \geq N_2$.

Hence, $\left\{\frac{1}{a_n}\right\} \to \frac{1}{a}$.

Therefore, $\lim_{n\to\infty}\frac{1}{a_n}=\frac{1}{a}$.

Hence part (iv) is proved
Hence the theorem.

Note 2.3.4

If $\{a_n\} \to a$ and $\{b_n\} \to b$, for $b_n \neq 0$, $b \neq 0$ and $n = 1,2,3,.....$, then $\left\{\frac{a_n}{b_n}\right\} \to \frac{a}{b}$.

Theorem 2.3.5

i) Suppose $\{x_n\} \in R^k$, $n = 1,2,3,......$, where $x_n = (\alpha_{1n}, \alpha_{2n}, ..., \alpha_{kn})$. Then, $x_n \to x = (\alpha_1, \alpha_2, ..., \alpha_k)$ if and only if $\alpha_{jn} \to \alpha_j$, for all $j = 1,2,....,k$.

ii) Suppose $\{x_n\},\{y_n\}$ are sequences in R^k and $\{\alpha_n\}$ is a sequence of real numbers and $x_n \to x,\ y_n \to y,\ \alpha_n \to \alpha$. Then, $x_n + y_n \to x + y$, $x_n y_n \to xy,\quad \alpha_n x_n \to \alpha x$

Proof

i) Given that $\{x_n\} \to x$.

We have to prove that, $\alpha_{jn} \to \alpha_j$, for all $j = 1,2,3,........,k$.

Since $\{x_n\} \to x$, for given $\varepsilon > 0$, there exists $N \in J$ such that $|x_n - x| < \varepsilon$, for all $n \geq N$.

That is, $|x_n - x|^2 < \varepsilon^2$, for all $n \geq N$.

Therefore, $\sum_{j=1}^{k} (\alpha_{jn} - \alpha_j)^2 < \varepsilon^2$, for all $n \geq N$.

That implies, $(\alpha_{jn} - \alpha_j)^2 < \varepsilon^2$, for all $j = 1,2,3,......,k$.

Hence, $\alpha_{jn} \to \alpha_j$, for all $j = 1,2,3,......,k$.

Conversely,

Given that $\{\alpha_{jn}\} \to \alpha_j$, for all $j = 1,2,.....,k$.

To prove that, $\{x_n\} \to x$.

Since $\{\alpha_{jn}\} \to \alpha_j$, for given $\frac{\varepsilon}{\sqrt{k}} > 0$, there exists

$N_j \in J$, for $j = 1,2,3,......,k$, such that

$$|\alpha_{jn} - \alpha_j| < \frac{\varepsilon}{\sqrt{k}},\quad \text{for all } n \geq N_j,\quad \text{for } j = 1, 2, 3, ..., k.$$

Let $N = \max.\ (N_1, N_2, ..., N_k)$.

Therefore, for all $n \geq N$, $|x_n - x| = \sqrt{\sum_{j=1}^{k} (\alpha_{jn} - \alpha_j)^2} < \sqrt{\sum_{j=1}^{k} \left(\frac{\varepsilon}{\sqrt{k}}\right)^2} = \varepsilon$.

Hence, $\{x_n\} \to x$.

Hence part (i) is proved.

ii) Let $\{x_n\} = (\alpha_{1n}, \alpha_{2n}, \ldots\ldots, \alpha_{kn})$ and $\{y_n\} = (\gamma_{1n}, \gamma_{2n}, \ldots\ldots, \gamma_{kn})$.

$$x = (\alpha_1, \alpha_2, \ldots\ldots, \alpha_k) \text{ and } y = (\gamma_1, \gamma_2, \ldots\ldots, \gamma_k).$$

Hence, $\{x_n + y_n\} = (\alpha_{1n} + \gamma_{1n}, \alpha_{2n} + \gamma_{2n}, \ldots\ldots, \alpha_{kn} + \gamma_{kn})$ and

$x + y = (\alpha_1 + \gamma_1, \alpha_2 + \gamma_2, \ldots\ldots, \alpha_k + \gamma_k)$.

$\beta_n x_n = (\beta_n \alpha_{1n}, \beta_n \alpha_{2n}, \ldots\ldots, \beta_n \alpha_{kn})$ and

$\beta x = (\beta\alpha_1, \beta\alpha_2, \ldots\ldots, \beta\alpha_k)$.

Since $\{x_n\} \to x$, $\alpha_{jn} \to \alpha_j$, for all $j = 1,2,3,\ldots\ldots, k$.

Since $\{y_n\} \to y$, $\gamma_{jn} \to \gamma_j$, for all $j = 1,2,3,\ldots\ldots, k$.

Therefore, $\alpha_{jn} + \gamma_{jn} \to \alpha_j + \gamma_j$, for all $j = 1,2,3,\ldots\ldots, k$.

$$\alpha_{jn}\gamma_{jn} \to \alpha_j\gamma_j, \quad \text{for all } j = 1,2,3,\ldots\ldots, k,$$

and $$\beta_n \alpha_{jn} \to \beta\alpha_j, \quad \text{for all } j = 1,2,3,\ldots\ldots, k.$$

Hence, $\{x_n + y_n\} \to x + y$ and $\beta_n x_n \to \beta x$.

Since, $\alpha_{jn}\gamma_{jn} \to \alpha_j\gamma_j$, for all $j = 1,2,3,\ldots\ldots, k$,

$$\sum_{j=1}^{k} \alpha_{jn}\gamma_{jn} \to \sum_{j=1}^{k} \alpha_j\gamma_j .$$

That is, $x_n y_n \to xy$, where $x_n y_n = \sum_{j=1}^{n} \alpha_{jn}\gamma_{jn}$ and $xy = \sum_{j=1}^{n} \alpha_j\gamma_j$.

Hence part (ii) is proved.

Hence the theorem.

2.4 Subsequences and Cauchy sequence

Definition 2.4.1

Let $\{x_n\}$ be a sequence and $\{n_i\}$ be a sequence of positve integers such that $n_1 < n_2 < n_3 < \ldots\ldots$, then $\{x_{n_i}\}$ is called a subsequence of the sequence $\{x_n\}$. If $\{x_{n_i}\} \to x$, then x is called a subsequential limit of $\{x_n\}$.

Example 2.4.2

Let us consider $\{a_n\} = (-1)^n$.

Here, the subsequence $\{p_1, p_3, \ldots\ldots\ldots\} = \{-1,-1,-1,\ldots\ldots\} \to -1$

and $\{p_2, p_4, p_6, \ldots\ldots\} = \{1,1,1,\ldots\ldots\ldots\} \to 1$.

Therefore, -1 and 1 are subsequential limits.

Note 2.4.3

i) $\{a_n\}$ converges to a, if and only if every subsequence of $\{a_n\}$ converges to a.

ii) Suppose a is a subsequential limit of sequence $\{a_n\}$, then every neighborhood of a contains infinitely many terms of $\{a_n\}$.

Definition 2.4.4

A sequence $\{a_n\}$ is said to be *bounded* if the range $\{a_n : n \in \mathbb{N}\}$ is a bounded set, that is, if there exists $M \geq 0$ such that $|a_n| \leq M$, for all $n \in \mathbb{N}$.

Theorem 2.4.5

i) If $\{a_n\}$ is a sequence in a compact metric space X, then some subsequence of $\{a_n\}$ converges to a point of X.

ii) Every bounded sequence in R^k has a convergent subsequence.

In this theorem, part (ii) is known as *Bolzano-Weierstrass theorem.*

Proof

i) Let E be the range of $\{a_n\}$.

Case i):

Suppose E is finite.

Therefore, there is a point $a \in E$ and a sequence $\{n_i\}$ of positive integers with $n_1 < n_2 < \ldots\ldots.$, $a = a_{n_1} = a_{n_2} = \ldots\ldots.$.

Hence, the constant subsequence $\{a_{n_i}\}$ converges to a.

Case ii):

Suppose E is infinite.

Since every infinite subset of a compact space has a limit point, E has a limit point a in X.

Since a is a limit point of E, $N_1(a) \cap E \neq \phi$.

Therefore, there exists $n_1 \in J$, such that $a_{n_1} \in N_1(a)$. That is, $d(a, a_{n_1}) < 1$.

Since a is a limit point of E, $N_{1/2}(a) \cap E$ is an infinite set.

Therefore, we can find a positive integer $n_2 > n_1$ such that $a_{n_2} \in N_{1/2}(a)$.

Therefore, $d(a_{n_2}, a) < 1/2$.

Having chosen $n_1, n_2, ..., n_{i-1}$, since a is a limit point of E, $N_{1/n}(a) \cap E$ is an infinite set.

Therefore, we can find a positive integer $n_i > n_{i-1}$ such that $a_{n_i} \in N_{1/i}(a)$.

Hence, $d(a_{n_i}, a) < 1/i$.

Thus, we get a subsequence $\{a_{n_i}\}$ of sequence $\{a_n\}$ such that $d(a_{n_i}, a) < 1/i$, for all $i \in J$.

We claim that, $\{a_{n_i}\} \to a$.

We choose $k \in J$, such that $1/k < \varepsilon$.

Therefore, $i \geq k$, $d(a_{n_i}, a) < \frac{1}{i} \leq \frac{1}{k} < \varepsilon$.

Hence, $d(a_{n_i}, a) < \varepsilon$, for all $i \geq k$.

That is, $\{a_{n_i}\} \to a$.

Thus part (i) is proved.

ii) Let $\{a_n\}$ be any bounded sequence in R^k.

Let E be the range of $\{a_n\}$.

Since $\{a_n\}$ is bounded, E is a bounded subset of R^k.

Therefore, $E \subset I$, for some k-cell I.

Since every k-cell is compact, I is compact and since $\{a_n\}$ is a sequence in I and I is compact, by (i), $\{a_n\}$ has a convergent subsequence.

Hence part (ii) is proved.

Hence the theorem.

Note 2.4.6

Any convergent sequence is bounded. But, the converse is not true.

Theorem 2.4.7

The subsequential limits of a sequence $\{a_n\}$ in a metric space X form a closed subset of X.

Proof

Let E^* be the set of all subsequential limits of $\{a_n\}$ in a metric space.

We have to prove that, E^* is closed in X.

Case (*i*)

Suppose E^* has no limit point in *X*.

Therefore, $(E^*)' = E^*$.

Hence, E^* is closed in *X*.

Case (*ii*)

Suppose E^* has limit point in *X*.

Let $b \in X$ be a limit point of E^*.

We have to prove that, there exists $n \in J$, such that $a_n \neq b$.

Suppose $a_n = b$, for all n.

Therefore, $\{a_n\} = \{b, b, b, ...\} \to b$.

Hence, every subsequence of $\{a_n\}$ is convergent to *b*.

That is, $E^* = \{a\}$. That implies, E^* has no limit point in *X*, which is a contradiction.

Therefore, there exists $n_1 \in J$, such that $a_{n_1} \neq b$.

Let $\delta = d(a_{n_1}, b) > 0$.

Since *b* is a limit point of E^*, $N_{2^{-2}\delta}(b) \cap E^* \neq \phi$.

Let us choose $x \in N_{2^{-2}\delta}(b) \cap E^*$.

Therefore, $d(x, b) < 2^{-2}\delta$ and $x \in E^*$. Since $x \in E^*$, *x* is a subsequential limit of $\{a_n\}$.

Therefore, every neighbourhood of *x* contains infinitely many terms of $\{a_n\}$.

Now, $N_{2^{-2}\delta}(x)$ is a neighbourhood of *x*.

Hence, there exists $n_2 > n_1$ such that $a_{n_2} \in N_{2^{-2}\delta}(x)$.

That implies, $d(a_{n_2}, x) < 2^{-2}\delta$.

Therefore, $d(a_{n_2}, b) \leq d(a_{n_2}, x) + d(x, b) < 2^{-1}\delta$.

Having chosen $n_1, n_2, ..., n_{i-1}$, since *b* is a limit point of E^*, $N_{2^{-i}\delta}(b) \cap E^* \neq \phi$.

That implies, $d(y, b) < 2^{-i}\delta$ and $y \in E^*$.

Since $y \in E^*$, every neighbourhood of y contains infinitely many terms of $\{a_n\}$.

Now, $N_{2^{-i}\delta}(y)$ is a neighbourhood of y. Therefore, there exists positive integer $n_i > n_{i-1}$, such that $a_{n_i} \in N_{2^{-i}\delta}(y)$. That is, $d(a_{n_i}, y) < 2^{-i}\delta$.

Hence, $d(a_{n_i}, b) \le d(a_{n_i}, y) + d(y, b) < 2^{1-i}\delta$.

Thus, we get a sequence $\{a_{n_i}\}$ of $\{a_n\}$ such that $d(a_{n_i}, a) < 2^{1-i}\delta$, for all $i \in J$.

We claim that, $\{a_{n_i}\} \to b$.

Let $\varepsilon > 0$ be given. We choose $k \in J$, such that $2^{1-k}\delta < \varepsilon$.

Now, for all $i \ge k$, $1-i \le 1-k$. That implies, $2^{1-i}\delta \le 2^{1-k}\delta$.

Hence, $d(a_{n_i}, b) < 2^{1-i}\delta \le 2^{1-k}\delta < \varepsilon$.

Therefore, $\{a_{n_i}\} \to b$. That is, b is a subsequential limit of $\{a_n\}$.

Hence, $b \in E^*$. We conclude that, E^* is closed in X.

Definition 2.4.8

A sequence $\{a_n\}$ in a metric space X is said to be a *Cauchy sequence*, if for every $\varepsilon > 0$, there exists $N \in J$ such that $d(a_n, a_m) < \varepsilon$, for all $n, m \ge N$.

Definition 2.4.9

Let $E \subset X$. The diameter of E is defined as $\text{diam.}E = \sup.\{d(x, y)/x, y \in E\}$.

Note 2.4.10

1. The diameter of a singleton set is zero.
2. The diameter of a set with atleast two elements is greater than zero.
3. The diameter of an unbounded set is $+\infty$.
4. The diameter of an empty set is $-\infty$.

Theorem 2.4.11

Every Cauchy sequence is bounded.

Proof

Let $\{a_n\}$ be a Cauchy sequence.

We have to prove that, $\{a_n\}$ is bounded.

Since $\{a_n\}$ is a Cauchy sequence, there exists $N \in J$, such that $d(a_n, a_m) < 1$, for all $n, m \geq N$.

That is, $d(a_n, a_N) < 1$, for all $n \geq N$.

Take $k = \max.\{1, d(a_1, a_N), d(a_2, a_N), ..., d(a_{N-1}, a_N)\}$.

Then, $d(a_n, a_N) \leq k$, for all $n \in J$.

Hence, $\{a_n\}$ is bounded.

Therefore, every Cauchy sequence is bounded.

Theorem 2.4.12

i) If $E \subset X$, then $\text{diam}.\overline{E} = \text{diam}.E$.

ii) If $\{K_n\}$ is a sequence of compact sets such that $K_n \supset K_{n+1}$,

for n = 1, 2, 3, ... and $\lim_{n\to\infty} \text{diam}.K_n = 0$, then $\bigcap_{n=1}^{\infty} K_n$ consists of exactly one point.

Proof

We know that, $E \subset \overline{E}$ and hence $\text{diam}.E \leq \text{diam}.\overline{E}$.

That is, $\text{diam}.\overline{E}\text{-diam}.E \geq 0$.

Let $\varepsilon > 0$ be given and $x, y \in \overline{E}$.

Hence, $N_{\varepsilon/3}(x) \cap E \neq \varphi$ and $N_{\varepsilon/3}(y) \cap E \neq \varphi$.

Let $x_1 \in N_{\varepsilon/3}(x) \cap E$ and $y_1 \in N_{\varepsilon/3}(y) \cap E$.

Therefore, $x_1, y_1 \in E$, also $x_1 \in N_{\varepsilon/3}(x)$ and $y_1 \in N_{\varepsilon/3}(y)$.

Hence, $d(x_1, x) < \dfrac{\varepsilon}{3}$ and $d(y_1, y) < \dfrac{\varepsilon}{3}$.

Now,

$$d(x, y) \leq d(x_1, x) + d(x_1, y_1) + d(y_1, y) < \frac{\varepsilon}{3} + d(x_1, y_1) + \frac{\varepsilon}{3} \leq \frac{2\varepsilon}{3} + \text{diam}.E.$$

Therefore, the distance between any two points of $\overline{E}$ is less than $\frac{2\varepsilon}{3} + diam.E$.

That is, $\text{diam}.\overline{E} < \frac{2\varepsilon}{3} + \text{diam}.E < \varepsilon + \text{diam}.E$.

That implies, $\text{diam}.\overline{E} - \text{diam}.E < \varepsilon$.

Since ε is arbitrary, $\text{diam}.\overline{E} = \text{diam}.E$.

Hence part (i) is proved.

ii) Given that, $\lim_{n\to\infty} \text{diam}.K_n = 0$ and $K_n \supset K_{n+1}$, for $n = 1,2,3,......$ and $\text{diam}\phi = -\infty$.

Hence, $K_n \neq \phi$, for n = 1, 2, 3, ……

Therefore, we get that, K_n is a sequence of non – empty compacts set, such that $K_n \supset K_{n+1}$, for $n = 1,2,3,......$

By the previous theorem, we have $\bigcap_{n=1}^{\infty} K_n \neq \phi$.

Let $K = \bigcap_{n=1}^{\infty} K_n$.

We claim that, K has exactly one point.

If possible, let K has two distinct elements x and y.

Therefore, $\delta = d(x, y) > 0$.

Now, since $K \subset K_n$, for all $n \in J$, we have,

$$\delta = d(x, y) \leq \text{diam}.K \leq \text{diam}.K_n.$$

That implies, $\lim_{n\to\infty} \text{diam}.K_n \geq \delta$. That is, $0 \geq \delta$ which is a contradiction to $\delta > 0$.

Therefore, K has exactly one point.

Hence part (ii) is proved.

Hence the theorem.

Theorem 2.4.13

i) In any metric space X, every convergent sequenc is a Cauchy sequence.

ii) If X is a compact space, then every Cauchy sequence in X converges to a point of X.

iii) Every Cauchy sequence converges in R^k.

Proof

i) Let $\{a_n\}$ be a convergent sequence in X, which converges to a.

Therefore, for given $\varepsilon > 0$, there exists $N \in J$ such that $d(a_n, a) < \frac{\varepsilon}{2}$, for all $n \geq N$.

Let $n, m \geq N$.

Hence, $d(a_n, a_m) \leq d(a_n, a) + d(a, a_m) < \varepsilon$.

Therefore, $\{a_n\}$ is a Cauchy sequence.

Hence part (i) is proved.

ii) Given that, X is compact and let $\{a_n\}$ be a Cauchy sequence in X.

For each $n \in J$, let $E_n = \{a_n, a_{n+1}, ...\}$.

Since $\{a_n\}$ is a Cauchy sequence, $\lim\limits_{n \to \infty} \text{diam.} E_n = 0$, and also we have, $\text{diam.} E_n = \text{diam.} \overline{E}_n$.

Therefore, $\lim\limits_{n \to \infty} \text{diam.} \overline{E}_n = 0$.

Since $E_n \supset E_{n+1}$, for $n = 1,2,3,......$, $\overline{E}_n \supset \overline{E}_{n+1}$, for $n = 1, 2, 3, ...$

Since $\overline{E}_n$ is a closed subset of a compact set X, $\overline{E}_n$ is compact.

Therefore, by theorem 2.4.12 (ii), there exists a unique point $a \in X$, such that $a \in \bigcap\limits_{n=1}^{\infty} \overline{E}_n$.

We claim that, $\{a_n\} \to a$.

Since $\lim\limits_{n \to \infty} \text{diam.} \overline{E}_n = 0$, for given $\varepsilon > 0$, there exists a positive integer $N \in J$, such that $\text{diam.} \overline{E}_n < \varepsilon$, for all $n \geq N$.

Hence, $\text{diam.} \overline{E}_N < \varepsilon$.

Let $n \geq N$. Hence $a_n \in E_n \subset \overline{E}_N$. That is, $a_n \in \overline{E}_N$. Also, $a \in \overline{E}_N$.

Therefore, $d(a_n, a) \leq \text{diam}.\overline{E}_N < \varepsilon$.

That is, $d(a_n, a) < \varepsilon$, for all $n \geq N$.

Hence, $\{a_n\} \rightarrow a$.

Hence part (ii) is proved.

iii) Let $\{a_n\}$ be a Cauchy sequence in R^k.
Since every Cauchy sequence is bounded, $\{a_n\}$ is bounded.
Let E be the range of $\{a_n\}$. Since $\{a_n\}$ is bounded, E is a bounded subset of R^k.
Therefore, $\overline{E}$ is a closed and bounded subset of R^k.
We know that, $\{a_n\}$ is a Cauchy sequence in $E \subset \overline{E}$.
Therefore, $\{a_n\}$ is a Cauchy sequence in $\overline{E}$, which is compact.
Hence, by part (ii), $\{a_n\}$ is convergent in $\overline{E} \subset R^k$.
That is, $\{a_n\}$is convergent in R^k.
Hence part (iii) is proved.
Hence the theorem.

Definition 2.4.13

A metric space X is said to be *complete*, if every Cauchy sequence in X converges to a point of X.

Examples 2.4.14

1. By theorem 2.4.12 (ii), every compact metric space is complete. Again, by theorem 2.4.12 (iii), R^k is complete.
2. (0, 1) is not complete. Now, $\left\{\frac{1}{n}\right\}_{n=2}^{\infty}$ is a Cauchy sequence in (0, 1), but it doesn't converge to a point of (0, 1).

Definition 2.4.15

A sequence $\{p_n\}$ of real numbers is said to be

i) *monotonically decreasing*, if $p_n \leq p_{n+1}$, for all $n \in J$.

ii) *monotonically increasing*, if $p_n \geq p_{n+1}$, for all $n \in J$.

iii) *monotonic*, if it is either monotonic increasing or monotonic decreasing.

Theorem 2.4.16

Suppose $\{p_n\}$ is monotonic. Then, $\{p_n\}$ is convergent if and only if it is bounded.

Proof

Direct part,

Given that $\{p_n\}$ is monotonic. Also, given that $\{p_n\}$ is convergent.

Since every convergent sequence is bounded, $\{p_n\}$ is bounded.

Converse part,

Given that $\{p_n\}$ is bounded.

We have to prove that, $\{p_n\}$ is convergent.

Case (i);

Let $\{p_n\}$ be monotonic increasing and E be the range of $\{p_n\}$.

Since $\{p_n\}$ is bounded, E is bounded.

Also, E is non-empty.

Therefore, E is a non-empty bounded subset of R.

Since R has least upper bound property, $p = \sup.E$ exists in R.

Hence, $p_n \leq p$, for all $n \in J$.

We claim that, $\{p_n\}$ converges to p.

Let $\varepsilon > 0$ be given. Then, $p - \varepsilon < p$.

Hence, $p - \varepsilon$ is not an upper bound. Therefore, there exists $N \in J$, such that $p_N > p - \varepsilon$.

Let $n \geq N$. Since, p_n is monotonic increasing, $p - \varepsilon < p_N \leq p_n \leq p < p + \varepsilon$.

Hence, $p - \varepsilon < p_n < p + \varepsilon$, for all $n \geq N$.

That is, $-\varepsilon < p_n - p < \varepsilon$, for all $n \geq N$, and hence

$$|p_n - p| < \varepsilon, \quad \text{for all } n \geq N.$$

Therefore, $\{p_n\} \to p$, that is, $\{p_n\}$ is convergent.

Case (ii);

Let $\{p_n\}$ be monotonic decreasing and E be the range of $\{p_n\}$.

Since $\{p_n\}$ is bounded, E is bounded.

Also, E is non-empty.

Therefore, E is a non-empty bounded subset of R.

Since R has greatest lower bound property, $p = \inf.E$ exists in R.

Hence, $p_n \geq p$, for all $n \in J$.

We claim that, $\{p_n\}$ converges to p.

Let $\varepsilon > 0$ be given. Then, $p+\varepsilon > p$.

Hence, $p+\varepsilon$ is not a lower bound. Therefore, there exists $N \in J$, such that $p_N < p+\varepsilon$.

Let $n \geq N$. Since, p_n is monotonic decreasing, $p+\varepsilon > p_N \geq p_n \geq p > p-\varepsilon$.

Hence, $p-\varepsilon < p_n < p+\varepsilon$, for all $n \geq N$.

That is, $-\varepsilon < p_n - p < \varepsilon$, for all $n \geq N$, and hence

$$|p_n - p| < \varepsilon, \quad \text{for all } n \geq N.$$

Therefore, $\{p_n\} \to p$, that is, $\{p_n\}$ is convergent.

Hence the theorem.

Theorem 2.4.17

i) If $p > 0$, then $\lim\limits_{n\to\infty} \dfrac{1}{n^p} = 0$,

ii) If $p > 0$, then $\lim\limits_{n\to\infty} p^{1/n} = 1$.

iii) $\lim\limits_{n\to\infty} n^{1/n} = 1$.

iv) If $p > 0$ and α is real, then $\lim\limits_{n\to\infty} \dfrac{n^\alpha}{(1+p)^n} = 0$.

v) If $|x| < 1$, then $\lim\limits_{n\to\infty} x^n = 0$.

Proof

i) Let $\varepsilon > 0$ be given and we choose a positive integer N, such that $N > \left(\dfrac{1}{\varepsilon}\right)^{1/p}$.

Therefore, $N^p > \frac{1}{\varepsilon}$. That is, $\frac{1}{N^p} < \varepsilon$. .... (2.4.17.1)

Let $n \geq N$.

Since $p > 0$, we have $n^p \geq N^p$. That implies, $\frac{1}{n^p} \leq \frac{1}{N^p} < \varepsilon$.

Therefore, $\frac{1}{n^p} < \varepsilon$, for all $n \geq N$.

Hence, $\lim_{n\to\infty} \frac{1}{n^p} = 0$.

Hence proved.

ii) Given that, $p > 0$.

We have to prove that, $\lim_{n\to\infty} p^{1/n} = 1$.

Case (*i*); When $p = 1$,

Then, $p^{1/n} = 1$, for all $n \in J$.

Therefore, $\lim_{n\to\infty} p^{1/n} = 1$.

Case (*ii*); When $p > 1$,

Then, $p^{1/n} > 1$, for all $n \in J$.

Therefore, $p^{1/n} = 1 + x_n$, where $x_n > 0$.

That is, $p = (1+x_n)^n = 1 + nx_n + \frac{n(n-1)}{2!} x_n^2 + \ldots\ldots + x_n^n$.

That implies, $p \geq 1 + nx_n$.

Therefore, $0 < x_n < \frac{p-1}{n}$.

We know that, $\lim_{n\to\infty} \frac{p-1}{n} = 0$.

Therefore, $\lim_{n\to\infty} x_n = 0$.

That implies, $\lim_{n\to\infty} p^{1/n} = \lim_{n\to\infty}(1+x_n) = 1$.

Case (*iii*); When $0 < p < 1$,

If $p < 1$, then $\frac{1}{p} > 1$.

Therefore, by case (ii), $\lim_{n \to \infty} \left(\frac{1}{p} \right)^{1/n} = 1$.

That is, $\lim_{n \to \infty} p^{1/n} = 1$.

Hence proved.

iii) We have to prove that, $\lim_{n \to \infty} n^{1/n} = 1$.

We know that, $n^{1/n} \geq 1$, for all $n \in J$.

Therefore, $n^{1/n} = 1 + x_n$, where $x_n \geq 0$.

That is,

$$n = (1 + x_n)^n = 1 + nx_n + \frac{n(n-1)}{2!} x_n^2 + \ldots\ldots + x_n^n > \frac{n(n-1)}{2!} x_n^2, \text{ if } n \geq 2.$$

Hence, $\frac{2}{n-1} > x_n^2$, if $n \geq 2$. That is, $0 \leq x_n < \frac{\sqrt{2}}{\sqrt{n-1}}$, if $n \geq 2$.

Since $\lim_{n \to \infty} \frac{\sqrt{2}}{\sqrt{n-1}} = 0$, we have $\lim_{n \to \infty} x_n = 0$.

Hence, $\lim_{n \to \infty} n^{1/n} = \lim_{n \to \infty} (1 + x_n) = 1$.

Hence proved.

iv) Given $p > 0$ and α is real.

We choose a positive integer k such that $k > \alpha$.

Let $n > 2k$.

Now,

$$(1+p)^n = 1 + np + \ldots\ldots + \binom{n}{k} p^k + \ldots\ldots + p^n > \binom{n}{k} p^k = \frac{n(n-1)(n-2)\ldots\ldots(n-k+1)}{k!}$$

Since $n > 2k$, we have, $n - 2k > 0$. That is, $2n - 2k > n$ implies $n - k > \frac{n}{2}$.

Hence, $n(n-1)\ .....\ (n-(k-1)) > n-k > \frac{n}{2}$.

Therefore, $(1+p)^n > \frac{\left(\frac{n}{2}\right)\cdot\left(\frac{n}{2}\right)\cdots\cdots\left(\frac{n}{2}\right)}{k!}p^k > \frac{n^k}{2^k k!}p^k$.

That is, $\frac{1}{(1+p)^n} < \frac{2^k k!}{n^k p^k}$ implies $0 < \frac{n^\alpha}{(1+p)^n} < \frac{2^k k!}{p^k}\frac{1}{n^{k-\alpha}}$.

Since $k > \alpha$, we have $k-\alpha > 0$.

Hence by (i), we get $\lim_{n\to\infty}\frac{1}{n^{k-\alpha}} = 0$. That is, $\frac{2^k k!}{p^k}\frac{1}{n^{k-\alpha}} = 0$.

Therefore, $\lim_{n\to\infty}\frac{n^\alpha}{(1+p)^n} = 0$.

Hence proved.

v) Given that $|x| < 1$.

We have to prove that, $\lim_{n\to\infty} x^n = 0$.

Case (*i*); When $x = 0$,

Then, $x^n = 0$, for all $n \in J$.

Hence, $\lim_{n\to\infty} x^n = 0$.

Case (*ii*); When $0 < x < 1$,

Then, $1-x > 0$ implies $\frac{1-x}{x} > 0$.

Let us take $p = \frac{1-x}{x} > 0$ and $\alpha = 0$.

Hence by (iv), we get $\lim_{n\to\infty}\frac{n^0}{\left(1+\frac{1-x}{x}\right)^n} = 0$.

Therefore, $\lim_{n\to\infty} x^n = 0$.

Case (*iii*); When $-1 < x < 0$,

Then, $1 > -x > 0$ or $0 < -x < 1$.

Now, by case (ii), $\lim_{n\to\infty}(-x)^n = 0$.

Therefore, for given $\varepsilon > 0$, there exists $N \in J$ such that $\left|(-x)^n - 0\right| < \varepsilon,$ for all $n \geq N$.

That is, $\left|(-1)^n\right|\left|x\right|^n < \varepsilon,$ for all $n \geq N$. Hence we have $\left|x\right|^n < \varepsilon,$ for all $n \geq N$.

Therefore, $\lim_{n\to\infty} x^n = 0$.

Hence proved.

2.5 Series

Definition 2.5.1

Let $\{a_n\}$ be a complex series. Then, $\sum_{n=1}^{\infty} a_n$ is called a series. Let $s_n = a_1 + a_2 + \ldots\ldots + a_n$, and $\{s_n\}$ is called the partial sums of $\sum_{n=1}^{\infty} a_n$. If $\{s_n\} \to s$, we say that $\sum_{n=1}^{\infty} a_n$ is convergent and s is the sum of $\sum_{n=1}^{\infty} a_n$. If $\{s_n\}$ is divergent, then $\sum_{n=1}^{\infty} a_n$ is divergent.

Theorem 2.5.2

This theorem is known as ***Cauchy's criteria for convergence of a series***.

$\sum a_n$ is convergent if and only if given $\varepsilon > 0$, there exists $N \in J$ such that $\left|\sum_{k=n}^{m} a_k\right| \leq \varepsilon$, for all $m \geq n \geq N$.

Proof

Let $\{s_n\}$ be the sequence of partial sums of $\sum a_n$.

Now, $\sum a_n$ is convergent if and only if $\{s_n\}$ is convergent.

If and only if $\{s_n\}$ is Cauchy, since C is complex.

If and only if, given $\varepsilon > 0$, there exists positive integer $N \in J$ such that $\left|\sum_{k=n}^{m} a_k\right| \leq \varepsilon$, for all $m \geq n \geq N$.

Hence the theorem.

Theorem 2.5.3

If $\sum a_n$ is convergent, then $\lim_{n\to\infty} a_n = 0$.

Proof

Given that $\sum a_n$ is convergent.

Therefore, by Cauchy's criteria for convergence of a series, for given $\varepsilon > 0$, there exists $N \in J$ such that $\left|\sum_{k=n}^{m} a_k\right| \leq \varepsilon$, for all $m \geq n \geq N$.

Put $m = n$.

Therefore, for given $\varepsilon > 0$, there exists $N \in J$ such that $|a_n| \leq \varepsilon$, for all $n \geq N$.

Hence $\lim_{n\to\infty} a_n = 0$.

Hence the theorem.

Note 2.5.4

i) The converse of the above theorem need not be true. That is, if $\lim_{n\to\infty} a_n = 0$, then $\sum a_n$ need not be convergent.

Let us give some examples. Consider $a_n = \frac{1}{n}$.

Now, $\lim_{n\to\infty} a_n = \lim_{n\to\infty} \frac{1}{n} = 0$.

But $\sum a_n = \sum \frac{1}{n}$ which is divergent.

ii) If $\lim_{n\to\infty} a_n \neq 0$, then $\sum a_n$ is divergent.

Suppose $\sum a_n$ is convergent, then by theorem 2.5.3, we have $\lim_{n\to\infty} a_n = 0$, which is a contradiction.

Hence, $\sum a_n$ is divergent.

Theorem 2.5.5

A series of non – negative terms is convergent if and only if its partial sums form a bounded sequence.

Proof

Let $\sum a_n$ be a series of non – negative terms and $\{s_n\}$ be the sequence of partial sums of $\sum a_n$.

Therefore, $s_{n+1} - s_n = a_{n+1} \geq 0$. That is, $s_{n+1} \geq s_n$.

Hence, $\{s_n\}$ is monotonic increasing.

That is, $\{s_n\}$ is convergent if and only if $\{s_n\}$ is bounded.

Therefore, $\sum a_n$ is convergent if and only if $\{s_n\}$ is bounded.

Hence the theorem

Theorem 2.5.6

This theorem is known as ***Comparison Test***.

i) If $|a_n| \leq c_n$, for all $n \geq N_0$, where N_0 is fixed and if $\sum c_n$ is convergent, then $\sum a_n$ is convergent.

ii) If $a_n \geq d_n \geq 0$, for all $n \geq N_0$, where N_0 is fixed and if $\sum d_n$ is divergent, then $\sum a_n$ is divergent.

Proof

i) Given that $\sum c_n$ is convergent.

Therefore, by Cauchy's criteria for convergence, we have, for given $\varepsilon > 0$, there exists $N > N_0 \in J$ such that $\left|\sum_{k=n}^{m} c_k\right| \leq \varepsilon$, for all $m \geq n \geq N$.

.... (2.5.6.1)

Now, for all $m \geq n \geq N$, we have $\left|\sum_{k=n}^{m} a_k\right| \leq \sum_{k=n}^{m} |a_k| \leq \sum_{k=n}^{m} c_k$, since $|a_n| \leq c_n$, for all $n \geq N$.

Since $c_k \geq |a_k| \geq 0$, by equation (2.5.6.1), we have $\left|\sum_{k=n}^{m} a_k\right| \leq \left|\sum_{k=n}^{m} c_k\right| \leq \varepsilon$.

Therefore, by Cauchy's criteria for convergence of series, we have, $\sum a_n$ is convergent.

Hence part (i) is proved.

ii) Given that $a_n \geq d_n \geq 0$, for all $n \geq N_0$, where N_0 is fixed and $\sum d_n$ is divergent.

We have to prove that, $\sum a_n$ is divergent.

Suppose $\sum a_n$ is convergent. Then by (i), we have $\sum d_n$ is convergent, which is a contradiction.

Therefore, $\sum a_n$ is divergent.

Hence part (ii) is proved.

Hence the theorem.

Theorem 2.5.7

i) If $0 \leq x < 1$, then $\sum_{n=0}^{\infty} x^n = \frac{1}{1-x}$.

ii) If $x \geq 1$, then $\sum_{n=0}^{\infty} x^n$ is divergent.

Proof

i) Given that, $0 \leq x < 1$.

Let $\{s_n\}$ be the sequence of partial sums of $\sum_{n=0}^{\infty} x^n$.

Therefore, $s_n = 1 + x + x^2 + \dots\dots + x^n = \frac{1(1-x^{n+1})}{1-x}$.

Since $0 \leq x < 1$, we have $x^{n+1} \to 0$ as $n \to \infty$.

Therefore, $\lim_{n\to\infty} s_n = \frac{1}{1-x}$. That is, $\{s_n\} \to \frac{1}{1-x}$.

Therefore, $\sum_{n=0}^{\infty} x^n$ is convergent and converges to $\frac{1}{1-x}$.

Hence, $\sum_{n=0}^{\infty} x^n = \frac{1}{1-x}$.

Hence part (i) is proved.

ii) Given that $x \geq 1$.

That implies, $x^n \geq 1$, for all n. Hence, $\lim_{n\to\infty} x^n \neq 0$.

Therefore, $\sum x^n$ is divergent.

Hence part (ii) is proved.
Hence the theorem.

Theorem 2.5.8

This theorem is known as ***Cauchy's condensation test***.

Suppose $a_1 \geq a_2 \geq .... \geq 0$. Then the series $\sum_{n=1}^{\infty} a_n$ converges if and only if

$\sum_{k=0}^{\infty} 2^k a_{2^k} = a_1 + 2a_2 + 4a_4 + ...$converges.

Proof
Direct part;

Let $s_n = a_1 + a_2 + ... + a_n$ and $t_k = a_1 + 2a_2 + 4a_4 + ... + 2^k a_{2^k}$.

Given that, $\{a_n\}$ is a sequence of non – negative terms.

Therefore, $\sum a_n$ converges if and only if $\{s_n\}$ is bounded and

$$\sum_{k=0}^{\infty} 2^k a_{2^k} \text{ converges if and only if } \{t_k\} \text{ is bounded.}$$

To prove the theorem, it is enough to prove that, $\{s_n\}$ is bounded if and only if $\{t_k\}$ is bounded.

Suppose $\{s_n\}$ is bounded. Therefore, there exists $l > 0$, such that $s_n \leq l$, for all $n \in J$. .... (2.5.8.1)

Let $n > 2^k$.
Since $\{s_n\}$ is monotonic increasing, we have

$s_n \geq s_{2^k} = a_1 + a_2 + a_3 + ... + a_{2^k}$.

$= a_1 + a_2 + (a_3 + a_4) + (a_5 + a_6 + a_7 + a_8) + ... + (a_{2^{k-1}+1} + ... + a_{2^k})$.

$\geq \dfrac{a_1}{2} + a_2 + 2a_4 + 4a_8 + ... + 2^{k-1} a_{2^k}$ [Since $\{a_n\}$ is monotonic decreasing]

$= \dfrac{1}{2}\left[a_1 + 2a_2 + 4a_4 + ... + 2^k a_{2^k}\right] = \dfrac{1}{2} t_k$.

Therefore, $s_n \geq \frac{1}{2}t_k$, if $n > 2^k$.

That implies, $t_k \leq 2s_n$, if $n > 2^k$. That is, $t_k \leq 2l$, for all $k \geq 0$.

Therefore, $\{t_n\}$ is bounded.

Converse part;

Suppose t_k is bounded. Therefore, there exists $l_1 > 0$ such that $t_k \leq l_1$, for all $k \geq 0$ .... (2.5.8.2)

Let $n < 2^k$. Since $\{s_n\}$ is monotonic increasing, we have

$$s_n \leq s_{2^k} = a_1 + a_2 + a_3 + \ldots\ldots + a_{2^k}.$$
$$\leq a_1 + (a_2 + a_3) + (a_4 + a_5 + a_6 + a_7) + \ldots\ldots. + (a_{2^k} + \ldots\ldots + a_{2^{k+1}-1}) = t_k.$$

Hence, $s_n \leq t_k$, for all $n < 2^k$. Then, by equation (2.5.8.2), we have $s_n \leq l_1$, for all $n \in J$.

Therefore, $\{s_n\}$ is bounded. Hence, $\sum a_n$ is convergent if and only if $\sum 2^k a_{2^k}$ is convergent.

Hence the theorem.

Theorem 2.5.9

$\sum \frac{1}{n^p}$ is convergent if $p > 1$ and divergent if $p \leq 1$.

Proof

Case (*i*); When $p \leq 0$

Then, $-p \geq 0$. That is, $n^{-p} \geq 1$, for all $n \in J$.

That implies, $\frac{1}{n^p} \geq 1$, for all $n \in J$ and hence $\lim_{n\to\infty} \frac{1}{n^p} \neq 0$.

Therefore, $\sum \frac{1}{n^p}$ is divergent.

Case (*ii*); When $p > 0$

Let $a_n = \frac{1}{n^p}$ and since $p > 0$, we have $a_n > 0$, for all n.

We know that, $\frac{1}{n^p} > \frac{1}{(n+1)^p}$. Hence, $a_n > a_{n+1}$. That is, $a_1 > a_2 > a_3 > .... > 0$.

By condensation test, we have, $\sum a_n$ is convergent if and only if $\sum_{k=0}^{\infty} 2^k a_{2^k}$ is convergent.

Now, $\sum_{k=0}^{\infty} 2^k a_{2^k} = \sum_{k=0}^{\infty} 2^k \frac{1}{(2^k)^p} = \sum_{k=0}^{\infty} (2^k)^{1-p} = \sum_{k=0}^{\infty} (2^{1-p})^k$.

Therefore, $\sum_{k=0}^{\infty} (2^{1-p})^k$ is convergent, if $2^{1-p} < 1$ and divergent, if $2^{1-p} \geq 1$.

That implies, $\sum_{k=0}^{\infty} 2^k a_{2^k}$ is convergent, if $1-p<0$ and divergent, if $1-p \geq 0$.

Hence, $\sum a_n$ is convergent, if $p>1$ and divergent, if $p \leq 1$.

Hence the theorem.

Theorem 2.5.10

$\sum_{n=2}^{\infty} \frac{1}{n(\log n)^p}$ is convergent, if $p>1$ and divergent, if $p \leq 1$.

Proof

Case (i); When $p \leq 0$,

Then, $-p \geq 0$.

Now, $\log_e n \geq 1$, if $n \geq 3$.

Hence, $(\log n)^{-p} \geq 1$, if $n \geq 3$. That is, $\frac{1}{n(\log n)^p} \geq \frac{1}{n} > 0$, if $n \geq 3$.

Since, $\sum \frac{1}{n}$ is divergent, by comparison test, $\sum \frac{1}{n(\log n)^p}$ is divergent.

Case (ii); When $p > 0$,

Let $a_n = \frac{1}{n(\log n)^p}$, for $n \geq 2$.

Hence, $a_n > 0$, for all $n \geq 2$.

We know that, $n < n+1$. That is, $n(\log n)^p < (n+1)(\log(n+1))^p$.

That implies, $\dfrac{1}{n(\log n)^p} > \dfrac{1}{(n+1)(\log(n+1))^p}$.

Hence, $a_n > a_{n+1}$, for all $n \geq 2$.

Therefore, $a_2 > a_3 > a_4 > \ldots\ldots > 0$.

Hence by Condensation test, we have $\sum a_n$ is convergent, if and only if $\sum_{k=1}^{\infty} 2^k a_{2^k}$ is convergent.

Now, $\sum_{k=1}^{\infty} 2^k a_{2^k} = \sum_{k=1}^{\infty} 2^k \dfrac{1}{2^k(\log 2^k)^p} = \sum_{k=1}^{\infty} \dfrac{1}{k^p(\log 2)^p} = \dfrac{1}{(\log 2)^p} \sum_{k=1}^{\infty} \dfrac{1}{k^p}$.

Therefore, $\sum_{p=1}^{\infty} \dfrac{1}{k^p}$ is convergent, if $p > 1$ and divergent, if $p \leq 1$.

Hence, $\dfrac{1}{(\log 2)^p} \sum_{k=1}^{\infty} \dfrac{1}{k^p}$ is convergent, if $p > 1$ and divergent, if $p \leq 1$.

That is, $\sum_{k=1}^{\infty} 2^k a_{2^k}$ is convergent, if $p > 1$ and divergent, if $p \leq 1$.

That implies, $\sum a_n = \sum \dfrac{1}{n(\log n)^p}$ is convergent, if $p > 1$ and divergent, if $p \leq 1$.

Hence the theorem.

Definition 2.5.11

We define $e = \sum_{n=0}^{\infty} \dfrac{1}{n!}$. Let $s_n = \sum_{k=0}^{n} \dfrac{1}{k!} = 1 + \dfrac{1}{1!} + \dfrac{1}{2!} + \ldots\ldots + \dfrac{1}{n!}$ be the partial sum of *e*.

Here, $\{s_n\}$ is monotonic increasing.

Now, $s_n = 1 + \dfrac{1}{1!} + \dfrac{1}{2!} + \dfrac{1}{3!} + \ldots\ldots. + \dfrac{1}{n!} < 1 + 1 + \dfrac{1}{2} + \dfrac{1}{2^2} + \ldots\ldots. + \dfrac{1}{2^{n-1}}$.

That is, $s_n < 1 + 1 + \dfrac{1}{2} + \dfrac{1}{2^2} + \dfrac{1}{2^3} + \ldots\ldots. = 1 + \dfrac{1}{1 - 1/2} = 3$.

Hence, $s_n < 3$, which implies that $\{s_n\}$ is bounded.

Therefore, $\{s_n\}$ is convergent.

Hence, $\sum_{n=0}^{\infty}\frac{1}{n!}$ is convergent.

Theorem 2.5.12

$\lim_{n\to\infty}\left(1+\frac{1}{n}\right)^n = e$.

Proof

Let $t_n = \left(1+\frac{1}{n}\right)^n$ and $s_n = \sum_{k=0}^{n}\frac{1}{k!}$.

We know that, $e = \sum_{n=0}^{\infty}\frac{1}{n!}$. Therefore, $\lim_{n\to\infty} s_n = e$.

That is, $\lim_{n\to\infty}\sup .s_n = \lim_{n\to\infty}\inf .s_n = e$. .... (2.5.12.1)

Consider, $t_n = \left(1+\frac{1}{n}\right)^n = 1 + n\frac{1}{n} + \frac{n(n-1)}{2!}\frac{1}{n^2} + \text{........} + \frac{1}{n^n}$.

That is, $t_n = 1 + 1 + \frac{(n-1)}{2!}\frac{1}{n} + \frac{(n-1)(n-2)}{3!}\frac{1}{n^2}\text{......} + \frac{n(n-1)\text{.......}1}{n^n}\frac{1}{n!}$.

That implies,

$$t_n = 1 + 1 + \frac{1}{2!}\left(1-\frac{1}{n}\right) + \frac{1}{3!}\left(1-\frac{1}{n}\right)\left(1-\frac{2}{n}\right) + \text{....} + \frac{1}{n!}\left(1-\frac{1}{n}\right)\left(1-\frac{2}{n}\right)\text{.....}\left(1-\frac{n-1}{n}\right).$$

Hence, $t_n < 1 + 1 + \frac{1}{2!} + \frac{1}{3!} + \text{.......} + \frac{1}{n!}$.

Therefore, $t_n < s_n$, for all $n \in J$.

Now, taking lim.sup. on both sides, we get $\lim_{n\to\infty}.\sup .t_n \le \lim_{n\to\infty}.\sup .s_n = e$.

That implies, $\lim_{n\to\infty}.\sup .t_n \le e$. .... (2.5.12.2)

Let us take $m < n$.

Then,

$$t_n = \left(1+\frac{1}{n}\right)^n = 1 + n\frac{1}{n} + \frac{n(n-1)}{2!}\frac{1}{n^2} + \frac{n(n-1)(n-2)}{3!}\frac{1}{n^3} + \ldots\ldots + \left(\frac{1}{n}\right)^n.$$

That is,

$$t_n = 1 + 1 + \frac{n-1}{2!}.\frac{1}{n} + \frac{n(n-1)(n-2)}{3!}.\frac{1}{n^3} + \ldots\ldots + \frac{n(n-1)(n-2)\ldots\ldots 1}{n^n}.\frac{1}{n!}.$$

Hence,

$$t_n = 1 + 1 + \frac{1}{2!}\left(1-\frac{1}{n}\right) + \frac{1}{3!}\left(\left(1-\frac{1}{n}\right)\left(1-\frac{2}{n}\right)\right) + \ldots.$$

$$\ldots. + \frac{1}{n!}\left(\left(1-\frac{1}{n}\right)\left(1-\frac{2}{n}\right)\ldots\ldots\left(1-\frac{n-1}{n}\right)\right).$$

That is, $t_n > 1 + 1 + \frac{1}{2!}\left(1-\frac{1}{n}\right) + \frac{1}{3!}\left(1-\frac{1}{n}\right)\left(1-\frac{2}{n}\right) + \ldots.$

$$\ldots.. + \frac{1}{m!}\left(1-\frac{1}{n}\right)\left(1-\frac{2}{n}\right)\ldots\ldots\left(1-\frac{m-1}{n}\right)$$

Let us fix *m*, and take lim.inf as $n \to \infty$.

Therefore, since $\lim_{n\to\infty}\frac{1}{n} = 0$, we have,

$$\liminf_{n\to\infty} t_n \geq 1 + \frac{1}{1!} + \frac{1}{2!} + \ldots\ldots + \frac{1}{m!} = s_m.$$

Taking limit as $m \to \infty$, we get $\liminf_{n\to\infty} t_n \geq \lim_{m\to\infty} s_m = e$.

We know that, $e \leq \liminf_{n\to\infty} t_n \leq \limsup_{n\to\infty} t_n \leq e$, by using equation (2.5.12.2).

Hence, $\lim_{n\to\infty} t_n = e$. That is, $\lim_{n\to\infty}\left(1+\frac{1}{n}\right)^n = e$.

Hence the theorem.

Note 2.5.13

We have, $e - s_n = \frac{1}{(n+1)!} + \frac{1}{(n+2)!} + \frac{1}{(n+3)!} + \ldots\ldots > 0$ always.

Therefore, $0 < (e - s_n)n! < \frac{1}{n}$.

Theorem 2.5.14

e is irrational.

Proof

Suppose e is rational.

Let $e = \frac{p}{q}$, where p and q are positive integers.

By note 2.5.13, we have, $0 < (e - s_q)q! < \frac{1}{q}$, where q is a positive integer.

Now, $eq! = e(1.2.3.......(q-1)q) = qe(1.2.3.......(q-1)) = p(1.2.3.......(q-1))$, which is an integer.

Again, $s_q q! = \left(1 + \frac{1}{1!} + \frac{1}{2!} + ... + \frac{1}{q!}\right)q!$, is an integer.

Hence, $eq! - s_q q!$ is an integer.

Since q is a positive integer, $q \geq 1$. That implies, $\frac{1}{q} \leq 1$.

Hence, we have, $0 < eq! - s_q q! < \frac{1}{q} \leq 1$. This implies that, there exists an integer lying between 0 and 1, which is a contradiction.

Therefore, e is irrational.

Hence the theorem.

Theorem 2.5.15

This theorem is known as ***Root test.***

Given $\sum a_n$. Take $a = \limsup_{n \to \infty} \sqrt[n]{|a_n|}$.

i) If $a < 1$, then $\sum a_n$ is convergent.

ii) If $a > 1$, then $\sum a_n$ is divergent.

iii) If $a = 1$, then the test gives no information.

Proof

i) Given that $a = \limsup_{n \to \infty} \sqrt[n]{|a_n|} < 1$.

Let us choose b such that $a < b < 1$. Hence, there exists $N \in J$ such that $|a_n|^{1/n} < b$, for all $n \geq N$. That is, $|a_n| < b^n$, for all $n \geq N$.

Since $0 < b < 1$, $\sum b_n$ is convergent. Hence, $\sum a_n$ is convergent.

Hence, part (i) is proved.

ii) Given that $a > 1$. We can find a sequence $\{n_k\}$ such that $\sqrt[n_k]{|a_{n_k}|} \to a$.

Therefore, $|a_n| > 1$, for finitely many values of n.

That implies, $\lim_{n\to\infty} a_n \neq 0$.

Therefore, $\sum a_n$ is divergent.

Hence part (ii) is proved.

iii) Given $a = 1$.

Let us consider the series $\sum \frac{1}{n^2}$.

$$\text{Now, } a = \limsup_{n\to\infty} |a_n|^{1/n} = \limsup_{n\to\infty} \left(\frac{1}{n^2}\right)^{1/n} = \limsup_{n\to\infty} \left(\frac{1}{n^{1/n}}\right)^2 = 1.$$

But, $\sum \frac{1}{n^2}$ is convergent.

Again, consider the series, $\sum \frac{1}{n}$.

$$\text{Now, } a = \limsup_{n\to\infty} |a_n|^{1/n} = \limsup_{n\to\infty} \left(\frac{1}{n}\right)^{1/n} = 1.$$

But, $\sum \frac{1}{n}$ is divergent.

Therefore, when $a = 1$, the test gives no information.

Hence part (iii) is proved.

Hence the theorem.

Theorem 2.5.16

This theorem as ***Ratio test***.

The series $\sum a_n$

i) converges, if $\limsup\limits_{n\to\infty}\left|\dfrac{a_{n+1}}{a_n}\right|<1$,

ii) diverges, if $\left|\dfrac{a_{n+1}}{a_n}\right|\geq 1$ for $n\geq n_0$, where n_0 is some fixed integer.

Proof

i) Let $a=\limsup\limits_{n\to\infty}\left|\dfrac{a_{n+1}}{a_n}\right|$.

Given that $a<1$. Hence, $0\leq a<1$.
Let us choose b such that $a<b<1$.

Since $a=\limsup\limits_{n\to\infty}\left|\dfrac{a_{n+1}}{a_n}\right|$, there exists $N\in J$ such that

$$\left|\frac{a_{n+1}}{a_n}\right|<b,\quad \text{for all } n\geq N.$$

That is, $|a_{n+1}|<b|a_n|$, for all $n\geq N$.
Hence, we have
$|a_{N+1}|<b|a_N|$, $|a_{N+2}|<b|a_{N+1}|$, $|a_{N+3}|<b|a_{N+2}|$, ..., $|a_{N+p}|<b|a_{N+(p-1)}|$,
where $p\in J$.
Multiplying all these inequalities, we get $|a_{N+p}|<b^p|a_N|$ .... (2.5.16.1)

Let $n>N$.
Hence by (2.5.16.1), we have $|a_n|=|a_{N+(n-N)}|<b^{n-N}|a_N|=b^n\left(|a_N|b^{-N}\right)$.

Since $0\leq b<1$, we have $\sum b^n$ is convergent.

Therefore, $\sum b^n|a_N|b^{-N}$ is convergent.

Hence, by comparison test, $\sum a_n$ is convergent.

Hence part (i) is proved.

ii) Given that $\left|\dfrac{a_{n+1}}{a_n}\right|\geq 1$, for $n\geq n_0$.

That is, $|a_{n+1}|\geq|a_n|$, for $n\geq n_0$.

Let $n > n_0$. Therefore, we have $|a_n| > |a_{n-1}| > |a_{n-2}| > .... > |a_{n_0}|$.

That is, $|a_n| > |a_{n_0}|$, for all $n \geq n_0$.

Since $|a_n|$ is fixed, $\lim_{n\to\infty} a_n \neq 0$.

Therefore, $\sum a_n$ is divergent.

Hence part (ii) is proved.
Hence the theorem.

Theorem 2.5.17

For any sequence $\{a_n\}$ of positive numbers,

i) $\liminf_{n\to\infty} \frac{a_{n+1}}{a_n} \leq \liminf_{n\to\infty} a_n^{1/n}$.

ii) $\limsup_{n\to\infty} a_n^{1/n} \leq \limsup_{n\to\infty} \frac{a_{n+1}}{a_n}$.

Proof

i) Now, $a_n > 0$, for all $n \in J$.

Let $\alpha = \liminf_{n\to\infty} \frac{a_{n+1}}{a_n}$. Let us assume that $\alpha = 0$.

Hence, $\liminf_{n\to\infty} a_n^{1/n} \geq 0 = \alpha = \liminf_{n\to\infty} \frac{a_{n+1}}{a_n}$.

Let $\alpha > 0$ and let us choose β such that $0 < \beta < \alpha$.

Since $\alpha = \liminf_{n\to\infty} \frac{a_{n+1}}{a_n}$ and $\beta < \alpha$, there exists $N \in J$ such that

$\frac{a_{n+1}}{a_n} > \beta$, for all $n \geq N$.

That is, $a_{n+1} > \beta a_n$, for all $n \geq N$.

Hence, we have

$a_{N+1} > \beta a_N$, $a_{N+2} > \beta a_{N+1}$, $a_{N+3} > \beta a_{N+2}$, ..., $a_{N+p} > \beta a_{N+(p-1)}$, for all $p \in J$.

Multiplying these inequalities, we get $a_{N+p} > \beta^p a_N$. .... (2.5.17.1)

Let $n > N$.

By equation (2.5.17.1), we have $a_n = a_{N+(n-N)} > \beta^{n-N} a_N = \beta^n (a_N \beta^{-N})$, for all $n \geq N$.

Taking n^{th} root, we have $a_n^{1/n} > \beta (a_N \beta^{-N})^{1/n}$, for all $n \geq N$ .... (2.5.17.2)

Now, $a_N \beta^{-N} > 0$ and hence $\lim_{n\to\infty} (a_N \beta^{-N})^{1/n} = 1$.

Taking limit infimum in equation (2.5.17.2), we get $\liminf_{n\to\infty} a_n^{1/n} > \beta$.

This is true for every $\beta < \alpha$.

Therefore, $\liminf_{n\to\infty} a_n^{1/n} \geq \alpha = \liminf_{n\to\infty} \frac{a_{n+1}}{a_n}$.

Hence part (i) is proved.

ii) Now, $a_n > 0$, for all $n \in J$.

Let $\alpha = \limsup_{n\to\infty} \frac{a_{n+1}}{a_n}$ and $\alpha = \infty$.

Hence, $\limsup_{n\to\infty} a_n^{1/n} \leq \infty = \alpha = \limsup_{n\to\infty} \frac{a_{n+1}}{a_n}$.

Let α be finite and let us choose β such that $0 < \alpha < \beta$.

Since $\alpha = \limsup_{n\to\infty} \frac{a_{n+1}}{a_n}$ and $\alpha < \beta$, there exists $N \in J$ such that $\frac{a_{n+1}}{a_n} < \beta$, for all $n \geq N$.

That is, $a_{n+1} < \beta a_n$, for all $n \geq N$.

Hence we have,

$a_{N+1} < \beta a_N$, $a_{N+2} < \beta a_{N+1}$, $a_{N+3} < \beta a_{N+2}$, ..., $a_{N+p} < \beta a_{N+(p-1)}$, for all $p \in J$.

Multiplying all these inequalities, we get $a_{N+p} < \beta^p a_N$. .... (2.5.17.3)

Let $n > N$.

By equation (2.5.17.3), we have $a_n = a_{N+(n-N)} < \beta^{n-N} a_N = \beta^n (a_N \beta^{-N})$, for all $n \geq N$.

Taking n^{th} root, we have $a_n^{1/n} < \beta\left(a_N\beta^{-N}\right)^{1/n}$, for all $n \geq N$. .... (2.5.17.4)

Now, $a_N\beta^{-N} > 0$.

Hence, by theorem 2.4.17 (ii), we get $\lim\limits_{n\to\infty}\left(a_n\beta^{-N}\right)^{1/n} = 1$.

Now, by taking limit supremum in equation (2.5.17.2), we get $\limsup\limits_{n\to\infty} a_n^{1/n} < \beta$.

This is true for all $\beta > \alpha$.

Hence, $\limsup\limits_{n\to\infty} a_n^{1/n} < \alpha = \limsup\limits_{n\to\infty} \frac{a_{n+1}}{a_n}$.

Hence part (ii) is proved.

Hence the theorem.

2.6 Power series

Definition 2.6.1

Given $\{c_n\}$ of complex numbers. The series $\sum c_n z^n$ is called a power series. The numbers c_n are called the coefficients of this power series.

Theorem 2.6.2

Given a power series $\sum c_n z^n$. Put $a = \limsup\limits_{n\to\infty} |c_n|^{1/n}$ and $R = \frac{1}{a}$, then $\sum c_n z^n$ is convergent, if $|z| < R$ and divergent, if $|z| > R$. R is called the radius of convergence.

Proof

Let $a_n = c_n z^n$.

Hence, by using the given data, we get,

$$\limsup_{n\to\infty} |a_n|^{1/n} = \limsup_{n\to\infty} \left|c_n z^n\right|^{1/n} = |z| \limsup_{n\to\infty} |c_n|^{1/n} = |z|a = \frac{|z|}{R}.$$

Hence, by root test, we get $\sum c_n z^n$ is convergent, if $\frac{|z|}{R} < 1$ and

$\sum c_n z^n$ is divergent, if $\frac{|z|}{R} > 1$.

That is, $\sum c_n z^n$ is convergent, if $|z| < R$ and $\sum c_n z^n$ is divergent, if $|z| > R$.

Hence the theorem.

Examples 2.6.3

1. Discuss the convergence and divergence of the power series $\sum_{n=0}^{\infty} n^n z^n$.

Solution

Given that, $c_n = n^n$.

Hence, $\lim_{n\to\infty} \sup|c_n|^{1/n} = \lim_{n\to\infty} \sup(n^n)^{1/n} = \lim_{n\to\infty} \sup n$.

Now, $R = \dfrac{1}{a} = \dfrac{1}{\lim\limits_{n\to\infty} \sup|c_n|^{1/n}} = \dfrac{1}{\lim\limits_{n\to\infty} \sup n} = \dfrac{1}{\infty} = 0$.

Hence, by theorem 2.6.2, the power series is convergent, if $|z| < 0$ and divergent, if $|z| > 0$. But, $|z| \geq 0$ always. Therefore, the power series is divergent, if $|z| > 0$.

When $|z| = 0$, then $\sum_{n=0}^{\infty} n^n z^n = 0$.

Therefore, $\sum_{n=0}^{\infty} n^n z^n$ is convergent.

2. Discuss the convergence and divergence of the power series $\sum_{n=0}^{\infty} \dfrac{z^n}{n!}$.

Solution

Given that, $c_n = \dfrac{1}{n!}$

Now, $\dfrac{c_{n+1}}{c_n} = \dfrac{n!}{(n+1)!} = \dfrac{n!}{n!(n+1)} = \dfrac{1}{n+1}$.

Therefore, $\lim_{n\to\infty} \sup \dfrac{c_{n+1}}{c_n} = 0$.

We know that, $0 \leq \lim_{n\to\infty} \sup c_n^{1/n} \leq \lim_{n\to\infty} \sup \dfrac{c_{n+1}}{c_n} = 0$.

Hence, $\limsup\limits_{n\to\infty} c_n^{1/n} = 0$ and $R = \dfrac{1}{a} = \dfrac{1}{\limsup\limits_{n\to\infty} c_n^{1/n}} = \dfrac{1}{0} = \infty$.

Therefore, the given power series is convergent in the whole complex plane.

3. Discuss the convergence and divergence of the power series $\sum z^n$.

Solution

Given that, $c_n = 1$.

Hence, $\limsup\limits_{n\to\infty} |c_n|^{1/n} = 1$.

Now, $R = \dfrac{1}{a} = \dfrac{1}{\limsup\limits_{n\to\infty} |c_n|^{1/n}} = 1$.

Therefore, $\sum\limits_{n=0}^{\infty} z^n$ is convergent, if $|z| < 1$ and divergent, if $|z| > 1$.

If $|z| = 1$, then $|z|^n = 1$, for all n.

Hence, $\lim\limits_{n\to\infty} |z^n| \neq 0$.

Therefore, $\sum\limits_{n=0}^{\infty} z^n$ is divergent.

Theorem 2.6.4

Given two sequences $\{x_n\}$ and $\{y_n\}$. Put $X_n = \sum\limits_{k=0}^{n} x_k$, and $X_{-1} = 0$.

If $0 \leq p \leq q$, then $\sum\limits_{n=p}^{q} x_n y_n = \sum\limits_{n=p}^{q-1} X_n (y_n - y_{n+1}) + X_q y_q - X_{p-1} y_p$. ...... (2.6.4.1)

Equation (2.6.4.1) is called ***partial summation formula.***

Proof

Let us take, $x_n = X_n - X_{n-1}$, for all $n \geq 0$.

Given that, $0 \leq p \leq q$.

Now, $\sum\limits_{n=p}^{q} x_n y_n = \sum\limits_{n=p}^{q} (X_n - X_{n-1}) y_n = \sum\limits_{n=p}^{q} X_n y_n - \sum\limits_{n=p}^{q} X_{n-1} y_n$.

$$= \sum_{n=p}^{q-1} X_n y_n + X_q y_q - \sum_{n=p-1}^{q-1} X_n y_{n+1}$$

$$= \sum_{n=p}^{q-1} X_n y_n + X_q y_q - \left[\sum_{n=p}^{q-1} X_n y_{n+1} + X_{p-1} y_p \right].$$

Therefore, $\sum_{n=p}^{q} x_n y_n = \sum_{n=p}^{q-1} X_n (y_n - y_{n+1}) + X_q y_q - X_{p-1} y_p$.

Hence the theorem.

Theorem 2.6.5

This theorem is known as ***Dirichlet's test***.

i) Suppose, the partial sums X_n of S_n is bounded,

ii) $y_0 \geq y_1 \geq y_2 \geq$........

iii) $\lim_{n \to \infty} y_n = 0$, then $\sum x_n y_n$ is convergent.

Proof

Given that, the partial sum X_n of S_n is bounded.

Therefore, there exists $M > 0$, such that $|X_n| \leq M$, for all $n \in J$ ..... (2.6.5.1)

Since $\lim_{n \to \infty} y_n = 0$, we have, for given $\varepsilon > 0$, there exists $N \in J$ such that

$|y_n| < \dfrac{\varepsilon}{4M}$, for all $n \geq N$. .... (2.6.5.2)

Let $q \geq p \geq 0$.

Hence, $\sum_{n=p}^{q} x_n y_n = \sum_{n=p}^{q-1} X_n (y_n - y_{n+1}) + X_q y_q - X_{p-1} y_p$.

That implies, $\left| \sum_{n=p}^{q} x_n y_n \right| \leq \sum_{n=p}^{q-1} |X_n| |y_n - y_{n+1}| + |X_q| |y_q| + |X_{p-1}| |y_p|$.

Since $y_n \geq y_{n+1}$ and also from equations (2.6.5.1) & (2.6.5.2), we get

$$\left| \sum_{n=p}^{q} x_n y_n \right| \leq M \sum_{n=p}^{q-1} (y_n - y_{n+1}) + M |y_q| + M |y_p|$$

$$\leq M\left(y_p - y_q\right) + M\left|y_q\right| + M\left|y_p\right| = M\left[\left|y_p\right| + \left|y_q\right| + \left|y_q\right| + \left|y_p\right|\right] < M\frac{4\varepsilon}{4M} = \varepsilon.$$

Hence, $\left|\sum_{n=p}^{q} x_n y_n\right| < \varepsilon$, for all $q \geq p \geq \mathrm{N}$.

Therefore, by Cauchy's criteria for convergence of a series, $\sum x_n y_n$ is convergent.

Hence the theorem.

Theorem 2.6.6

This theorem is known as ***Leibnitz's theorem***.

Suppose (i) $|z_1| \geq |z_2| \geq |z_3| \geq \ldots\ldots.$

(ii) $z_{2n-1} \geq 0$ and $z_{2n} \leq 0$, for $n = 1,2,3,\ldots\ldots$

(iii) $\lim_{n\to\infty} z_n = 0$,

then $\sum_{n=1}^{\infty} z_n$ is convergent.

Proof

Let us assume that $x_n = (-1)^{n+1}$ and $y_n = |z_n|$.

Let X_n be the sequence of partial sums of $\sum_{n=1}^{\infty} x_n$.

From our assumption, we have $X_n = \begin{cases} 1, & \text{if } n \text{ is odd} \\ 0, & \text{if } n \text{ is even} \end{cases}$

Hence, $|X_n| \leq 1$, for all n. Therefore, $\{X_n\}$ is bounded.

Given that $|z_1| \geq |z_2| \geq |z_3| \geq \ldots\ldots$, thus, we have $y_1 \geq y_2 \geq y_3 \geq \ldots\ldots$

Also, given that $\lim_{n\to\infty} z_n = 0$, which implies $\lim_{n\to\infty} |z_n| = 0$. That is, $\lim_{n\to\infty} y_n = 0$.

Hence by Dirichlet's test, we have $\sum_{n=1}^{\infty} x_n y_n$ is convergent.

Now, $\sum_{n=1}^{\infty} x_n y_n = x_1 y_1 + x_2 y_2 + \ldots\ldots + x_n y_n$.

Since $x_n = (-1)^{n+1}$ and $y_n = |z_n|$, we have

$$\sum_{n=1}^{\infty} x_n y_n = y_1 - y_2 + y_3 - y_4 + \ldots\ldots = |z_1| - |z_2| + |z_3| - |z_4| + \ldots\ldots$$

Again, since $|z_1| = z_1, |z_2| = -z_2, \ldots\ldots$, we get

$$\sum_{n=1}^{\infty} x_n y_n = z_1 + z_2 + z_3 + z_4 + \ldots\ldots = \sum_{n=1}^{\infty} z_n .$$

Therefore, $\sum_{n=1}^{\infty} z_n$ is convergent.

Hence the theorem.

Definition 2.6.7

The series $\sum_{n=1}^{\infty} (-1)^{n+1} a_n$ where $a_n \geq 0$ is called an *alternating series*. By Leibnitz theorem, if $\sum_{n=1}^{\infty} (-1)^{n+1} a_n$ is an alternating series with $\{a_n\}$ is monotonic decreasing and converging to 0, then $\sum_{n=1}^{\infty} (-1)^{n+1} a_n$ is convergent.

Example 2.6.8

Discuss the convergence of the series $\sum_{n=1}^{\infty} (-1)^{n+1} \frac{1}{n}$.

Solution

Let $a_n = \frac{1}{n}$.

Hence, $\lim_{n \to \infty} a_n = \lim_{n \to \infty} \frac{1}{n} = 0$.

We have to prove that, a_n is monotonic decreasing. That is, to prove that $a_n \geq a_{n+1}$.

Now, $a_n - a_{n+1} = \frac{1}{n} - \frac{1}{n+1} > 0$ and hence $a_n > a_{n+1}$.

Therefore, a_n is monotonic decreasing.

Hence, by Leibnitz test, $\sum_{n=1}^{\infty}(-1)^{n+1}a_n$ is convergent and hence $\sum_{n=1}^{\infty}(-1)^{n+1}\frac{1}{n}$ is convergent.

Hence the theorem.

Theorem 2.6.9

Suppose the radius of convergence of the power series $\sum_{n=0}^{\infty}c_n z^n$ is 1 and $c_0 \geq c_1 \geq c_2 \geq ......$ and $\lim_{n\to\infty} c_n = 0$. Then, $\sum c_n z^n$ is convergent at all points on the circle $|z| = 1$ except possibly at $z = 1$.

Proof

Let us assume $a_n = z^n$ and $b_n = c_n$.

Since c_n is monotonic decreasing and $\lim_{n\to\infty} c_n = 0$, we have b_n is monotonic decreasing and $\lim_{n\to\infty} b_n = 0$.

Let A_n be the sequence of partial sum of $\sum a_n$.

Therefore, $|A_n| = |1 + z + z^2 + z^3 + ... + z^{n-1}| = \left|\frac{1-z^n}{1-z}\right|$, if $z \neq 1$.

That is, $|A_n| \leq \frac{1+|z|^n}{|1-z|}$, if $z \neq 1$, which implies that $|A_n| \leq \frac{2}{|1-z|}$, if $|z| = 1$ and $z \neq 1$.

Therefore, A_n is bounded. Hence, by Dirichlet's test, $\sum_{n=0}^{\infty} a_n b_n$ is convergent, if $|z| = 1$ and $z \neq 1$.

Therefore, $\sum c_n z^n$ is convergent at all points on the circle $|z| = 1$, except possibly at $z = 1$.

Hence the theorem.

Definition 2.6.10

$\sum a_n$ is said to be *absolute convergence*, if $\sum |a_n|$ is convergent.

Theorem 2.6.11

Every absolutely convergent series is convergent.

Proof

Given that $\sum a_n$ is absolutely convergent and hence $\sum |a_n|$ is convergent. Therefore, for given $\varepsilon > 0$, there exists $N \in J$, such that $\left|\sum_{k=n}^{m} |a_k|\right| \leq \varepsilon$, for all $m \geq n \geq N$.

That is, $\sum_{k=n}^{m} |a_k| \leq \varepsilon$, for all $m \geq n \geq N$. .... (2.6.11.1)

Let $m \geq n \geq N$. Hence, by equation (2.6.11.1), we have $\left|\sum_{k=n}^{m} a_k\right| \leq \sum_{k=n}^{m} |a_k| \leq \varepsilon$.

Therefore, by Cauchy's criteria for convergence of a series, $\sum a_n$ is convergent.

Hence the theorem.

Note 2.6.12

The converse of theorem 2.6.11 need not be true. That is, every convergent series need not be absolutely convergent.

Theorem 2.6.13

If $\sum_{n=0}^{\infty} x_n = X$ and $\sum_{n=0}^{\infty} y_n = Y$, then

i) $\sum_{n=0}^{\infty} (x_n + y_n) = X + Y$

ii) $\sum_{n=0}^{\infty} z x_n = zX$, for some scalar z.

Proof

Let $X_n = \sum_{k=0}^{n} x_k$ and $Y_n = \sum_{k=0}^{n} y_k$.

Hence, $X_n + Y_n = \sum_{k=0}^{n}(x_k + y_k)$ and $zX_n = \sum_{k=0}^{n} zx_k$.

Since $\sum_{n=0}^{\infty} x_n = X$ and $\sum_{n=0}^{\infty} y_n = Y$, we have $X_n \to X$ and $Y_n \to Y$.

Hence, $X_n + Y_n \to X + Y$ and $zX_n \to zX$.

Therefore, $\sum_{n=0}^{\infty}(x_n + y_n) = X + Y$ and $\sum_{n=0}^{\infty} zx_n = zX$.

Hence the theorem.

Definition 2.6.14

The Cauchy product of two series $\sum x_n$ and $\sum y_n$ is $\sum z_n$, where

$$z_n = x_0 y_n + x_1 y_{n-1} + ... + x_n y_0 = \sum_{k=0}^{n} x_k y_{n-k}.$$

Theorem 2.6.15

The product of two convergent series may diverge.

Proof

Let $x_n = \frac{(-1)^n}{\sqrt{n+1}}$ and $y_n = \frac{(-1)^n}{\sqrt{n+1}}$.

Now, $\sum_{n=0}^{\infty} x_n = \sum_{n=0}^{\infty} \frac{(-1)^n}{\sqrt{n+1}}$, which is an alternating series.

Here, $\frac{1}{\sqrt{n+1}}$ is monotonic decreasing and converges to 0.

Hence, by Leibnitz test, $\sum a_n$ is convergent. Similarly, we can prove that $\sum b_n$ is convergent.

Let the product of $\sum x_n$ and $\sum y_n$ be $\sum z_n$.

Now, $z_n = \sum_{k=0}^{n} x_k y_{n-k} = \sum_{k=0}^{n} \frac{(-1)^k}{\sqrt{k+1}} \frac{(-1)^{n-k}}{\sqrt{n-k+1}} = (-1)^n \sum_{k=0}^{n} \frac{1}{\sqrt{k+1}\sqrt{n-k+1}}$.

Consider, $(k+1)(n-k+1) = \left(\frac{n}{2}+1\right)^2 - \left(\frac{n}{2}-k\right)^2 \leq \left(\frac{n}{2}+1\right)^2$.

That implies, $\dfrac{1}{\sqrt{k+1}\sqrt{n-k+1}} \geq \dfrac{1}{(n/2)+1} = \dfrac{2}{n+2}$.

Therefore, $|z_n| = \left|\sum_{k=0}^{n} \dfrac{1}{\sqrt{k+1}\sqrt{n-k+1}}\right|$

$$= \sum_{k=0}^{n} \frac{1}{\sqrt{k+1}\sqrt{n-k+1}} \geq \sum_{k=0}^{n} \frac{2}{n+2} = \frac{2}{n+2}(n+1)$$

Hence, $\lim_{n\to\infty} |z_n| \geq \lim_{n\to\infty} \dfrac{2(1+(1/n))}{1+(2/n)} \geq 2$.

Therefore, $\lim_{n\to\infty} |z_n| \neq 0$ and hence $\sum z_n$ is divergent.

Hence the theorem.

Theorem 2.6.16

This theorem is known as ***Mertern's theorem***.

Suppose (i) $\sum_{n=0}^{\infty} x_n$ is absolutely continuous,

(ii) $\sum_{n=0}^{\infty} x_n = X$,

(iii) $\sum_{n=0}^{\infty} y_n = Y$,

(iv) $z_n = \sum_{k=0}^{n} x_k y_{n-k}$, then $\sum_{n=0}^{\infty} z_n = XY$.

Proof

Let $X_n = \sum_{k=0}^{n} x_k$, $\quad Y_n = \sum_{k=0}^{n} y_k$ and $Z_n = \sum_{k=0}^{n} z_k$.

Since, $\sum_{n=0}^{\infty} x_n = X$, then $X_n \to X$ as $n \to \infty$ and

$\sum_{n=0}^{\infty} y_n = Y$, then $Y_n \to Y$ as $n \to \infty$.

Let us take $\beta_n = Y_n - Y$. .... (2.6.16.1)

Now, $Z_n = z_0 + z_1 + z_2 + ... + z_n$.

$= x_0 y_0 + (x_0 y_1 + x_1 y_0) + (x_0 y_2 + x_1 y_1 + x_2 y_0) + ... + (x_0 y_n + x_1 y_{n-1} + ... + x_n y_0).$

$= x_0 (y_0 + y_1 + ... + y_n) + x_1 (y_0 + y_1 + ... + y_{n-1}) + ... + x_n y_0.$

$= x_0 Y_n + x_1 Y_{n-1} + ... + x_n Y_0.$

Hence, by equation (2.6.16.1), we have

$$Z_n = x_0(\beta_n + Y) + x_1(\beta_{n-1} + Y) + ... + x_n(\beta_0 + Y).$$

$$= Y(x_0 + x_1 + ... + x_n) + x_0\beta_n + x_1\beta_{n-1} + ... + x_n\beta_0.$$

Let $Z_n = YX_n + \gamma_n$, where $\gamma_n = x_0\beta_n + x_1\beta_{n-1} + ... + x_n\beta_0$.

Since $X_n \to X$, as $n \to \infty$, we have $YX_n \to YX$ as $n \to \infty$.

To prove that, $Z_n \to XY$, as $n \to \infty$, it is enough if we prove that $\gamma_n \to 0$ as $n \to \infty$.

Since $Y_n \to Y$, we have $Y_n - Y \to 0$.

Hence $\beta_n \to 0$ as $n \to \infty$.

That is, for given $\varepsilon > 0$, there exists $N \in J$ such that $|\beta_n| \le \varepsilon$, for all $n \ge N$ .... (2.6.16.2)

Since $\sum X_n$ converges absolutely, $\sum_{n=0}^{\infty} |x_n| = \alpha < \infty$. .... (2.6.16.3)

Let $n > N$.

Now,

$$|\gamma_n| = |x_n\beta_0 + x_{n-1}\beta_1 + ... + x_0\beta_n|$$

$$= |(x_n\beta_0 + x_{n-1}\beta_1 + ... + x_{n-N}\beta_N) + (x_{n-N-1}\beta_{N+1} + ... + x_0\beta_n)|.$$

$$\le |x_n\beta_0 + x_{n-1}\beta_1 + ... + x_{n-N}\beta_N| + |x_{n-N-1}||\beta_{N+1}| + ... + |x_0||\beta_n|.$$

Hence by equation (2.6.16.2). we have

$$|\gamma_n| \le |x_n\beta_0 + x_{n-1}\beta_1 + ... + x_{n-N}\beta_N| + \varepsilon[|x_{n-N-1}| + ... + |x_0|].$$

By using equation (2.6.16.3), we get

$$|\gamma_n| \le |x_n\beta_0 + x_{n-1}\beta_1 + ... + x_{n-N}\beta_N| + \varepsilon\alpha. \quad .... (2.6.16.4)$$

Since $\sum x_n$ is convergent, $\lim_{n \to \infty} x_n = 0$.

Let us fix N and take $\limsup_{n\to\infty}$. in the equation (2.6.16.4).

Hence we get $\limsup_{n\to\infty}|\gamma_n| \le \varepsilon\alpha$.

Since ε is arbitrary, we have $\limsup_{n\to\infty}|\gamma_n| = 0$.

Therefore, $0 \le \liminf_{n\to\infty}|\gamma_n| \le \limsup_{n\to\infty}|\gamma_n| = 0$.

That is, $\liminf_{n\to\infty}|\gamma_n| = \limsup_{n\to\infty}|\gamma_n| = 0$, which implies $\lim_{n\to\infty}|\gamma_n| = 0$.

Hence, $\lim_{n\to\infty}\gamma_n = 0$.

Therefore, $Z_n \to XY$. That implies, $\sum_{n=0}^{\infty} Z_n = XY$.

Hence the theorem.

Theorem 2.6.17

This theorem is known as ***Cauchy's First limit theorem***.

If $x_n \to x$, then $\dfrac{x_0 + x_1 + \ldots\ldots. + x_n}{n+1} \to x$.

Theorem 2.6.18

This theorem is known as ***Cesara's theorem***.

If $x_n \to x$ and $y_n \to y$, then $\dfrac{x_0 y_n + x_1 y_{n-1} + \ldots.. + x_n y_0}{n+1} \to xy$.

Theorem 2.6.19

This theorem is known as ***Abel's theorem***.

Suppose (i) $\sum_{n=0}^{\infty} x_n = X$, (ii) $\sum_{n=0}^{\infty} y_n = Y$,

(iii) $z_n = \sum_{k=0}^{n} x_k y_{n-k}$, (iv) $\sum_{n=0}^{\infty} z_n = z$, then $Z = XY$.

Proof

Let $X_n = \sum_{k=0}^{n} x_k$, $Y_n = \sum_{k=0}^{n} y_k$ and $Z_n = \sum_{k=0}^{n} z_k$.

Since $\sum_{n=0}^{\infty} x_n = X$, $\sum_{n=0}^{\infty} y_n = Y$, $\sum_{n=0}^{\infty} z_n = Z$,

we have $X_n \to X$, $Y_n \to Y$ and $Z_n \to Z$.

Now, $Z_n = z_0 + z_1 + \ldots\ldots + z_n$.

$$= x_0 y_0 + (x_0 y_1 + x_1 y_0) + (x_0 y_2 + x_1 y_1 + x_2 y_0) + \ldots\ldots$$

$$\ldots\ldots + (x_o y_n + x_1 y_{n-1} + \ldots\ldots + x_n y_0).$$

$$= x_0(y_0 + y_1 + \ldots\ldots + y_n) + x_1(y_0 + y_1 + \ldots\ldots + y_{n-1}) + \ldots\ldots + x_n y_0.$$

$$= x_0 Y_n + x_1 Y_{n-1} + \ldots\ldots + x_n Y_0.$$

Hence $Z_0 + \ldots\ldots + Z_n = x_0 Y_0 + (x_0 Y_1 + x_1 Y_0) + (x_0 Y_2 + x_1 Y_1 + x_2 Y_0) + \ldots\ldots$

$$\ldots\ldots + (x_0 Y_n + \ldots\ldots + x_n Y_0).$$

$$= Y_0(x_0 + x_1 + \ldots\ldots + x_n) + Y_1(x_0 + x_1 + \ldots\ldots + x_{n-1}) + \ldots\ldots + Y_n x_0.$$

$$= Y_0 X_n + Y_1 X_{n-1} + \ldots\ldots + Y_n X_0.$$

Therefore, $\dfrac{Z_0 + Z_1 + \ldots\ldots + Z_n}{n+1} = \dfrac{Y_0 X_n + Y_1 X_{n-1} + \ldots\ldots + Y_n X_0}{n+1}$. .... (2.6.19.1)

Since $Z_n \to Z$, then by Cauchy's first limit theorem,

$$\frac{Z_0 + Z_1 + \ldots\ldots + Z_n}{n+1} \to Z$$

Since $X_n \to X$ and $Y_n \to Y$, by Cesara's theorem, we have

$$\frac{Y_0 X_n + \ldots\ldots + Y_n X_0}{n+1} \to XY.$$

From equation (2.6.19.1), taking limit on both sides, we get $Z = XY$.

Hence the theorem.

Theorem 2.6.20

The product of two absolutely convergent series is absolutely convergent.

Proof

Let $\sum_{n=0}^{\infty} x_n$ and $\sum_{n=0}^{\infty} y_n$ be two absolutely convergent series.

Hence, $\sum_{n=0}^{\infty} |x_n|$ and $\sum_{n=0}^{\infty} |y_n|$ are convergent.

Let $\sum_{n=0}^{\infty}|x_n| = \alpha < \infty$, $\sum_{n=0}^{\infty}|y_n| = \beta$ and $z_n = \sum_{n=0}^{\infty} x_k y_{n-k}$.

Then, $\sum_{n=0}^{\infty} z_n$ is the Cauchy product of $\sum x_n$ and $\sum y_n$.

We have to prove that, $\sum_{n=0}^{\infty} z_n$ is absolutely convergent.

That is, we have to prove that, $\sum_{n=0}^{\infty}|z_n|$ is convergent.

Let $Z_n = |z_0| + |z_1| + \ldots.. + |z_n|$.

Hence, $\sum_{n=0}^{\infty}|Z_n|$ is a series of non-negative terms.

In order to prove that, $\sum_{n=0}^{\infty}|Z_n|$ is convergent, it is enough if we prove its sequence of partial sums $\{z_n\}$ is bounded.

Now, $Z_n = |z_0| + |z_1| + \ldots\ldots + |z_n| = |x_0 y_0| + |x_0 y_1 + x_1 y_0| + \ldots..$

$$\ldots.. + |x_0 y_n + x_1 y_{n-1} + \ldots.. + x_n y_0|.$$

$$\leq |x_0||y_0| + |x_0||y_1| + |x_1||y_0| + \ldots.. + |x_0||y_n| + |x_1||y_{n-1}| + \ldots.. + |x_n||y_0|.$$

$$\leq (|x_0| + |x_1| + \ldots.. + |x_n|)\ (|y_0| + |y_1| + \ldots.. + |y_n|) \leq \alpha\beta.$$

Hence, $\{z_n\}$ is bounded and therefore $\sum_{n=0}^{\infty}|z_n|$ is convergent.

That is, $\sum_{n=0}^{\infty} z_n$ is absolutely convergent.

Hence the theorem.

Definition 2.6.21

Rearrangement of a series

Let $\sum a_n$ be a series and $\{k_n\}$ be a sequence of positive terms in which every positive integer appears exactly once. Let $a_n' = a_{k_n}$, for $n = 1, 2, 3, \ldots\ldots.$ Then $\sum a_n'$ is called a rearrangement of $\sum a_n$.

Example 2.6.22

Consider the series $\sum_{n=1}^{\infty}(-1)^{n+1}\frac{1}{n}$.

Now, $\sum_{n=1}^{\infty}(-1)^{n+1}\frac{1}{n}=1-\frac{1}{2}+\frac{1}{3}-\frac{1}{4}+\frac{1}{5}-.......$ .... (2.6.22.1)

is convergent.

Now, $\sum_{n=1}^{\infty}(-1)^{n+1}\frac{1}{n}=1-\frac{1}{2}+\frac{1}{3}-\left(\frac{1}{4}-\frac{1}{5}\right)-\left(\frac{1}{6}-\frac{1}{7}\right)-......<1-\frac{1}{2}+\frac{1}{3}=\frac{5}{6}$.

Consider the rearrangement,

$$\sum_{n=1}^{\infty}(-1)^{n+1}\frac{1}{n}=\left(1+\frac{1}{3}-\frac{1}{2}\right)+\left(\frac{1}{5}+\frac{1}{7}-\frac{1}{4}\right)+\left(\frac{1}{9}+\frac{1}{11}-\frac{1}{6}\right)+......$$

.... (2.6.22.2)

Let s_n' be the sequence of partial sums of equation (2.6.22.2).

Now, for $k\geq 1$, we have

$$\frac{1}{4k-3}+\frac{1}{4k-1}-\frac{1}{2k}=\frac{(4k-1)2k+(4k-3)2k-(4k-1)(4k-3)}{(4k-3)(4k-1)2k}>0.$$

Therefore, $s_3'<s_6'<s_9'<.......$

Now, $\limsup_{n\to\infty} s_n' > s_3' = 1-\frac{1}{2}+\frac{1}{3}=\frac{5}{6}$.

Therefore, $\{s_n'\}$ does not converge to $\sum_{n=1}^{\infty}(-1)^{n+1}\frac{1}{n}$.

Theorem 2.6.23

This theorem is known as ***Riemann's theorem***.

Let $\sum a_n$ be a series of real numbers which converges but not absolutely. If $-\infty\leq\alpha\leq\beta\leq+\infty$, then there is a rearrangement $\sum a_n'$ of $\sum a_n$ with partial sums s_n' such that $\liminf_{n\to\infty} s_n'=\alpha$ and $\limsup_{n\to\infty} s_n'=\beta$.

Proof

Given that, $\sum a_n$ is convergent and $\sum |a_n|$ is divergent.

Also, given that, $-\infty \leq \alpha \leq \beta \leq +\infty$.

Let us assume that, $p_n = \dfrac{|a_n| + a_n}{2}$ and $q_n = \dfrac{|a_n| - a_n}{2}$, for $n = 1, 2, 3, \ldots\ldots$

Therefore, $p_n = \begin{cases} a_n, & \text{if } a_n > 0 \\ 0, & \text{if } a_n \leq 0 \end{cases}$ and $q_n = \begin{cases} 0, & \text{if } a_n > 0 \\ -a_n, & \text{if } a_n \leq 0 \end{cases}$

Hence, $p_n \geq 0$ and $q_n \geq 0$.

Now, $p_n + q_n = |a_n|$ and $p_n + q_n = a_n$.

We have to prove that, $\sum p_n$ and $\sum q_n$ diverge.

Suppose, both $\sum p_n$ and $\sum q_n$ converge.

Hence, $\sum (p_n + q_n)$ is convergent and therefore, $\sum |a_n|$ is convergent, which is a contradiction, since $\sum |a_n|$ is divergent.

Suppose $\sum p_n$ is convergent and $\sum q_n$ is divergent.

Now, $p_n - q_n = a_n$ implies $p_n - a_n = q_n$.

Since $\sum p_n$ is convergent and $\sum a_n$ is convergent, we have $\sum (p_n - a_n)$ is convergent.

That is, $\sum q_n$ is convergent, which is a contradiction to $\sum q_n$ is divergent.

Suppose $\sum p_n$ is divergent and $\sum q_n$ is convergent.

Now, $p_n - q_n = a_n$ implies $p_n = q_n + a_n$.

Since $\sum q_n$ is convergent and $\sum a_n$ is convergent, we have $\sum (q_n + a_n)$ is convergent.

That is, $\sum p_n$ is convergent, which is a contradiction to $\sum p_n$ is divergent.

Therefore, both $\sum p_n$ and $\sum q_n$ diverge to $+\infty$.

Let P_n denote the non-negative terms of $\sum a_n$ in the order in which they occur.

Let Q_n denote the absolute values of the negative terms of $\sum a_n$ in the order in which they occur.

Now, the series $\sum P_n$ and $\sum Q_n$ differ from the series $\sum p_n$ and $\sum q_n$ only by 0 – terms and hence $\sum P_n$ and $\sum Q_n$ are divergent.

Choose real sequences $\{\alpha_n\}$ and $\{\beta_n\}$ such that $\alpha_n \to \alpha$ and $\beta_n \to \beta$.

Here, $\alpha_n < \beta_n$ and $\beta_1 > 0$.

Since $\sum P_n$ diverges to ∞, we can choose a least positive integer m_1, such that $P_1 + P_2 + ... + P_{m_1} > \beta_1$.

Since $\sum Q_n$ diverges to ∞, we can choose a least positive integer k_1, such that $P_1 + P_2 + ... + P_{m_1} - Q_1 - Q_2 - ...... - Q_{k_1} < \alpha_1$.

Choose a least positive integer m_2, such that

$P_1 + P_2 + ... + p_{m_1} - Q_1 - Q_2 - ..... - Q_{k_1} + P_{m_1+1} + ... + P_{m_2} > \beta_2$.

Choose a least positive integer k_2, such that

$P_1 + P_2 + ... + P_{m_1} - Q_1 - Q_2 - .... - Q_{k_1} + P_{m_1+1} + ... + P_{m_2} - Q_{k_1+1} + ... + Q_{k_2} < \alpha_2$.

Since $\sum P_n$ and $\sum Q_n$ are divergent, the above process will not stop at a finite stage.

Therefore, we get the sequences $\{m_n\}$ and $\{k_n\}$ of positive integers and a rearrangement

$$P_1 + P_2 + ... + P_{m_1} - Q_1 - Q_2 - ...... - Q_{k_1} + P_{m_1+1} + ...$$
$$.... + P_{m_2} - Q_{k_1+1} - ...... - Q_{k_2} + ... \qquad .... (2.6.23.1)$$

of $\sum a_n$.

Let $\{s'_n\}$ be the sequence of partial sums of equation (2.6.23.1).

Let x_n and y_n be the partial sums of equation (2.6.23.1) whose last terms are P_{m_n} and $-Q_{k_n}$.

Then, $\{x_n\}$ and $\{y_n\}$ are subsequences of $\{s'_n\}$.

We claim that, $|x_n - \beta_n| \leq P_{m_n}$.

Now, $x_n = P_1 + P_2 + ... + P_{m_1} - Q_1 - Q_2 - ...... - Q_{k_1} + P_{m_1+1} + ... + P_{m_n} > \beta_n$.

Hence, $x_n > \beta_n$ implies $x_n - \beta_n > 0$.

Suppose $|x_n - \beta_n| > P_{m_n}$.

Since $x_n - \beta_n > 0$, from the above equation, we have $x_n - \beta_n > P_{m_n}$.

That is, $x_n - P_{m_n} > \beta_n$.

Hence, $$P_1 + \ldots\ldots + P_{m_1} - Q_1 - \ldots\ldots - Q_{k_1} + P_{m_1+1} + \ldots\ldots + P_{m_n-1} + P_{m_n} - P_{m_n} > \beta_n,$$

which is a contradiction to the choice of m_n.

Therefore, $|x_n - \beta_n| \leq P_{m_n}$. .... (2.6.23.2)

Next, we claim that, $|y_n - \alpha_n| \leq + Q_{k_n}$.

Now, $$y_n = P_1 + P_2 + \ldots\ldots + P_{m_1} - Q_1 - \ldots\ldots - Q_{k_1} + P_{m_1+1} + \ldots$$

$$\ldots + P_{m_2} - Q_{k_1+1} - \ldots\ldots - Q_{k_n} < \alpha_n.$$

Hence, $y_n < \alpha_n$ implies $\alpha_n - y_n > 0$ or $y_n - \alpha_n < 0$.

Suppose $|y_n - \alpha_n| > Q_{k_n}$. That is, $\alpha_n - y_n > Q_{k_n}$ which implies $y_n + Q_{k_n} < \alpha_n$.

Hence, $$P_1 + \ldots\ldots + P_{m_1} - Q_1 - \ldots\ldots - Q_{k_1} + P_{m_1+1} + \ldots\ldots + P_{m_2} - Q_{k_1+1} - \ldots$$

$$\ldots - Q_{k_n-1} - Q_{k_n} + Q_{k_n} < \alpha_n.$$

which is a contradiction to the choice of k_n.

Therefore, $|y_n - \alpha_n| \leq Q_{k_n}$. .... (2.6.23.3)

Since $\sum a_n$ is convergent, $\lim_{n\to\infty} a_n = 0$.

Hence, $\lim_{n\to\infty} P_n = 0$ and $\lim_{n\to\infty} Q_n = 0$.

When $n \to \infty$, $m_n \to \infty$ and $k_n \to \infty$.

This implies that, $\lim_{n\to\infty} P_{m_n} = 0$ and $\lim_{n\to\infty} Q_{k_n} = 0$.

Therefore, taking limit as $n \to \infty$ in equations (2.6.23.2) and (2.6.23.3), we get

$$\lim_{n\to\infty} x_n = \lim_{n\to\infty} \beta_n = \beta \quad \text{and} \quad \lim_{n\to\infty} y_n = \lim_{n\to\infty} \alpha_n = \alpha.$$

Finally, it is clear that no number less than α or greater than β can be a sub-sequential limit of (2.6.23.1).

Therefore, $\limsup_{n\to\infty} s'_n = \beta$ and $\liminf_{n\to\infty} s'_n = \alpha$.

Hence the theorem.

Theorem 2.6.24

If $\sum a_n$ is a series of complex numbers which converge absolutely, then every rearrangement of $\sum a_n$ converges, and they converge to the same sum.

Proof

Given that $\sum a_n$ converges absolutely.

Hence, $\sum |a_n|$ is convergent.

Let $\{k_n\}$ be a sequence of positive integers in which each positive integer appears exactly once.

Let $a'_n = a_{k_n}$, for $n = 1,2,3,.....$

Then, $\sum a'_n$ is called a rearrangement of $\sum a_n$.

Let $\{s_n\}$ and $\{s'_n\}$ be the sequence of partial sums of $\sum a_n$ and $\sum a'_n$.

Hence, we have to prove that, $\lim_{n\to\infty} s_n = \lim_{n\to\infty} s'_n$.

Since $\sum |a_n|$ is convergent, there exists $N \in J$ such that

$$\sum_{k=n}^{m} |a_k| \leq \varepsilon, \text{ for all } m \geq n \geq N . \qquad .... (2.6.24.1)$$

Let p be a positive integer such that $\{1,2,3,....., N\} \subseteq \{k_1, k_2,....., k_p\}$.

If $n > p$, then the terms $a_1, a_2,....., a_N$ will cancel in the difference $s_n - s'_n$.

Therefore, by using equation (2.6.24.1), we have $|s_n - s'_n| \leq \varepsilon$.

Now, taking limit as $n \to \infty$, we get $\lim_{n\to\infty} s_n - s'_n = 0$.

That is, $\lim_{n\to\infty} s_n = \lim_{n\to\infty} s'_n$.

Therefore, $\sum a'_n$ converges and also converge to the same sum as $\sum a_n$.

Hence the theorem.

Unit	
3	**LIMITS AND CONTINUITY**

3.1 Introduction

The key features that distinguishes real analysis from algebraic and combinatorial branches of mathematics, is the concept of a limit. We can say that, limit is the heart of real analysis. There are various types of limits, like limits of functions and limits of sequences. However, all these types of limits have similar features and when we gain familiarity in one type of limit will make learning about the other types much easier. Limits take an important role in real analysis than in calculus.

The intuitive idea of a limit of a function f as x goes to a number c in the domain of f is that the value of $f(x)$ gets closer and closer to a number L as the value of x gets closer and closer to c. It is not necessary that every function has a limit at each number c. It is important that, when we look at the values of $f(x)$ as x gets closer and closer to c, we never consider the value of f at c. In fact, the function f need not be defined at c for the limit of f to exist as x goes to c.

3.2 Limit of functions

Definition 3.2.1

Let X and Y be metric spaces. Let $E \subset X$ and $p \in X$ be a limit point of E. Let $f: E \rightarrow Y$ be any function. We say, $\lim_{x \rightarrow p} f(x) = q$, or write $f(x) \rightarrow q$ as $x \rightarrow p$, if there is a point q in Y, with the property that, for a given $\varepsilon > 0$, there exists $\delta > 0$ such that $x \in E$ and $0 < d_X(x, p) < \delta \Rightarrow d_Y(f(x), q) < \varepsilon$.

Here, p need not be a point of E and if $p \in E$, then $\lim_{x \rightarrow p} f(x)$ need not be equal to $f(p)$.

Theorem 3.2.2

Let X and Y be metric spaces. Let $E \subset X$ and $p \in X$ be a limit point of E. Let $f : X \to Y$ be any function. Then, $\lim_{x \to p} f(x) = q$ if and only if $\lim_{n \to \infty} f(p_n) = q$ for every sequence $\{p_n\}$ in E, such that

$$p_n \neq p \text{ and } \lim_{n \to \infty} p_n = p.$$

Proof

Direct part,

Given that, $\lim_{x \to p} f(x) = q$ and $\{p_n\}$ is a sequence in E, such that $p_n \neq p$ and $\lim_{n \to \infty} p_n = p$.

We have to prove that, $\lim_{n \to \infty} f(p_n) = q$.

Since $\lim_{x \to p} f(x) = q$, we have, for given $\varepsilon > 0$, there exists $\delta > 0$ such that

$x \in E$ and $0 < d_X(x, p) < \delta \Rightarrow d_Y(f(x), q) < \varepsilon$. .... (3.2.2.1)

Since $p_n \neq p$, we have, $d_X(p_n, p) > 0$. .... (3.2.2.2)

Again, since $\lim_{n \to \infty} p_n = p$, there exists a positive integer $N \in J$ such that

$d_X(p_n, p) < \delta$ for all $n \geq N$. .... (3.2.2.3)

Hence from equations (3.2.2.2) and (3.2.2.3), we get $0 < d_Y(f(p_n), q) < \varepsilon$, for all $n \geq N$.

Therefore, $\lim_{n \to \infty} f(p_n) = q$.

Converse part,

Given that, for every sequence $\{p_n\}$ in E, such that $p_n \neq p$ and $\lim_{n \to \infty} p_n = p$, we have

$$\lim_{n \to \infty} f(p_n) = q.$$

We have to prove that, $\lim_{x \to p} f(x) = q$.

Suppose $\lim_{x \to p} f(x) \neq q$.

Therefore, for given $\varepsilon > 0$ and for all $\delta > 0$ there exists $x \in E$ and $0 < d_X(x, p) < \delta \Rightarrow d_Y(f(x), q) \geq \varepsilon$. .... (3.2.2.4)

Let us choose $\delta = \frac{1}{n}$.

Hence, there exists a point $p_n \in E$ such that

$$0 < d_X(p_n, p) < \frac{1}{n} \Rightarrow d_Y(f(p_n), q) \geq \varepsilon .$$

When $n = 1,2,3,....$, we have a sequence of points p_n in E, such that $p_n \neq p$ and $\lim_{n\to\infty} p_n = p$.

But, $\lim_{n\to\infty} f(p_n) \neq q$, which is a contradiction.

Hence our assumption is wrong and therefore $\lim_{x\to p} f(x) = q$.

Hence the theorem.

Theorem 3.2.3

Let X be a metric space, $E \subset X$ and p be a limit point of E. If f and g are complex functions on E such that $\lim_{x\to p} f(x) = A$ and $\lim_{x\to p} g(x) = B$, then

i) $\lim_{x\to p}(f+g)(x) = A + B$,

ii) $\lim_{x\to p}(fg)(x) = AB$,

iii) $\lim_{x\to p}\left(\frac{f}{g}\right)(x) = \frac{A}{B}$, provided $B \neq 0$.

Proof

i) Let $\{p_n\}$ be a sequence in E such that $p_n \neq p$ and $\lim_{n\to\infty} p_n = p$.

Given that $\lim_{x\to p} f(x) = A$ and $\lim_{x\to p} g(x) = B$.

Hence, by theorem 3.2.2, $\lim_{n\to\infty} f(p_n) = A$, ... (3.2.3.1)

and $\lim_{n\to\infty} g(p_n) = B$. .... (3.2.3.2)

Since $\{f(p_n)\}$ and $\{g(p_n)\}$ are complex sequences,

$$\lim_{n\to\infty}(f+g)(p_n) = \lim_{n\to\infty}[f(p_n) + g(p_n)] = \lim_{n\to\infty} f(p_n) + \lim_{n\to\infty} g(p_n).$$

Hence, by equations (3.2.3.1) and (3.2.3.2), we have

$$\lim_{x\to p}(f+g)(x) = A+B.$$

Hence part (i) is proved.

ii) $\lim_{n\to\infty}(fg)(p_n) = \lim_{n\to\infty} f(p_n)g(p_n) = AB$.

Hence, by theorem 3.2.2, $\lim_{x\to p}(fg)(x) = AB$.

Hence part (ii) is proved.

iii) $\lim_{n\to\infty}\left(\dfrac{f}{g}\right)(p_n) = \lim_{n\to\infty}\dfrac{f(p_n)}{g(p_n)}$, if $g(p_n) \neq 0$.

That is, $\lim_{n\to\infty}\left(\dfrac{f}{g}\right)(p_n) = \dfrac{A}{B}$, if $B \neq 0$.

Therefore, by theorem 3.2.2, we have $\lim_{x\to p}\left(\dfrac{f}{g}\right)(x) = \dfrac{A}{B}$, provided $B \neq 0$.

Hence part (iii) is proved.
Hence the theorem.

Note 3.2.4

1) Let $\bar{f},\bar{g} : E \to \mathbb{R}^k$, $\lim_{x\to p}\bar{f}(x) = \bar{A}$ and $\lim_{x\to p}\bar{g}(x) = \bar{B}$, then

i) $\lim_{x\to p}(\bar{f}+\bar{g})(x) = \bar{A}+\bar{B}$ and ii) $\lim_{x\to p}(\bar{f}\bar{g})(x) = \bar{A}\bar{B}$.

Proof

i) Let

$$\bar{f} = (f_1, f_2, \ldots\ldots, f_k),\ \bar{g} = (g_1, g_2, \ldots\ldots, g_k),$$
$$\bar{A} = (A_1, A_2, \ldots\ldots, A_k) \text{ and } \bar{B} = (B_1, B_2, \ldots\ldots, B_k).$$

Here, $f_i : E \to \mathbb{R}$ and $g_i : E \to \mathbb{R}$, for all $i = 1,2,\ldots.,k$.

Now, $\lim_{x\to p}\bar{f}(x) = \bar{A}$ if and only if $\lim_{x\to p} f_i(x) = A_i$, for all $i = 1,2,3,\ldots\ldots,k$,

and $\lim_{x\to p}\bar{g}(x) = \bar{B}$ if and only if $\lim_{x\to p} g_i(x) = B_i$, for all $i = 1,2,3,\ldots\ldots,k$.

Therefore, $\lim_{x\to p} f_i(x) + g_i(x) = A_i + B_i$, for all $i = 1,2,\ldots.,k$.

That is, $\lim_{x \to p}(f_i + g_i)(x) = A_i + B_i,$ for all $i = 1,2,....,k$.

That implies, $\lim_{x \to p}\sum_{i=1}^{k}(f_i + g_i)(x) = \sum_{i=1}^{k}(A_i + B_i)$.

Hence, $\lim_{x \to p}(\bar{f} + \bar{g})(x) = \overline{A} + \overline{B}$.

Hence part (i) is proved.

ii) Also, $\lim_{x \to p} f_i(x) g_i(x) = A_i B_i,$ for $i = 1,2,.....,k$.

That is, $\lim_{x \to p}(f_i g_i)(x) = A_i B_i,$ for $i = 1,2,.....,k$.

That implies, $\lim_{x \to p}\sum_{i=1}^{k}(f_i g_i)(x) = \sum_{i=1}^{k} A_i B_i$.

Therefore, $\lim_{x \to p}(\bar{f}\bar{g})(x) = \overline{A}\overline{B}$.

Hence part (ii) is proved.

2. If f has a limit at p, then this is unique.

Proof

Suppose $\lim_{x \to p} f(x) = q$ and $\lim_{x \to p} f(x) = q'$.

We have to prove that, $q = q'$.

Since f is a limit point of E, there exists a sequence p_n in E such that $\lim_{n \to \infty} p_n = p$.

Since $\lim_{x \to p} f(x) = q$ and $\lim_{x \to p} f(x) = q'$, by theorem (2.3.2.(ii)) and theorem (3.2.2), we get

$$\lim_{n \to \infty} f(p_n) = q \text{ and } \lim_{n \to \infty} f(p_n) = q'.$$

Since convergent sequences have unique limit, $q = q'$.

Hence proved.

3.3 Continuity of functions

Definition 3.3.1

Let X and Y be metric spaces, $E \subset X$ and $p \in E$. Let $f : E \to Y$. Then, f is said to be *continuous* at p, if for given $\varepsilon > 0$, there exists $\delta > 0$ such that

$x \in E$ and

$$d_X(x, p) < \delta \Rightarrow d_Y(f(x), f(p)) < \varepsilon .$$

Note 3.3.2

1. If *f* is continuous at all points in *E*, then *f* is said to be continuous on *E*.
2. Every function is continuous at every isolated point of its domain.

Proof

Let *p* be an isolated point of *E*.

Then, *p* is not limit point of *E*.

Hence, there exists $\delta > 0$ such that $N_\delta(p) \cap E = \{p\}$.

Therefore, $x \in E$ and $d_X(x, p) < \delta \Rightarrow x = p$.

We have to prove that *f* is continuous at *p*.

For given $\varepsilon > 0$, there exists $\delta > 0$ such that $x \in E$ and $d_X(x, p) < \delta$.

Therefore, $x = p$.

Hence, $d_Y(f(p), f(p)) = 0 < \varepsilon$ and thus, *f* is continuous at *p*.

Since *p* is arbitrary, *f* is continuous at every isolated point of its domain.

Hence proved.

Theorem 3.3.3

Suppose $p \in E$ and *p* is a limit point of *E*. Then, *f* is continuous at *p* if and only if $\lim_{x \to p} f(x) = f(p)$.

Proof

Direct part,

Given that *f* is continuous at *p*. Therefore, for given $\varepsilon > 0$, there exists $\delta > 0$ such that $x \in E$ and

$$d_X(x, p) < \delta \Rightarrow d_Y(f(x), f(p)) < \varepsilon .$$

Hence, there exists $\delta > 0$ such that $x \in E$ and $0 < d_X(x, p) < \delta$.

That implies, $d_Y(f(x), f(p)) < \varepsilon$.

Therefore, $\lim_{x \to p} f(x) = f(p)$.

Converse part,

Given that $\lim_{n \to \infty} f(x) = f(p)$.

Therefore, for given $\varepsilon > 0$, there exists $\delta > 0$ such that $x \in E$ and

$$0 < d_X(x,p) < \delta \Rightarrow d_Y(f(x), f(p)) < \varepsilon. \quad \text{.............. (3.3.3.1)}$$

When $d_X(x,p) = 0$, then $x = p$. In this case,

$$d_Y(f(x), f(p)) = d_Y(f(p), f(p)) = 0 < \varepsilon.$$

Therefore, by equation (3.3.3.1), there exists $\delta > 0$ such that

$$0 \leq d_X(x,p) < \delta \Rightarrow d_Y(f(x), f(p)) < \varepsilon.$$

Therefore, *f* is continuous at *p*.

Hence the theorem.

Theorem 3.3.4

Suppose *X*, *Y* and *Z* are metric spaces. Let $E \subset X$. If *f* maps *E* into *Y* and *g* maps *f* (*E*) into *Z*. Define $h(x) = g(f(x), x \in E$. If *f* is continuous at *p* and *g* is continuous at *f* (*p*). Then, *h* is continuous at *p*. This theorem can also be stated as, "Composition of two continuous is continuous."

Proof

Let $\varepsilon > 0$ be given.

Since *g* is continuous at *f* (*p*), there exists $\eta > 0$ such that $y \in f(E)$ and

$$d_Y(y, f(p)) < \eta \Rightarrow d_Z(g(y), g(f(p))) < \varepsilon \quad \text{................ (3.3.4.1)}$$

Since *f* is continuous at *p*, there exists $\delta > 0$ such that

$$d_X(x,p) < \delta \Rightarrow d_Y(f(x), f(p)) < \eta. \quad \text{................ (3.3.4.2)}$$

Now, there exists $\delta > 0$ such that $x \in E$ and $d_X(x,p) < \delta$.

Hence, by equation (3.3.4.2), we have $d_Y(f(x), f(p)) < \eta$.

Since $f(x) \in f(E)$, then by equation (3.3.4.1), we have

$$d_Z(g(f(x)), g(f(p))) < \varepsilon.$$

That is, $d_Z(h(x), h(p)) < \varepsilon$.

Therefore, *h* is continuous at *p*.

Hence the theorem.

Theorem 3.3.5

A function $f: X \to Y$ is continuous on *X* if and only if $f^{-1}(V)$ is open in *X*, for every open subset *V* of *Y*. This theorem can also be stated as, "Under a continuous map, inverse image of every open set is open".

Proof

Direct part,

Given that, $f: X \to Y$ is continuous.

Let V be open in Y.

We have to prove that, $f^{-1}(V)$ is open in X.

Let $p \in f^{-1}(V)$, then $f(p) \in V$.

Since V is open, there exists $\varepsilon > 0$ such that $N_\varepsilon(f(p)) \subset V$. ...... (3.3.5.1)

Again, since f is continuous at p, there exists $\delta > 0$ such that $x \in X$ and

$$d_X(x, p) < \delta \Rightarrow d_Y(f(x), f(p)) < \varepsilon.$$

Therefore, by equation (3.3.5.1) we get

$$x \in N_\delta(p) \Rightarrow f(x) \in N_\varepsilon(f(p)) \subset V.$$

That implies, $f(x) \in V \Rightarrow x \in f^{-1}(V)$.

Hence, $x \in N_\delta(p) \subset f^{-1}(V)$.

Therefore, $f^{-1}(V)$ is open in X.

Converse part,

Given that $f^{-1}(V)$ is open in X for every open subset V of Y.

We have to prove that, $f: X \to Y$ is continuous on X.

Let $\varepsilon > 0$ be given and $p \in X$.

Hence, $f(p) \in f(X)$.

Therefore, $N_\varepsilon(f(p))$ is an open subset of Y.

By the given data, we have $f^{-1}(N_\varepsilon(f(p)))$ is an open subset of X.

Now, $f(p) \in N_\varepsilon(f(p))$.

That implies, $p \in f^{-1}(N_\varepsilon(f(p)))$.

Since $p \in f^{-1}(N_\varepsilon(f(p)))$ and $f^{-1}(N_\varepsilon(f(p)))$ is open, there exists $\delta > 0$ such that

$$N_\delta(p) \subset f^{-1}(N_\varepsilon(f(p))). \qquad \text{................ (3.3.5.2)}$$

Let $x \in X$, then $d_X(x, p) < \delta \Rightarrow x \in N_\delta(p)$.

Then by equation (3.3.5.2), we have $x \in N_\delta(p) \subset f^{-1}(N_\varepsilon(f(p)))$.

Therefore, $x \in f^{-1}(N_\varepsilon(f(p))) \Rightarrow f(x) \in N_\varepsilon(f(p)) \Rightarrow d_Y(f(x), f(p)) < \varepsilon$.

Hence f is continuous at p. Since p is arbitrary, f is continuous on X.

Hence the theorem.

Theorem 3.3.6

A mapping f of a metric space X into a metric space Y is continuous if and only if $f^{-1}(F)$ is closed in X, for every closed set F in Y. This theorem can also be stated as,

"Under a continuous map, inverse image of every closed set is closed."

Proof

Given that $f: X \to Y$ is continuous.

That implies, $f^{-1}(F)$ is open in X, whenever F^c is open in Y.

That is, $\left[f^{-1}(F)\right]^c$ is open in X, whenever F is closed in Y.

Hence, $f^{-1}(F)$ is closed in X, whenever F is closed in Y.

Hence the theorem.

Theorem 3.3.7

Let f and g be complex continuous functions on a metric space X, then $f+g,\ fg,\ f/g$ are continuous on X.

Proof

Let $p \in X$.

Case (i)

Let p be an isolated point of X.

Since every function is continuous at every isolated point p of its domain, then

$$f+g, \quad fg \quad \text{and} \quad \frac{f}{g} \text{ are continuous at } p.$$

Hence, $f+g$, fg and $\dfrac{f}{g}$ are continuous on X.

Case (ii)

Let p be a limit point of X.

Since f and g are continuous on X, f and g are continuous at p.

Therefore, $\lim\limits_{x \to p} f(x) = f(p)$ .... (3.3.7.1)

and $\lim\limits_{x \to p} g(x) = g(p)$. .... (3.3.7.2)

Now, by theorem (3.2.3), we have,

$$\lim_{x \to p}(f+g)(x) = f(p) + g(p) = (f+g)(p).$$

Therefore, $f+g$ is continuous at p.

Since p is arbitrary, $f+g$ is continuous on *X*.

Again, by theorem (3.2.3), we have, $\lim_{x\to p}(fg)(x)=f(p)g(p)=(fg)(x)$.

Therefore, *fg* is continuous at *p*.
Since *p* is arbitrary, *fg* is continuous on *X*.
Also by theorem (3.2.3), we have,

$$\lim_{x\to p}\left(\frac{f}{g}\right)(x)=\frac{f(p)}{g(p)}=\left(\frac{f}{g}\right)(p), \text{ if } g(p)\neq 0.$$

Therefore, $\frac{f}{g}$ is continuous at p.

Since p is arbitrary, $\frac{f}{g}$ is continuous on *X*.

Hence the theorem.

Theorem 3.3.8

Let $f_1, f_2, \ldots\ldots, f_k$ be real functions on *X*, and $\bar{f}$ be a function of *X* into R^k denoted by $\bar{f}(x)=(f_1(x), f_2(x), \ldots\ldots, f_k(x))$, where $x\in X$. Then,

i) $\bar{f}$ is continuous if and only if $f_1, f_2, \ldots\ldots, f_k$ are continuous.

ii) If $\bar{f}$ and $\bar{g}$ are continuous functions of *X* into R^k, then $\bar{f}+\bar{g}$ and $\bar{f}\bar{g}$ are continuous on *X*.

Proof

Direct part,

Given that, $\bar{f}: X\to \mathbb{R}^k$ is continuous.

We have to prove that, $f_i: X\to\mathbb{R}$ is continuous, for all $i=1,2,\ldots,k$.

Let $\varepsilon>0$ be given.

Since $\bar{f}: X\to\mathbb{R}^k$ is continuous, there exists $\delta>0$ such that $x\in X$ and

$$d_X(x,p)<\delta \Rightarrow \left|\bar{f}(x)-\bar{f}(p)\right|<\varepsilon, . \qquad \ldots\ldots\ldots\ldots\ldots (3.3.8.1)$$

Now, for all $i=1,2,\ldots\ldots,k$, $x\in X$ and $d_X(x,p)<\delta$, by equation (3.3.8.1), we have

$$\left|f_i(x)-f_i(p)\right|\leq\left|\bar{f}(x)-\bar{f}(p)\right|<\varepsilon.$$

Therefore, f_i is continuous at *p*, for all $i=1,2,\ldots\ldots,k$.

Hence, f_i is continuous on X, for all $i = 1,2,.....,k$.

Converse part,

Given that, $f_i : X \to \mathbb{R}$ is continuous on X, for all $i = 1,2,.....,k$.

We have to prove that, $\bar{f} : X \to \mathbb{R}^k$ is continuous.

Let $\varepsilon > 0$ be given.

Since $f_i : X \to \mathbb{R}$ is continuous, for all $i = 1,2,.......,k$, there exists $\delta_i > 0$ such that $x \in X$ and

$$d_X(x, p) < \delta_i \Rightarrow |f_i(x) - f_i(p)| < \frac{\varepsilon}{\sqrt{k}}, \text{ for all } i = 1,2,.....,k. \quad .... (3.3.8.2)$$

Let $\delta = \min.\{\delta_1, \delta_2, ..., \delta_k\}$.

Hence, $\delta > 0$.

Now, $x \in X$, $d_X(x, p) < \delta$ implies

$$|\bar{f}(x) - \bar{f}(p)| = \sqrt{\sum_{i=1}^{k} (f_i(x) - f_i(p))^2} < \sqrt{\sum_{i=1}^{k} \frac{\varepsilon^2}{k}} = \sqrt{k\frac{\varepsilon^2}{k}} = \varepsilon.$$

Therefore, $\bar{f}$ is continuous at p and hence $\bar{f}$ is continuous on X.

Hence part (i) is proved.

ii) Given that, $\bar{f} : X \to \mathbb{R}^k$ and $\bar{g} : X \to \mathbb{R}^k$ are continuous.

Let $\bar{f} = (f_1, f_2, ..., f_k)$ and $\bar{g} = (g_1, g_2, ..., g_k)$.

Since $\bar{f}$ and $\bar{g}$ are continuous, by part (i), $f_1, f_2, ..., f_k$ and $g_1, g_2, ..., g_k$ are continuous.

Therefore, $f_1 + g_1, f_2 + g_2, ..., f_k + g_k$ are continuous and also $f_1 g_1, f_2 g_2, \; ....., f_k g_k$ are continuous.

Now, $\bar{f} + \bar{g} = (f_1 + g_1, f_2 + g_2, ..., f_k + g_k)$ and

$$\bar{f} . \bar{g} = (f_1 . g_1, f_2 . g_2, ..., f_k . g_k).$$

Hence by part (i), we have $\bar{f} + \bar{g}$ and $\bar{f} . \bar{g}$ are continuous.

Hence part (ii) is proved.

Hence the theorem.

3.4 Compactness and Continuity

Theorem 3.4.1

Suppose f is a continuous mapping of a compact metric space X into a metric space Y. Then, $f(X)$ is compact. This theorem can also be stated as, "Continuous image of a compact set is compact".

Proof

Given that $f : X \to Y$ is continuous and X is compact.

We have to prove that, $f(X)$ is compact.

Let $\{G_\alpha\}_{\alpha\in\wedge}$ be a collection of open subsets of Y, such that $f(X) \subset \bigcup_{\alpha\in\wedge} G_\alpha$, where $\wedge$ is the index set.

Therefore, $X \subset f^{-1}\left(\bigcup_{\alpha\in\wedge} G_\alpha\right) = \bigcup_{\alpha\in\wedge}\left[f^{-1}(G_\alpha)\right]$.

That is, $X \subset \bigcup_{\alpha\in\wedge}\left[f^{-1}(G_\alpha)\right]$. .... (3.4.1.1)

Since, $f : X \to Y$ is continuous, and G_α is open in Y, $f^{-1}(G_\alpha)$ is open in X. Therefore, by equation (3.4.1.1), we have $\left\{f^{-1}(G_\alpha)\right\}_{\alpha\in\wedge}$ is an open cover for X.

Since X is compact, there exists a finite subcollection of open sets,

$$\left\{f^{-1}(G_{\alpha_1}), f^{-1}(G_{\alpha_2}), \ldots\ldots, f^{-1}(G_{\alpha_n})\right\} \text{ such that } X \subset \bigcup_{i=1}^{n}\left\{f^{-1}(G_{\alpha_i})\right\}.$$

Therefore, $f(X) \subset f\left(\bigcup_{i=1}^{n} f^{-1}(G_{\alpha_i})\right) = f\left(f^{-1}\left(\bigcup_{i=1}^{n}(G_{\alpha_i})\right)\right) \subset \bigcup_{i=1}^{n} G_{\alpha_i}$.

Hence, $f(X)$ is compact.

Definition 3.4.2

A mapping $\bar{f}$ of a set E into R^k is said to be *bounded* if there is a real number M, such that $\left|\bar{f}(x)\right| \le M$, for all $x \in E$. Also, $\bar{f} : E \to \mathrm{R}^k$ is bounded, if $\bar{f}(E)$ is bounded.

Theorem 3.4.3

If $\bar{f}$ is a continuous mapping of a compact metric space X into R^k, then $\bar{f}(E)$ is closed and bounded. Also, $\bar{f}$ is bounded.

Proof

Given that $\bar{f}: X \to \mathbb{R}^k$ is continuous and X is compact.

Since $\bar{f}$ is continuous and X is compact, $\bar{f}(X)$ is a compact subset of R^k.

Therefore, $\bar{f}(X)$ is closed and bounded subset of R^k.

Since $\bar{f}(X)$ is bounded, $\bar{f}$ is bounded.

Hence the theorem.

Theorem 3.4.4

Suppose f is a continuous real function on a compact metric space X and $M = \sup_{x\in X} f(x)$ and $m = \inf_{x\in X} f(x)$, then there exists $p, q \in X$ such that $f(p) = M$ and $f(q) = m$. This theorem can also be stated as, "A continuous real valued function defined on a compact set is bounded and attains its bound".

Proof

Given that $f: X \to \mathbb{R}$ is continuous and X is compact.

Therefore, $f(X)$ is a compact subset of $\mathbb{R}$.

Hence, $f(X)$ is a closed and bounded subset of $\mathbb{R}$.

Since $f(X)$ is bounded, $M = \sup_{x\in X} f(x)$ and $m = \inf_{x\in X} f(x)$ exists in $\mathbb{R}$.

Therefore, $M, m \in \bar{f}(X) = f(X)$, since $f(X)$ is closed.

Hence, there exists $p, q \in X$ such that $f(p) = M$ and $f(q) = m$.

Hence the theorem.

Theorem 3.4.5

Suppose f is a continuous 1 – 1 mapping of a compact metric space X onto a metric space Y. Then, the inverse mapping f^{-1} defined in Y by $f^{-1}(f(x)) = x$, where $x \in X$ is a continuous mapping of Y into X.

Proof

Given that $f: X \to Y$ is 1 – 1, onto and continuous, X is compact. Since f is 1 – 1 and onto, f^{-1} exists.

We have to prove that, $f^{-1}: Y \to X$ is continuous.

Let V be a closed subset of X.

Since X is compact and V is a closed subset of X, V is compact.

Since f is continuous and V is compact, $f(V)$ is a compact subset of Y.
Since every compact subset of a metric space is closed, $f(V)$ is closed in Y.
That is, $(f^{-1})^{-1}(V)$ is closed in Y.

Therefore, V is closed in X implies $(f^{-1})^{-1}(V)$ is closed in Y.

Hence under f^{-1}, inverse image of closed set is closed.

Therefore, f^{-1} is continuous.

Hence the theorem.

Definition 3.4.6

A function $f: X \to Y$ is said to be *uniformly continuous*, if, for given $\varepsilon > 0$ there exists $\delta > 0$ such that $p, q \in X$ and

$$d_X(p,q) < \delta \Rightarrow d_Y(f(p), f(q)) < \varepsilon.$$

Note 3.4.7

1. Continuity can be defined at a point but uniform continuity can be defined only on sets.
2. When we define continuity at p, δ depends on both ε and p. But in uniform continuity δ depends only on ε.
3. Every uniformly continuous function is continuous but the converse need not be true.

Proof

Let us consider the following example;

Define $f:(0,1) \to \mathbb{R}$ by $f(x) = \dfrac{1}{x}$.

Here, f is continuous.
We prove that, f is not uniformly continuous.
Let $\varepsilon > 0$ be given. Suppose there exists $\delta > 0$ such that $|x - y| < \delta \Rightarrow |f(x) - f(y)| < \varepsilon$.

Let $x = y + \dfrac{1}{2}\delta$.

That is, $|x - y| = (1/2)\delta < \delta$.

Therefore, $|f(x)-f(y)|<\varepsilon \Rightarrow \left|\frac{1}{x}-\frac{1}{y}\right|<\varepsilon$.

That implies, $\left|\frac{1}{y+(1/2)\delta}-\frac{1}{y}\right|<\varepsilon \Rightarrow \frac{\delta}{(2y+\delta)y}<\varepsilon$.

This inequality cannot be true for all $y\in(0,1)$, since $\frac{\delta}{(2y+\delta)y}$ becomes arbitrarily large as y approaches zero.

Therefore, *f* is not uniformly continuous.

Hence proved.

Theorem 3.4.8

Let *f* be a continuous mapping of a compact metric space *X* into a metric space *Y*. Then, f is uniformly continuous on *X*.

Proof

Given that $f: X \rightarrow Y$ is continuous and *X* is compact.

We have to prove that, *f* is uniformly continuous.

Let $\varepsilon>0$ be given and let $p\in X$.

Then, *f* is continuous at *p*.

Therefore, there exists $\delta_p>0$ such that $x\in X$ and

$$d_X(x,p)<\delta_p \Rightarrow d_Y(f(x),f(p))<\frac{\varepsilon}{2}. \qquad \text{................ (3.4.8.1)}$$

Now, the collection of open sets $\left\{N_{\frac{\delta p}{2}}(p)/\,p\in X\right\}$ forms a cover for *X*. Since *X* is compact, there exists a finite sub collection $\left\{N_{\frac{\delta p_1}{2}}(p_1), N_{\frac{\delta p_2}{2}}(p_2), \ldots\ldots, N_{\frac{\delta p_n}{2}}(p_n)\right\}$ such that $X\subset\bigcup_{i=1}^{n}N_{\frac{\delta p_i}{2}}(p_i)$. ... (3.4.8.2)

Let $\delta=\min\left\{\frac{\delta p_1}{2},\frac{\delta p_2}{2},\ldots\ldots,\frac{\delta p_n}{2}\right\}$.

Then, $\delta>0$.

Let $p,q\in X$ and $d_X(x,p)<\delta$.

Now, by equation (3.4.8.2), we have $p \in X \subset \bigcup_{i=1}^{n} N_{\frac{\delta p_i}{2}}(p_i)$.

That is, $p \in N_{\frac{\delta p_i}{2}}(p_i)$, for some p_i, where $1 \le i \le n$.

That implies, $d_X(p, p_i) < \frac{\delta p_i}{2} < \delta p_i$.

Hence, by equation (3.4.8.1), we have, $d_Y(f(p), f(p_i)) < \varepsilon / 2$. .. (3.4.8.3)

Now, $d_X(q, p_i) \le d_X(q, p) + d_X(p, p_i) < \delta + \frac{\delta p_i}{2} \le \frac{\delta p_i}{2} + \frac{\delta p_i}{2} = \delta p_i$.

Hence, by equation (3.4.8.1), we have,

$d_Y(f(q), f(p_i)) < \varepsilon / 2$. .... (3.4.8.4)

Therefore, by equations (3.4.8.3) and (3.4.8.4), we have

$$d_Y(f(p), f(q)) \le d_Y(f(p), f(p_i)) + d_Y(f(p_i), f(q)) < \varepsilon.$$

Hence f is uniformly continuous.
Hence the theorem.

Note 3.4.9

In the above theorem, if X is not compact, then f need not be uniformly continuous.

Theorem 3.4.10

Let E be a non compact set in $\mathbb{R}$. Then,

i) There exists a continuous function on E, which is not bounded,

ii) There exists a continuous and bounded function on E which has no maximum,

iii) If in addition, E is bounded, then there exists a continuous function on E which is not uniformly continuous.

Proof

Case (i)

Given that E is non compact.

Suppose E is bounded. Then, E is not closed in $\mathbb{R}$.

Therefore, there exists a limit point x_0 of E, such that $x_0 \notin E$.

Let us define a function $f:E \to \mathbb{R}$ as $f(x)=\dfrac{1}{x-x_0}$, for all $x\in E$.

Therefore, f is continuous on E.

Let M be any positive real number. Hence, $\dfrac{1}{M}>0$.

Since x_0 is a limit point of E, then there exists $x\in E$ such that $|x-x_0|<M$.

Now, $|f(x)|=\dfrac{1}{|x-x_0|}>\dfrac{1}{M}$.

Therefore, f is not bounded.

Case (ii)

Suppose E is not bounded.

Define $f:E\to\mathbb{R}$ as $f(x)=x$.

Then, f is continuous on E.

Now, $f(E)=E$, which is not bounded.

Therefore, f is not bounded.

Hence part (i) is proved.

ii) Given that E is not compact.

Case (i)

Suppose E is bounded.

Then, E is not closed in $\mathbb{R}$.

Therefore, there exists a limit point x_0 of E such that $x_0\notin E$.

Define $g:E\to R$ as $g(x)=\dfrac{1}{1+(x-x_0)^2}$, for all $x\in E$.

Then, g is continuous on E.

Since $0<g(x)<1$, $\sup\limits_{x\in E} g(x)\le 1$.

We have to prove that, $\sup\limits_{x\in E} g(x)=1$.

Let $0<\delta<1$.

Therefore, $\dfrac{\sqrt{1-\delta}}{\sqrt{\delta}}>0$.

Since x_0 is a limit point of *E*, there exists $x \in E$ such that $|x - x_0| < \dfrac{\sqrt{1-\delta}}{\sqrt{\delta}}$.

Now, $1+(x-x_0)^2 < 1+\dfrac{1-\delta}{\delta} = \dfrac{1}{\delta}$.

That is, $\dfrac{1}{1+(x-x_0)^2} > \delta$, which implies $g(x) > \delta$.

Therefore, $\sup\limits_{x\in E} g(x) = 1$.

But, there exists no $x \in E$ such that $g(x) = 1$.

Therefore, *g* has no maximum in *E*.

Case (ii)

Suppose *E* is not bounded.

Define $h : E \to \mathbb{R}$ as $h(x) = \dfrac{x^2}{1+x^2}$, for all $x \in E$.

Then, *h* is continuous on *E*.

Since $0 \le h(x) < 1$, *h* is bounded.

Hence, $\sup\limits_{x\in E} h(x) \le 1$.

We have to prove that, $\sup\limits_{x\in E} h(x) = 1$.

Let $0 < \delta < 1$.

Then, $\dfrac{\sqrt{\delta}}{\sqrt{1-\delta}} > 0$.

Since *E* is not bounded, there exist $x \in E$ such that $|x| > \dfrac{\sqrt{\delta}}{\sqrt{1-\delta}}$.

That is, $x^2 - x^2\delta > \delta \quad \Rightarrow x^2 > (1+x^2)\delta \;\Rightarrow \dfrac{x^2}{1+x^2} > \delta$.

Therefore, $h(x) > \delta$ implies $\sup\limits_{x\in E} h(x) = 1$.

But there exists no $x \in E$ such that $h(x) = 1$.

Therefore, *h* has no maximum in *E*.

Hence part (ii) is proved.

iii) Given that E is not compact and E is bounded.

Therefore, E is not closed in $\mathbb{R}$. Hence, there exists a limit point x_0 of E such that $x_0 \notin E$.

Define $f : E \to R$ as $f(x) = \dfrac{1}{x - x_0}$, for all $x \in E$.

Then, f is continuous on E.

We have to prove that, f is not uniformly continuous on E.

Let $\varepsilon > 0$ be given and $\delta > 0$ be any arbitrary positive real number.

Since x_0 is a limit point of E, there exists $x \in E$ such that $x_0 - \delta < x < x_0 < x_0 + \delta$.

Now, for all $t \in E$, $|f(t) - f(x)| = \left| \dfrac{1}{t - x_0} - \dfrac{1}{x - x_0} \right| = \dfrac{|x - t|}{|t - x_0||x - x_0|}$.

Since x_0 is a limit point of E, we can choose $t \in E$ such that $|t - x| < \delta$, and

$$|x - t| > \frac{|x - x_0|}{2} \text{ and } |t - x_0| < \frac{1}{2\varepsilon}.$$

Now, $|x - t| < \delta$, but $|f(t) - f(x)| > \dfrac{|x - x_0|/2}{|t - x_0||x - x_0|} > \varepsilon$.

Therefore, f is not uniformly continuous.

Hence part (iii) is proved.

Hence the theorem.

3.5 Connectedness and Continuity

Theorem 3.5.1

If f is a continuous mapping of a metric space X into a metric space Y and if E is a connected subset of X, then f (E) is connected. This theorem can otherwise be stated as,

"Under a continuous map, image of a connected set is connected."

Proof

Given that $f : X \to Y$ is continuous and $E \subset X$. Also, given that E is connected.

Now, $f(E) \subset Y$.

We have to prove that, f (E) is connected.

Suppose $f(E)$ is not connected.

Then, $f(\bar{E}) = A \cup B$, where $A \neq \varphi$, $B \neq \varphi$, $\bar{A} \cap B = \varphi$ and $A \cap \bar{B} = \varphi$.

Let $G = E \cap f^{-1}(A)$ and $H = E \cap f^{-1}(B)$.

We have to prove that, $G \neq \phi$.

Since $A \neq \phi$, there exists some x in A.

Since $A \subset f(E)$, $x \in f(E)$. Therefore, $x = f(y)$, for some $y \in E$.

Now, $f(y) = x \in A \Rightarrow y \in f^{-1}(A)$.

Therefore, $y \in E$ and $y \in f^{-1}(A) \Rightarrow y \in E \cap f^{-1}(A) \Rightarrow y \in G$.

Hence, $G \neq \phi$.

Similarly, we can prove that $E = G \cup H$.

$$\text{Now, } G \cup H = \left(E \cap f^{-1}(A)\right) \cup \left(E \cap f^{-1}(B)\right) = E \cap \left(f^{-1}(A) \cup f^{-1}(B)\right)$$
$$= E \cap f^{-1}(A \cup B).$$

That is, $G \cup H = E \cap f^{-1}(f(E)) = E$.

Next,, we claim that, $\bar{G} \cap H = \phi$.

Now, $G = E \cap f^{-1}(A)$.

We know that, $A \subset \bar{A} \Rightarrow f^{-1}(A) \subset f^{-1}(\bar{A})$. .... (3.5.1.1)

Since $\bar{A}$ is closed in Y and f is continuous, $f^{-1}(\bar{A})$ is closed in X.

Therefore, $\overline{f^{-1}(\bar{A})} = f^{-1}(\bar{A})$. .... (3.5.1.2)

From equation (3.5.1.1) and (3.5.1.2), we get

$$\overline{f^{-1}(A)} \subset \overline{f^{-1}(\bar{A})} = f^{-1}(\bar{A}). \quad \text{........... (3.5.1.3)}$$

Now, by equation (3.5.1.1),

we have, $G = E \cap f^{-1}(A) \subset f^{-1}(A) \subset f^{-1}(\bar{A})$

and $H = E \cap f^{-1}(B) \subset f^{-1}(B)$.

Therefore, by equation (3.5.1.2), we have, $\bar{G} \subset \overline{f^{-1}(\bar{A})} = f^{-1}(\bar{A})$.

Since $H = E \cap f^{-1}(B) \subset f^{-1}(B)$, we have

$$\bar{G} \cap H \subset f^{-1}(\bar{A}) \cap f^{-1}(B) = f^{-1}(\bar{A} \cap B) = f^{-1}(\phi) = \phi.$$

Similarly, we can proved that, $G \cap \bar{H} = \phi$.

Therefore, $E = G \cup H$, where $G \neq \phi$, $H \neq \phi$, $\bar{G} \cap H = \varphi$ and $G \cap \bar{H} = \varphi$

This is a contradiction to E is connected.
Therefore, $f(E)$ is connected.
Hence the theorem.

Theorem 3.5.2

This theorem is known as ***Intermediate value theorem***.
Let f be a continuous function on the interval $[a, b]$. If $f(a) < f(b)$ and if c is a number such that $f(a) < c < f(b)$, then there exists a point $x \in (a,b)$ such that $f(x) = c$.

Proof

Given $f : X \to \mathbb{R}$ is continuous.

Now, $[a,b]$ is connected.

Since f is continuous, $f([a,b])$ is a connected subset of $\mathbb{R}$.

Again, $f(a), f(b) \in f([a,b])$ and $f(a) < c < f(b)$.

We know that, "A subset E of a real line R^1 is connected, if and only if, it has the following property: If $x, y \in E$ and $x < z < y$, then $z \in E$."

Hence, by the above theorem, $c \in f([a,b])$.

Therefore, $c = f(x)$, where $x \in (a,b)$.

Hence the theorem.

3.6 Discontinuities

Definition 3.6.1

Let $f : (a,b) \to \mathbb{R}$ and $a \le x < b$. We say $f(x+) = q$, if $f(t_n) \to q$ as $n \to \infty$ whenever $\{t_n\}$ is a sequence in (x, b) such that $\{t_n\} \to x$.

Let $a < x \le b$. We say $f(x-) = q$, if $f(t_n) \to q$ as $n \to \infty$ whenever $\{t_n\}$ is a sequence in (a, x) such that $\{t_n\} \to x$.

Note 3.6.2

1. $f(x+) = q$ if and only if, for given $\varepsilon > 0$, there exists $\delta > 0$ such that $x < t < x + \delta < b \quad \Rightarrow |f(t) - q| < \varepsilon$.
2. $f(x-) = q$ if and only if, for given $\varepsilon > 0$ there exists $\delta > 0$ such that

$a < x - \delta < t < x \quad \Rightarrow |f(t) - q| < \varepsilon .$

3. $\lim_{t \to x} f(t) = q$ if and only if $f(x+) = f(x-) = q$.

4. The function f is continuous at x, if and only if $f(x+) = f(x-) = f(x)$.

Definition 3.6.3

If f is not continuous at x, then f is *discontinuous at x* or f has *discontinuity at x*. Suppose f has a discontinuity at x, if $f(x+)$ and $f(x-)$ exists, then the discontinuity of f at x is of the *first kind* or f has a *simple discontinuity at x*. Otherwise, the discontinuity of f at x is of the *second kind*. Suppose f has a simple discontinuity at x, then either

$$f(x+) \neq f(x-) \text{ or } f(x+) = f(x-) \neq f(x).$$

Examples 3.6.4

1. Define $f : (a,b) \to \mathbb{R}$ by $f(x) = \begin{cases} 1, & \text{if } x \text{ is rational} \\ 0, & \text{if } x \text{ is irrational} \end{cases}$

Let $x \in (a,b)$.

We have to prove that, $f(x+)$ does not exist.

Suppose $f(x+) = q$. Let $\{x_n\}$ be a sequence of rational numbers in (x,b) such that $\{x_n\} \to x$.

Therefore, $f(x_n) \to q$.

Now, since x_n is rational, $f(x_n) = 1$, for all n.

Hence, $q = 1$.

Let $\{y_n\}$ be a sequence of rational numbers in (x, b) such that $\{y_n\} \to x$.

Therefore, $f(y_n) \to q$.

Now, since y_n is irrational, $f(y_n) = 0$, for all n.

Hence, $q = 0$, which is a contradiction to $q = 1$.

Therefore, $f(x+) = q$ is impossible.

Hence, $f(x+)$ does not exist.

Similarly, we can prove that $f(x-)$ does not exist.

Therefore, f has a discontinuity of the second kind in (a, b).

2. Define $f : \mathbb{R} \to \mathbb{R}$ by $f(x) = \begin{cases} 0, \text{ if } x \text{ is irrational} \\ x, \text{ if } x \text{ is rational} \end{cases}$

We have to prove that, f is continuous at 0.

That is, to prove that, for given $\varepsilon > 0$, there exists $\delta > 0$ such that $|x - 0| < \delta \Rightarrow |f(x) - f(0)| < \varepsilon$.

Choose $\delta = \varepsilon$.

Now, we have $|x| = \delta$.

When x is irrational, $|f(x)| = |0| = 0 < \varepsilon$.

When x is rational, $|f(x)| = |x| < \delta = \varepsilon$.

Therefore, f is continuous at 0.

When $x > 0$, we have to prove that, $f(x+)$ does not exist.

Suppose, $f(x+) = q$.

Let $\{x_n\}$ be a sequence of irrational numbers such that $x_n > x$ and $\{x_n\} \to x$.

Therefore, $f(x_n) = q$.

Since x_n is irrational, $f(x_n) = 0$, for all n.

Hence, $q = 0$. .... (3.6.4.1)

Let $\{y_n\}$ be a sequence of rational numbers such that $y_n > x$ and $\{y_n\} \to x$.

Therefore, $f(y_n) = q$.

Since y_n is rational, $f(y_n) = y_n$, for all n.

Hence, by equation (3.6.4.1), $y_n = q = 0$.

Since $y_n > x$, $\lim_{n \to \infty} y_n \geq x$.

That is, $x \leq 0$, which is a contradiction to $x > 0$.

Therefore, $f(x+)$ does not exist.

Similarly, we can prove that, when $x < 0$, $f(x-)$ does not exist.

Therefore, f is continuous at 0, and has a discontinuity of the second kind in all the other points in $\mathbb{R}$.

3. Define $f(x)=\begin{cases} x+2, & -3<x<-2 \\ -x-2, & -2\le x<0 \\ x+2, & 0\le x<1 \end{cases}$

To test the continuity at -2;

$f(-2+)=\lim_{h\to 0} f(-2+h)=\lim_{h\to 0} f[(-2+h)-2]=0,$ where $h>0$.

$f(-2-)=\lim_{h\to 0} f(-2-h)=\lim_{h\to 0} f[(-2-h)+2]=0$.

$f(-2)=0$.

Therefore, $f(-2+)=f(-2-)=f(-2)$.

Hence, *f* is continuous at -2.

To test the continuity at 0;

$f(0+)=\lim_{h\to 0} f(0+h)=\lim_{h\to 0} h+2=2,$ where $h>0$.

$f(0-)=\lim_{h\to 0} f(0-h)=\lim_{h\to 0} -(-h)=-2,$ where $h>0$.

Therefore, $f(0+)$ and $f(0-)$ exist, but $f(0+)\neq f(0-)$.

Hence, *f* has a simple discontinuity at $x=0$ and is continuous at every other point of (-3, 1).

4. Define $f(x)=\begin{cases} \sin(1/x), & x\neq 0 \\ 0, & x=0 \end{cases}$

To test the continuity at 0;

$f(0+)=\lim_{h\to 0} f(0+h)=\lim_{h\to 0} \sin\frac{1}{h},$ where $h>0$.

Since $\lim_{h\to 0} \sin\frac{1}{h}$ lies between +1 and -1, $\lim_{h\to 0} \sin\frac{1}{h}$ does not exist.

Therefore, $f(0+)$ does not exist.

Now, $f(0-)=\lim_{h\to 0} f(0-h)=\lim_{h\to 0} \sin\frac{1}{-h}=-\lim_{h\to 0} \sin\frac{1}{h}$.

Since $\lim_{h\to 0} \sin\frac{1}{h}$ lies between -1 and +1, $-\lim_{h\to 0} \sin\frac{1}{h}$ does not exist.

Therefore, $f(0-)$ does not exist.

Hence, *f* has discontinuity of the second kind at $x=0$ and *f* is continuous at every other point.

Definition 3.6.5

Let $f:(a,b)\to\mathbb{R}$. We say *f* is *monotonically increasing* if and only if

$$a<x<y<b\Rightarrow f(x)\le f(y).$$

and if *f* is *monotonically decreasing* if and only if

$$a<x<y<b\Rightarrow f(x)\ge f(y).$$

Hence, *f* is *monotonic*, if it is either monotonic increasing or monotonic decreasing.

Theorem 3.6.6

Suppose *f* is monotonic increasing on (*a*, *b*). Then, $f(x+)$ and $f(x-)$ exist at all points x in (a,b). More precisely, $\sup\limits_{a<t<x} f(t)=f(x-)\le f(x)\le f(x+)=\inf\limits_{x<t<b} f(t)$ Also, if $a<x<y<b$, then $f(x+)\le f(y-)$.

Proof

Let $x\in(a,b)$.

Let $E=\{f(t)/a<t<x\}$.

Since *f* is monotonic increasing, $f(t)\le f(x)$, for all $t\in(a,x)$.

Therefore, $f(x)$ is an upper bound for *E*.

Since $\mathbb{R}$ has least upper bound property, $A=\sup.E$ exists in $\mathbb{R}$.

We have to prove that, $A=f(x-)$.

Let $\varepsilon>0$ be given.

Now, $A-\varepsilon<A$.

Therefore, $A-\varepsilon$ is not an upper bound for *E*. Hence, there exists $p\in(a,x)$ such that $A-\varepsilon<f(p)$. .... (3.6.6.1)

Now, $a<p<x\Rightarrow x-p>0$.

Let $\delta=x-p>0$. Hence, $p=x-\delta$.

We have to prove that, $x-\delta<t<x\Rightarrow |f(t)-A|<\varepsilon$.

Suppose $x-\delta<t<x$.

Since *f* is monotonic increasing, from equation (3.6.6.1), we have,

$$A-\varepsilon<f(p)\le f(t)\le A<A+\varepsilon.$$

That implies, $|f(t)-A|<\varepsilon$.

Therefore, $\sup_{a<t<x} f(t) = f(x-)$. .... (3.6.6.2)

Since A is the least upper bound of E and $f(x)$ is an upper bound of E,

$A \le f(x)$. .... (3.6.6.3)

Combining equations (3.6.6.2) and (3.6.6.3), we have

$$\sup_{a<t<x} f(t) = f(x-) \le f(x).$$

Similarly, we can prove that

$$f(x) \le f(x+) = \inf_{x<t<b} f(t).$$

Now, we will prove also part;

Given that $a < x < y < b$.

Now, $f(x+) = \inf_{x<t<b} f(t)$.

Since f is monotonic increasing, $f(x+) = \inf_{x<t<y} f(t)$.

Again, $f(y-) = \sup_{a<t<y} f(t)$.

Since f is monotonic increasing, $f(y-) = \sup_{x<t<y} f(t)$.

Therefore, $f(x+) = \inf_{x<t<y} f(t) \le \sup_{x<t<y} f(t) = f(y-)$.

That is, $f(x+) \le f(y-)$.

Hence the theorem.

Note 3.6.7

1. If f is monotonic decreasing on (a, b), then $f(x+)$ and $f(x-)$ exist at every point x in (a, b).
2. Monotone functions have no discontinuity of the second kind.

Proof

Since f is monotonic on (a, b), by theorem 3.6.6, $f(x+)$ and $f(x-)$ exist at every point x in (a,b).

Therefore, f has no discontinuity of the second kind.

Hence proved.

Theorem 3.6.8

Let f be monotonic on (a, b), then the set of all points of (a, b) at which f is discontinuous is atmost countable.

Proof

Let f be monotonic increasing on (a, b).

Let $E = \{x \in (a,b) / f \text{ is discontinuous at } x\}$.

We have to prove that, E is atmost countable.

Let $x \in E$. Then, f is discontinuous at x.

Since f is monotonic increasing, f has no discontinuity of the second kind at x.

Therefore, $f(x+) \neq f(x-)$ or $f(x+) = f(x-) \neq f(x)$.

By theorem 3.6.6, we have $f(x-) \leq f(x) \leq f(x+)$.

Suppose $f(x+) = f(x-)$. That implies, $f(x+) = f(x-) = f(x)$.

Therefore, $f(x+) = f(x-) \neq f(x)$ will not occur.

Hence, we have $f(x+) \neq f(x-)$. That is, $f(x-) < f(x+)$.

Let us choose a rational number $r(x)$, such that $f(x-) < r(x) < f(x+)$.

Define, $g : E \to \mathbb{Q}$ as $g(x) = r(x)$.

We have to prove that, g is one-to-one.

Let $x \neq y$ in E.

Hence, $a < x < y < b$. Therefore, $f(x+) \leq f(y-)$.

Now, $f(x-) < r(x) < f(x+) \leq f(-y) < r(y) < f(y+)$.

That is, $r(x) < r(y) \Rightarrow g(x) < g(y)$.

Therefore, $g(x) \neq g(y)$.

Hence, g is $1-1$. Therefore, $E \sim g(E) \subset \mathbb{Q}$.

Since $\mathbb{Q}$ is countable, $g(E)$ is atmost countable. Hence, E is atmost countable.

Similarly, E is atmost countable, if f is monotonic decreasing.

Unit	
4	**SEQUENCES AND SERIES OF FUNCTIONS**

4.1 Introduction

The concept of a sequence provides a common theme unifying the concepts of series, continuity, and differentiation. In this chapter, we discuss the most natural generalization of a sequence of real numbers. The basic notion of a sequence is to use behavior of the terms in order to identify a limit. Here we will take a collection of functions which are related in some fashion, the most typical relation being a natural number index such as $x, x^2, x^3, \ldots\ldots$; the natural concept of a limit here will take the form of a function. With such a definition in place, we hope to determine when properties of the sequence functions, e.g. continuity, differentiability, etc., pass on to the limit function.

4.2 Pointwise convergence and Uniform convergence

Definition 4.2.1

Pointwise convergence of a sequence of functions

Suppose $\{f_n\}$, $n = 1, 2, \ldots\ldots$ is a sequence of functions defined on a set E, and suppose that the series $\sum_{n=1}^{\infty} f_n(x)$ converges for each x in E and $f(x) = \sum_{n=1}^{\infty} f_n(x)$ converges for each x in E. Then, f is called the sum of the series $\sum f_n$.

Problems 4.2.2

1. Find $\lim_{n\to\infty} x^n$, where $0 < x < 1$ and $x = 1$.

Solution

Let $f_n(x) = x^n$, $0 < x < 1$, for $n = 1, 2, 3, \ldots\ldots$

Then, $\lim_{n\to\infty} f_n(x) = \lim_{n\to\infty} x^n = 0$, if $0 < x < 1$.

And, $\lim_{n\to\infty} f_n(x) = \lim_{n\to\infty} x^n = 1$, if $x = 1$.

Hence the limit function f is

$$f(x) = \begin{cases} 0, & \text{if } 0 < x < 1 \\ 1, & \text{if } x = 1 \end{cases}$$

2. Find $\sum_{n=0}^{\infty} f_n(x)$, if $f_n(x) = \dfrac{x^2}{(1+x^2)^n}$, x is real.

Solution

Let $f_n(x) = \dfrac{x^2}{(1+x^2)^n}$, x is real and $n = 0, 1, 2, \ldots\ldots$

Then,
$$\sum_{n=0}^{\infty} f_n(x) = \sum_{n=0}^{\infty} \frac{x^2}{(1+x^2)^n}$$
$$= x^2 \left[1 - \frac{1}{1+x^2}\right]^{-1}$$
$$= x^2 \left[\frac{1+x^2-1}{1+x^2}\right]^{-1}$$
$$= 1 + x^2, \text{ if } x > 0.$$

If $x = 0$, $\sum_{n=0}^{\infty} f_n(0) = 0$.

Hence sum of the series is $f(x) = \begin{cases} 0, & \text{if } x = 0 \\ 1, & \text{if } x > 0. \end{cases}$

Definition 4.2.3

Uniform convergence of a sequence of functions

A sequence of functions $\{f_n\}$, $n = 1, 2, 3, \ldots\ldots.$ is said to *converge uniformly* on E to a function f, if for every $\varepsilon > 0$, there exists an integer N such that $n \geq N$ implies $|f_n(x) - f(x)| \leq \varepsilon$, for all $x \in E$. We write $f_n \to f$ uniformly on E.

Note 4.2.4

Here N depends only on ε and not on x. Hence the convergence is said to be *uniform*. Otherwise the convergence is only *pointwise convergence*.

Theorem 4.2.5

This theorem is known as ***Cauchy criterion for uniform convergence***.
The sequence of functions $\{f_n\}$, defined on E converges uniformly on E if and only if there exists an integer N such that $m \geq N, n \geq N, x \in E$ implies

$$|f_n(x) - f_m(x)| \leq \varepsilon.$$

Proof

Direct part,

Suppose that $\{f_n\}$ converges uniformly on E to a limit function f.
Then by definition of uniform convergence,
For given $\varepsilon > 0$, there exists an N such that $n \geq N$ implies

$$|f_n(x) - f(x)| \leq \varepsilon/2 \text{ for all } x \in E. \qquad \text{..........} (4.2.5.1)$$

Therefore, if $m \geq N, n \geq N, x \in E$, then

$$\begin{aligned} |f_n(x) - f_m(x)| &= |[f_n(x) - f(x)] - [f_m(x) - f(x)]| \\ &\leq |f_n(x) - f(x)| + |f_m(x) - f(x)| \\ &\leq \varepsilon/2 + \varepsilon/2 = \varepsilon. \end{aligned}$$

Converse part,

Assume that for given $\varepsilon > 0$, there exists an integer N such that $m \geq N, n \geq N, x \in E$ implies $|f_n(x) - f_m(x)| \leq \varepsilon$. .... (4.2.5.2)

Then, for every x in E, $\{f_n(x)\}$ is a Cauchy sequence of numbers.

Since a Cauchy sequence of real or complex numbers converges, $\{f_n(x)\}$ converges for every x in E.

Define f on E as

$$f(x) = \lim_{n \to \infty} f_n(x) \text{ for each } x \text{ in } E.$$

Then, $\{f_n\}$ converges pointwise on E.

We have to show that the convergence is uniform.

In (4.2.5.2), let n be fixed and let $m \to \infty$, then

$$|f_n(x) - f(x)| \leq \varepsilon, \text{ for all } x \text{ in } E \text{ and for all } n \geq N.$$

Hence the theorem.

Theorem 4.2.6

Suppose $\lim_{n\to\infty} f_n(x) = f(x)$. Put $M_n = \sup\{|f_n(x) - f(x)|, x \in E\}$. Then, $f_n \to f$ uniformly on E if and only if $M_n \to 0$ as $n \to \infty$.

Proof

By definition $f_n \to f$ uniformly on *E*.

$\Leftrightarrow$ for given $\varepsilon > 0$, there exists N such that $n \geq N$ implies $|f_n(x) - f(x)| \leq \varepsilon$, for all $x \in E$.

$\Leftrightarrow M_n = \sup\{|f_n(x) - f(x)|, x \in E\} \leq \varepsilon$.

$\Leftrightarrow \lim_{n\to\infty} M_n = 0$.

Hence the theorem.

Theorem 4.2.7

Suppose $\{f_n\}$ is a sequence of functions defined on E and suppose that $|f_n(x)| \leq M_n, x \in E, n = 1,2,3,\ldots\ldots$. Then, $\sum f_n$ converges uniformly on E if $\sum M_n$ converges.

Proof

Let $t_n = M_1 + M_2 + \ldots\ldots + M_n$.

and $s_n(x) = \sum_{k=1}^{n} f_k(x)$.

Assume that $\sum M_n$ converges.

Hence $\{t_n\}$ is a Cauchy sequence.

Hence, for given $\varepsilon > 0$, there exists N such that $m \geq N, n \geq N$ implies

$$|t_m - t_n| < \varepsilon. \qquad \ldots\ldots\ldots\ldots (4.2.7.1)$$

Let $m > n \geq N$.

Then (4.2.7.1) implies that $\sum_{k=n+1}^{m} M_k < \varepsilon$. $\qquad \ldots\ldots\ldots\ldots$ (4.2.7.2)

Therefore, for $m > n \geq N$,

$$|s_m(x) - s_n(x)| = \left|\sum_{k=n+1}^{m} f_k(x)\right|$$

$$\leq \sum_{k=n+1}^{m} |f_k(x)|$$

$$< \sum_{k=n+1}^{m} M_k$$

$< \varepsilon$ for all x in E.

Hence $\{s_n\}$ satisfies the Cauchy criterion for uniform convergence.

Hence $\{s_n\}$ converges uniformly on E.

Hence $\sum f_n$ converges uniformly on E.

Hence the theorem.

4.3 Uniform convergence and continuity

Definition 4.3.1

Continuity of a function

A function f defined on E is said to be *continuous* at $x \in E$, if

$$\lim_{t \to x} f(t) = f(x).$$

Theorem 4.3.2

Suppose $f_n \to f$ uniformly on a set E in a metric space.

Let x be a point of E and suppose that

$$\lim_{t \to x} f_n(t) = A_n,\ n = 1, 2, 3, \qquad \text{.......... (4.3.2.1)}$$

Then, $\{A_n\}$ converges and $\lim_{t \to x} f(t) = \lim_{n \to \infty} A_n$. .... (4.3.2.2)

In other words, $\lim_{t \to x} \lim_{n \to \infty} f_n(t) = \lim_{n \to \infty} \lim_{t \to x} f_n(t)$. .... (4.3.2.3)

Proof

Since $f_n \to f$ uniformly on E, by definition

for given $\varepsilon > 0$, there exists an integer N_1 such that $n \geq N_1$ implies

$$|f_n(t) - f(t)| \leq \varepsilon / 3 \qquad \text{for all } t \in E. \qquad \text{.......... (4.3.2.4)}$$

By Cauchy's criterion for uniform convergence,

for the same $\varepsilon > 0$, there exists an integer N_2 such that $m \geq N_2, n \geq N_2, t \in E$ implies that

$$|f_n(t) - f_m(t)| \leq \varepsilon / 3 \qquad \text{for all } t \in E. \qquad \text{.......... (4.3.2.5)}$$

Letting $t \to x$ in (4.3.2.5), we get

$$|A_n - A_m| \le \varepsilon/3 \qquad \text{for } m \ge N_2, n \ge N_2. \qquad \text{.......... (4.3.2.6)}$$

Therefore, $\{A_n\}$ is a Cauchy sequence.
Hence, $\{A_n\}$ converges, say to A.
Therefore, for the same $\varepsilon > 0$ taken above, there exists an integer N_3 such that $n \ge N_3$ implies

$$|A_n - A| \le \varepsilon. \qquad \text{.......... (4.3.2.7)}$$

Let $N = \max.\ (N_1, N_2, N_3)$.
Then, from (4.3.2.4),

$$|f_N - f(t)| \le \varepsilon/3. \qquad \text{.......... (4.3.2.8)}$$

From (4.3.2.7), $|A_N - A| \le \varepsilon/3$ .... (4.3.2.9)

By assumption (1),

$$\lim_{t \to x} f_N(t) = A_N.$$

Therefore, there exists a neighbourhood V of x such that

$$|f_N(t) - A_N| \le \varepsilon/3 \text{ for all } t \in V \cap E, t \ne x \qquad \text{.......... (4.3.2.10)}$$

From (4.3.2.8), (4.3.2.9) and (4.3.2.10),

$$\begin{aligned} |f(t) - A| &= |f(t) - f_N(t) + f_N(t) - A_N + A_N - A| \\ &\le |f(t) - f_N(t)| + |f_N(t) - A_N| + |A_N - A| \\ &\le \varepsilon/3 + \varepsilon/3 + \varepsilon/3 \\ &= \varepsilon \text{ for all } t \in V \cap E, t \ne x. \end{aligned}$$

Hence, $\lim_{t \to x} f(t) = A$.

$$\Rightarrow \lim_{t \to x} f(t) = \lim_{n \to \infty} A_n.$$

$$\lim_{t \to x} \lim_{n \to \infty} f_n(t) = \lim_{n \to \infty} \lim_{t \to x} f_n(t).$$

Hence the theorem.

Note 4.3.3

The above theorem says that if $f_n \to f$ uniformly and x is a limit point of E, then the limit processes can be interchanged.

Theorem 4.3.4

If $\{f_n\}$ is a sequence of continuous functions on E and if $f_n \to f$ uniformly on E, then f is continuous.

Proof

Let $\{f_n\}$ be sequence of continuous functions on E.

Let x be any point of E.

Then, each f_n is continuous at x.

Therefore,

$$\lim_{t\to x} f_n(t) = f_n(x), \qquad n = 1, 2, 3, \ldots\ldots\ldots\ldots$$

Applying the above theorem 4.3.2 and replacing A_n by $f_n(x)$, we get

$$\lim_{t\to x} f(t) = \lim_{n\to\infty} f_n(x).$$

Since $f_n \to f$ on E,

$$\lim_{n\to\infty} f_n(x) = f(x)$$

Hence $\lim_{t\to x} f(t) = f(x)$.

Hence the theorem.

Problem 4.3.5

1. Is the converse of the above theorem true?

Solution

No.

A sequence of continuous functions may converge to a continuous function although the convergence is not uniform.

Example

Let $f_n(x) = n^2x(1-x^2)^n,\ 0 \le x \le 1, \quad n = 1, 2, 3, \ldots\ldots\ldots$

Each f_n is a continuous function on [0, 1].

And $\lim_{n\to\infty} f_n(x) = 0,\ 0 < x \le 1$.

Since $f_n(0) = 0$, for all n,

$$\lim_{n\to\infty} f_n(x) = 0, \text{ for } x = 0.$$

Hence the limit function is

$$f(x) = 0,\ 0 \le x \le 1.$$

Here, $f(x)$ is also a continuous function on [0, 1].

But, $\{f_n\}$ does not converge uniformly to f on [0, 1].

Theorem 4.3.6

Suppose K is compact and

i) $\{f_n\}$ is a sequence of continuous functions on K

ii) $\{f_n\}$ converges pointwise to a continuous function f on K

iii) $f_n(x) \ge f_{n+1}(x)$ for all $x \in E$, $n = 1, 2, 3, \ldots\ldots..$

Then, $f_n \to f$ uniformly on K.

Proof

Let $g_n = f_n - f$, $n = 1, 2, 3, \ldots\ldots.$

Since f_n and f are continuous functions on K, g_n is also continuous on K, $n = 1, 2, 3, \ldots\ldots..$

Since $f_n \to f$ pointwise on K, $g_n \to 0$ pointwise on K.

Since $f_n(x) \ge f_{n+1}(x)$ for all $x \in K$, $n = 1, 2, 3, \ldots\ldots$

$$f_n(x) - f(x) \ge f_{n+1}(x) - f(x), \text{ for all } x \in K,\ n = 1, 2, 3, \ldots\ldots..$$

Hence, $g_n(x) \ge g_{n+1}(x)$, for all $x \in K$, $n = 1, 2, 3,$ .... (4.3.6.1)

Claim: $g_n \to 0$ uniformly on K.

Let $\varepsilon > 0$ be given,

$$\text{Let } K_n = \{x \in K / g_n(x) \ge \varepsilon\}$$

$$= g_n^{-1}([\varepsilon, \infty)) \qquad \ldots\ldots\ldots.. (4.3.6.2)$$

Since each g_n is continuous and inverse image of a closed set under a continuous function is closed, each K_n is a closed subset of K.

Since K is compact and a closed subset of a compact set is closed, each K_n is compact.

If $x \in K_{n+1}$, then $g_{n+1}(x) \ge \varepsilon$.

Therefore,

$$g_n(x) \geq g_{n+1}(x) \geq \varepsilon.$$

Hence $x \in K_n$.

Hence $K_{n+1} \subseteq K_n$.

Fix $x \in K$.

Since $g_n(x) \to 0$, for given $\varepsilon > 0$, there exists an integer N_1 such that $n \geq N_1$ implies

$$g_n(x) < \varepsilon.$$

This implies that

$$x \notin K_n \text{ for all } n \geq N_1.$$

$\Rightarrow K_n$ is empty.

Hence, there exists an integer N such that K_N is empty.

This implies that

$$g_N(x) < \varepsilon \text{ for all } x \in K.$$

From (4.3.6.1), it is clear that, $\{g_n\}$ is monotonically decreasing sequence.

Therefore,

$$g_n(x) \geq g_N(x) \text{ for all } n \geq N.$$

Hence

$$0 \leq g_n(x) - g_N(x) < \varepsilon, \text{ for all } x \in K, n \geq N.$$

Hence, $g_n \to 0$ uniformly on K.

Hence, $f_n - f \to 0$ uniformly on K.

Hence, $f_n \to f$ uniformly on K.

Hence the theorem.

Problem 4.3.7

1. Is the compactness necessary in the above theorem?

Solution

Yes.

Example:

Let $f_n(x) = \dfrac{1}{nx+1}$, $0 < x < 1$, $n = 1, 2, 3, \ldots\ldots\ldots\ldots$

Then, $f_n(x) \to 0$ monotonically in (0, 1), but the convergence is not uniform.

4.4 Uniform convergence and Integration

Theorem 4.4.1

Let α be a monotonically increasing function on $[a, b]$. Suppose $f_n \in R(\alpha)$ on $[a, b]$, for $n = 1, 2, 3, \ldots\ldots\ldots$ and suppose $f_n \to f$ uniformly on $[a, b]$, then $f \in R(\alpha)$ on $[a, b]$ and $\int_a^b f d\alpha = \lim_{n\to\infty} \int_a^b f_n d\alpha$.

Proof

Let $\{f_n\}$ be a sequence of real valued functions on $[a, b]$ such that

i) $f_n \in R(\alpha)$ on $[a, b]$, for $n = 1, 2, 3, \ldots\ldots\ldots$

ii) $f_n \to f$ uniformly on $[a, b]$.

Let $\varepsilon_n = \sup_{a \le x \le b} |f_n(x) - f(x)|$.

Then, $\varepsilon_n \to 0$ as $n \to \infty$.

This implies that

$$|f_n(x) - f(x)| \le \varepsilon_n,\ a \le x \le b$$

$$\Rightarrow |f(x) - f_n(x)| \le \varepsilon_n,\ a \le x \le b$$

$$\Rightarrow -\varepsilon_n \le f(x) - f_n(x) \le \varepsilon_n,\ a \le x \le b$$

$$\Rightarrow f_n(x) - \varepsilon_n \le f(x) \le f_n(x) + \varepsilon_n,\ a \le x \le b \qquad \ldots\ldots\ldots\ldots (4.4.1.1)$$

$$\Rightarrow f_n(x) - \varepsilon_n \le f(x),\ a \le x \le b$$

$$\Rightarrow \underline{\int_a^b} (f_n - \varepsilon_n) d\alpha \le \underline{\int_a^b} f d\alpha \qquad \ldots\ldots\ldots\ldots (4.4.1.2)$$

Similarly from (4.4.1.1), we have

$$\Rightarrow f(x) \le f_n(x) + \varepsilon_n, \; a \le x \le b$$

$$\Rightarrow \overline{\int_a^b} f d\alpha \le \overline{\int_a^b} (f_n + \varepsilon_n) d\alpha \qquad \text{..........} (4.4.1.3)$$

Since $f_n \in R(\alpha)$ on $[a, b]$, $n = 1, 2, 3, \ldots\ldots\ldots$

$f_n - \varepsilon_n \in R(\alpha)$ on $[a, b]$ and

$f_n + \varepsilon_n \in R(\alpha)$ on $[a, b]$.

Hence,

$$\Rightarrow \underline{\int_a^b} (f_n - \varepsilon_n) d\alpha = \int_a^b (f_n - \varepsilon_n) d\alpha \qquad \text{..........} (4.4.1.4)$$

and

$$\Rightarrow \overline{\int_a^b} (f_n + \varepsilon_n) d\alpha = \int_a^b (f_n + \varepsilon_n) d\alpha \qquad \text{..........} (4.4.1.5)$$

We know that,

$$\underline{\int_a^b} f d\alpha \le \overline{\int_a^b} f d\alpha \qquad \text{..........} (4.4.1.6)$$

From (4.4.1.2), (4.4.1.3), (4.4.1.4), (4.4.1.5) and (4.4.1.6), we get that

$$\int_a^b (f_n - \varepsilon_n) d\alpha \le \underline{\int_a^b} f d\alpha \le \overline{\int_a^b} f d\alpha \le \int_a^b (f_n + \varepsilon_n) d\alpha .$$

Therefore,

$$\int_a^b f_n d\alpha - \varepsilon_n \int_a^b d\alpha \le \overline{\int_a^b} f d\alpha - \underline{\int_a^b} f d\alpha \le \int_a^b f_n d\alpha + \varepsilon_n \int_a^b d\alpha$$

$$\Rightarrow 0 \le \overline{\int_a^b} f d\alpha - \underline{\int_a^b} f d\alpha \le 2\varepsilon_n [\alpha(b) - \alpha(a)]$$

Since $\varepsilon_n \to 0$, as $n \to \infty$, the upper and lower integrals of *f* are equal.

Thus, $f \in R(\alpha)$.

Therefore,

$$\int_a^b f_n d\alpha - \varepsilon_n \int_a^b d\alpha \le \int_a^b f d\alpha \le \int_a^b f_n d\alpha + \varepsilon_n \int_a^b d\alpha$$

$$\Rightarrow -\varepsilon_n \int_a^b d\alpha \le \int_a^b f d\alpha - \int_a^b f_n d\alpha \le \varepsilon_n \int_a^b d\alpha$$

$$\Rightarrow \left| \int_a^b f d\alpha - \int_a^b f_n d\alpha \right| \le \varepsilon_n [\alpha(b) - \alpha(a)] > 0 \text{ as } n \to \infty.$$

Therefore, $\lim_{n\to\infty} \int_a^b f_n d\alpha = \int_a^b f d\alpha$.

$$\Rightarrow \lim_{n\to\infty} \int_a^b f_n d\alpha = \int_a^b \left(\lim_{n\to\infty} f_n \right) d\alpha.$$

Hence the theorem.

Theorem 4.4.2

If $f_n \in R(\alpha)$ on $[a, b]$, for $n = 1, 2, 3, \ldots\ldots$ and if $f(x) = \sum_{n=1}^{\infty} f_n(x),\ a \le x \le b$, and the series converges uniformly on $[a, b]$, then $\int_a^b f d\alpha = \sum_{n=1}^{\infty} \int_a^b f_n d\alpha$.

In other words, the series can be integrated term by term.

Proof

Let $s_n(x) = \sum_{k=1}^{n} f_k(x)$, $\quad n = 1, 2, 3, \ldots\ldots\ldots$

Since $\sum f_n$ converges uniformly to f on $[a, b]$, $s_n \to f$ uniformly on $[a, b]$. Therefore, by the above theorem 4.4.1,

$$\int_a^b f d\alpha = \lim_{n\to\infty} \int_a^b s_n d\alpha$$

$$= \lim_{n\to\infty} \int_a^b \left(\sum_{k=1}^{n} f_k \right) d\alpha$$

$$= \lim_{n\to\infty}\left(\sum_{k=1}^{n}\int_a^b f_k\,d\alpha\right)$$

$$= \sum_{k=1}^{\infty}\int_a^b f_k\,d\alpha .$$

Hence the theorem.

4.5 Uniform convergence and differentiation

Uniform convergence of a sequence of differentiable functions $\{f_n\}$ implies nothing about the sequence $\{f_n'\}$. Consider the following example.

Problems 4.5.1

If $f_n(x) = \sin nx$, x is real and n = 1, 2, 3, ... and $f_n \to f$, then check whether $f_n' \to f'$.

Solution

Let $f_n(x) = \sin nx$, x is real and n = 1, 2, 3, ...

Then, $f(x) = \lim\limits_{n\to\infty} f_n(x) = 0$ for all real x.

$\Rightarrow f'(x) = 0$.

Now, $f_n'(x) = n\cos nx$

For $x = 0$, $f_n'(0) = n \to \infty$, as $n \to \infty$

Hence, $\{f_n'\}$ does not converge to $\{f'\}$.

Theorem 4.5.2

Suppose $\{f_n\}$is sequence of functions differentiable on $[a, b]$ and such that $\{f_n(x_0)\}$ converges for some point x_0 in $[a, b]$. If $\{f_n'\}$ converges uniformly on $[a, b]$, then $\{f_n\}$ converges uniformly on $[a, b]$, to a function f and $f'(x) = \lim\limits_{n\to\infty} f_n'(x)$, $a \le x \le b$.

Proof

Since $\{f_n(x_0)\}$ converges for some x_0 in $[a, b]$,

for given ε > 0, there exists an integer N_1 such that $n \ge N_1, m \ge N_1$ implies

$$|f_n(x_0) - f_m(x_0)| < \varepsilon/2 \quad \text{...(4.5.2.1)}$$

Since $\{f_n'\}$ converges uniformly on $[a, b]$, $\{f_n'\}$ satisfies Cauchy criterion for uniform convergence.

Therefore, for the same $\varepsilon > 0$, there exists N_2 such that $n \geq N_2, m \geq N_2$ implies

$$|f_n'(t) - f_m'(t)| < \frac{\varepsilon}{2(b-a)}, \quad a \leq t \leq b \quad \text{.........................(4.5.2.2)}$$

Let $N = \max.(N_1, N_2)$.

Let $x, t \in [a, b]$.

Applying mean Value theorem to the function $(f_n - f_m)$, we get

$(f_n - f_m)(x) - (f_n - f_m)(t) = (f_n - f_m)'(\xi)(x - t)$, where ξ lies between x and t.

$\Rightarrow |(f_n - f_m)(x) - (f_n - f_m)(t)| = |(f_n - f_m)'(\xi)(x - t)|$.

Therefore, for any x, t in $[a, b]$, $m \geq N, n \geq N$ implies

$$\begin{aligned} |f_n(x) - f_m(x) - f_n(t) + f_m(t)| &= |(f_n - f_m)'(\xi)||(x - t)| \\ &< \frac{\varepsilon}{2(b-a)}|x - t| \\ &< \varepsilon/2 \quad \text{.........................(4.5.2.3)} \end{aligned}$$

Therefore by choosing $t = x_0$,

for any $x \in [a, b]$ and $m \geq N, n \geq N$, we get

$$\begin{aligned} |f_n(x) - f_m(x)| &= |(f_n(x) - f_m(x)) - (f_n(x_0) - f_m(x_0)) + (f_n(x_0) - f_m(x_0))| \\ &\leq |(f_n(x) - f_m(x)) - (f_n(x_0) - f_m(x_0))| + |(f_n(x_0) - f_m(x_0))| \\ &< \varepsilon/2 + \varepsilon/2 = \varepsilon. \end{aligned}$$

Therefore, $\{f_n\}$ satisfies Cauchy criterion for uniform convergence.

$\Rightarrow \{f_n\}$ converges uniformly on $[a, b]$.

Let $f(x) = \lim_{n \to \infty} f_n(x)$, $a \leq x \leq b$

Fix a point x in $[a, b]$.

Define $\phi_n(t) = \dfrac{f_n(t) - f_n(x)}{t - x}$, $\quad a \leq t \leq b, \quad t \neq x$. .... (4.5.2.4)

And $\varphi(t) = \dfrac{f(t) - f(x)}{t - x}$, $\quad a \leq t \leq b, \ t \neq x$. .... (4.5.2.5)

Then, $\lim_{t \to x} \varphi_n(t) = \lim_{t \to x} \dfrac{f_n(t) - f_n(x)}{t - x}$

$$= f_n'(x), \qquad n = 1,2,3,\dots\dots\dots \qquad \dots\dots\dots\dots (4.5.2.6)$$

and $\lim\limits_{t\to x}\varphi(t) = \lim\limits_{t\to x}\dfrac{f(t)-f(x)}{t-x}$

$$= f'(x), \qquad n = 1,2,3,\dots\dots\dots \qquad \dots\dots\dots\dots (4.5.2.7)$$

From (4.5.2.3), for any $x,t \in [a,b]$, $m \geq N, n \geq N$ implies

$$\left|\left(f_n(x)-f_n(t)\right)-\left(f_m(x)-f_m(t)\right)\right| \leq \frac{\varepsilon}{2(b-a)}|x-t|$$

Therefore, $|\varphi_n(t)-\varphi_m(t)| = \left|\dfrac{f_n(t)-f_n(x)}{t-x} - \dfrac{f_m(t)-f_m(x)}{t-x}\right|$

$< \dfrac{\varepsilon}{2(b-a)}$, for all $t \in [a,b]$, $t \neq x, m \geq N, n \geq N$.

Hence $\{\varphi_n\}$ satisfies Cauchy criterion for uniform convergence, for $t \neq x$.

Therefore, $\{\varphi_n\}$ converges uniformly for $t \neq x$.

Also for $t \neq x$,

$$\lim_{n\to\infty}\varphi_n(t) = \lim_{n\to\infty}\frac{f_n(t)-f_n(x)}{t-x}$$

$$= \frac{f(t)-f(x)}{t-x}$$

$$= \varphi(t).$$

Therefore, $\lim\limits_{t\to x}\varphi(t) = \lim\limits_{t\to x}\left[\lim\limits_{n\to\infty}\varphi_n(t)\right]$

$= \lim\limits_{n\to\infty}\left[\lim\limits_{t\to x}\varphi_n(t)\right]$ [by theorem 4.3.2]

$= \lim\limits_{n\to\infty} f_n'(x)$ [from (4.5.2.6)]

$$\Rightarrow \lim_{t\to x}\varphi(t) = \lim_{n\to\infty} f_n'(x) \qquad \dots\dots\dots\dots (4.5.2.8)$$

From (4.5.2.7) & (4.5.2.8), we get that

$$f'(x) = \lim_{t\to x}\varphi(t)$$

$$= \lim_{n\to\infty} f_n'(x).$$

Hence the theorem.

Problem 4.5.3

Is there exist a real continuous function on the line which is nowhere differentiable.

Solution

Yes.

Define $\phi(x) = |x|, \quad (-1 \le x \le 1)$

Extend φ to all real x by defining

$$\varphi(x+2) = \varphi(x).$$

Then, φ is continuous on R^1.

Define f on R^1 as

$$f(x) = \sum_{n=0}^{\infty} \left(\frac{3}{4}\right)^n \varphi(4^n x).$$

f is the required function.

4.6 Equicontinuous families of functions

Definition 4.6.1

Let $\{f_n\}$ be a sequence of functions defined on a set E. $\{f_n\}$ is said to be pointwise bounded on E, if the sequence $\{f_n(x)\}$ is bounded for every $x \in E$. In other words, $\{f_n\}$ is said to be pointwise bounded on E, if there exists a finite valued function φ defined on E such that

$$|f_n(x)| < \varphi(x) \qquad x \in E,\ n = 1,2,3,.........$$

Theorem 4.6.2

If $\{f_n\}$ is a pointwise bounded sequence of complex functions on a countable set E, then $\{f_n\}$ has a subsequence $\{f_{n_k}\}$ such that $\{f_{n_k}(x)\}$ converges for every $x \in E$.

Proof

Let $\{f_n\}$ be a pointwise bounded sequence of complex functions on a countable set E.

Since E is countable,

the points of E can be arranged in a sequence as $\{x_i\}$, $i = 1, 2, 3, \ldots\ldots$

Since $\{f_n\}$ is a pointwise bounded on E,

$\{f_n(x_1)\}$ is a bounded sequence of complex numbers.

Therefore, $\{f_n(x_1)\}$ contains a convergent subsequence.

Denote it by $\{f_{1,k}\}$, that is, $\{f_{1,k}(x_1)\}$ converges as $k \to \infty$.

Now consider $\{f_{1,k}(x_2)\}$.

Since $\{f_{1,k}\}$ is a subsequence of $\{f_n\}$,

$\{f_{1,k}\}$ is also pointwise bounded on E.

Therefore, $\{f_{1,k}(x_2)\}$ contains a convergent subsequence, say $\{f_{2,k}(x_2)\}$.

Proceeding in this way, we get the sequences represented by the following array:

F_1: $f_{1,1}\, f_{1,2}\, f_{1,3}$ ....

F_2: $f_{2,1}\, f_{2,2}\, f_{2,3}$ ....

F_3: $f_{3,1}\, f_{3,2}\, f_{3,3}$ ....

....

....

which have the following properties:

i) F_n is a subsequence of F_{n-1}, for $n = 2, 3, 4,$ ....

ii) $\{f_{n,k}(x_2)\}$ converges as $k \to \infty$.

iii) the order in which the functions appear is the same in each sequence. That is, if one function precedes another in F_1, then they are in same relation in every F_n, until one or the other is deleted.

Hence, when going from one row in the above array to the next row below, functions may move left but never to the right.

Now consider the sequence

F: $f_{1,1}\, f_{2,2}\, f_{3,3}\, f_{4,4}$ ....

obtained by going along the diagonal of the array.

By (iii), the sequence F is a subsequence of F_n (except possibly its first $n-1$ terms), for $n = 1,2,3,\ldots\ldots$

Hence (ii) implies that, $\{f_{n,n}(x_i)\}$ converges for every $x_i \in E$.

Hence $\{f_n\}$ has a subsequence that converges for every $x_i \in E$.

Hence the theorem.

Definition 4.6.3

Let $\{f_n\}$ be a sequence of functions defined on a set E. $\{f_n\}$ is said to be uniformly bounded on E, if there exists a number M such that $|f_n(x)| < M, \qquad (x \in E,\ n = 1,2,3,..........)$.

Problems 4.6.4

1. If pointwise bounded sequence of functions contains a convergent subsequence, can you say that it will be true for uniformly bounded functions also?

Solution

No.

Example:

Let $f_n(x) = \sin nx, \qquad 0 \le x \le 2\pi, \quad n = 1,2,3,........$

Then, $|f_n(x)| = |\sin nx| \le 1, \qquad 0 \le x \le 2\pi, \quad n = 1,2,3,........$

Therefore, $\{f_n\}$ is uniformly bounded on the compact set $[0, 2\pi]$.

Suppose there exists a sequence $\{n_k\}$ such that $\{\sin n_k x\}$ converges for every $x \in [0,2\pi]$.

Then,

$$\lim_{k\to\infty}(\sin n_k x - \sin n_{k+1} x) = 0, \qquad 0 \le x \le 2\pi.$$

$$\Rightarrow \lim_{k\to\infty}(\sin n_k x - \sin n_{k+1} x)^2 = 0, \qquad 0 \le x \le 2\pi.$$

$$\Rightarrow \lim_{k\to\infty}\int_0^{2\pi}(\sin n_k x - \sin n_{k+1} x)^2\, dx = 0. \qquad .... (4.6.4.1)$$

But, if we evaluate $\int_0^{2\pi}(\sin n_k x - \sin n_{k+1} x)^2\, dx$, we get

$$\int_0^{2\pi}(\sin n_k x - \sin n_{k+1} x)^2\, dx = \int_0^{2\pi}\left(\sin^2 n_k x + \sin^2 n_{k+1} x - 2\sin n_k x.\sin n_{k+1} x\right)dx$$

$$= 2\pi.$$

Therefore,

$$\lim_{k\to\infty}\int_0^{2\pi}(\sin n_k x - \sin n_{k+1} x)^2\,dx = 2\pi$$, which contradicts (4.6.4.1).

Therefore, $\{f_n\}$ does not have any convergent subsequence.

2. Define $f_n(x) = \dfrac{x^2}{x^2 + (1-nx)^2}$, $\quad (0 \le x \le 1, \quad n = 1,2,3,......)$ Show that $\{f_n\}$ is uniformly bounded on [0, 1].

Solution

Let $f_n(x) = \dfrac{x^2}{x^2 + (1-nx)^2}$, $\quad (0 \le x \le 1, \quad n = 1,2,3,......)$.

Then, $|f_n(x)| \le 1$.

Hence, $\{f_n\}$ is uniformly bounded on [0, 1].

3. Does every convergent sequence contain a uniformly convergent subsequence?

Solution

Need not be so.

Let $f_n(x) = \dfrac{x^2}{x^2 + (1-nx)^2}$, $\quad (0 \le x \le 1, \quad n = 1,2,3,......)$.

$\Rightarrow \lim_{n\to\infty} f_n(x) = 0, \qquad 0 \le x \le 1$

That is, $\{f_n\}$ converges pointwise to $f(x) = 0$ on [0, 1].

But, $\{f_n\}$ does not contain any uniformly convergent subsequence, because

$$f_n(1/n) = 1, \quad \forall n = 1,2,3,.......$$

Definition 4.6.5

A family F of complex functions f defined on a set E in a metric space X is said to be equicontinuous on E, if for every $\varepsilon > 0$, there exists a $\delta > 0$, such that

$$|f(x) - f(y)| < \varepsilon, \text{ whenever } d(x, y) < \delta,\ x \in E,\ y \in E.$$

Here, d denotes the metric of X.

Problem 4.6.6

1. Is every member of an equicontinuous family of functions is uniformly continuous or not?

Solution

Yes.

2. Let $f_n(x) = \dfrac{x^2}{x^2 + (1-nx)^2}$, $(0 \le x \le 1,\quad n = 1,2,3,......)$. Is $\{f_n\}$ an equicontinuous family of functions?

Solution

No.

Theorem 4.6.7

If K is a compact metric space, if $f_n \in C(K)$ for $n = 1, 2, 3, ...$ and if $\{f_n\}$ converges uniformly on K, then $\{f_n\}$ is equicontinuous on K. ($C(K)$ denotes the set of all continuous functions on K).

Proof

Let $\varepsilon > 0$ be given.

Since $\{f_n\}$ converges uniformly on K, $\{f_n\}$ satisfies the Cauchy's condition for uniform convergence.

Hence for the given $\varepsilon > 0$, there exists an integer N such that $m \ge N, n \ge N, x \in K$ implies

$$|f_n(x) - f_m(x)| < \varepsilon/3$$

$\Rightarrow |f_n(x) - f_N(x)| < \varepsilon/3$, whenever $n > N$ and $x \in K$ .... (4.6.7.1)

$\Rightarrow \sup\{|f_n(x) - f_N(x)|,\ x \in K\} < \varepsilon/3$, whenever $n > N$.

$\Rightarrow \|f_n - f_N\| < \varepsilon/3$, $(n > N)$ ....(4.6.7.2)

Since continuous functions on a compact metric space are uniformly continuous and K is compact, each f_i is a uniformly continuous function on K.

In particular, $f_1, f_2, ..., f_N$ are uniformly continuous on K.

Therefore for the same $\varepsilon > 0$, there exists a $\delta > 0$ such that

$|f_i(x) - f_i(y)| < \varepsilon/3$ whenever $d(x, y) < \delta$ and $i = 1, 2, ..., N$.

.... (4.6.7.3)

From (4.6.7.1) and (4.6.7.3), we get that

If $n > N$ and $d(x, y) < \delta$,

$$|f_n(x) - f_n(y)| = |f_n(x) - f_N(x) + f_N(x) - f_N(y) + f_N(y) - f_n(y)|$$

$$\leq |f_n(x) - f_N(x)| + |f_N(x) - f_N(y)| + |f_N(y) - f_n(y)|$$

$$< \varepsilon/3 + \varepsilon/3 + \varepsilon/3 = \varepsilon \qquad \text{.......... (4.6.7.4)}$$

From (4.6.7.2) and (4.6.7.3),

$$|f_n(x) - f_n(y)| < \varepsilon \text{ whenever } d(x, y) < \delta, x, y \in K, \; n = 1,2,3,.....$$

Hence $\{f_n\}$ is equicontinuous on K.

Hence the theorem.

Theorem 4.6.8

If K is compact, if $f_n \in C(K)$ for n = 1, 2, 3, ... and if $\{f_n\}$ is pointwise bounded and equicontinuous on K, then

i) $\{f_n\}$ is uniformly bounded on K,

ii) $\{f_n\}$ contains a uniformly convergent subsequence.

Proof

Let K be compact and $f_n \in C(K)$ for n = 1, 2, 3, ...

Let $\{f_n\}$ be pointwise bounded and equicontinuous on K.

i) To prove that $\{f_n\}$ is uniformly bounded on K.

Since $\{f_n\}$ is equicontinuous on K, for given $\varepsilon > 0$, there exists a $\delta > 0$ such that

$$|f_n(x) - f_n(y)| < \varepsilon, \text{ whenever } d(x, y) < \delta, \; n = 1, 2, 3, \qquad \text{.......... (4.6.8.1)}$$

Since K is compact, for this $\delta > 0$, there exists a finite subset $\{p_1, p_2, ..., p_r\}$ of K which is δ-dense in K.

That is, for every $x \in K$, there correspondence atleast one p_i with $d(x, p_i) < \delta$ .... (4.6.8.2)

Since $\{f_n\}$ is pointwise bounded on K, $\{f_n(p_i)\}$ is bounded for all p_i.

That is, there exist M_i such that $|f_n(p_i)| < M_i$, i = 1, 2, ..., r.

Let M = Max. $(M_1, M_2, ...,M_r)$.

Let x be any element of K.

Then, by (4.6.8.2), there exists one p_i such that $d(x, p_i) < \delta$.

Therefore, by (4.6.8.1), we have, $|f_n(x) - f_n(p_i)| < \varepsilon$, $\forall n$.

Hence, $|f_n(x)| = |f_n(x) - f_n(p_i) + f_n(p_i)|$

$$\leq |f_n(x) - f_n(p_i)| + |f_n(p_i)|$$

$$< \varepsilon + M_i \quad \forall n.$$

Therefore, $\{f_n\}$ is uniformly bounded on K.

ii) To prove that $\{f_n\}$ contains a uniformly convergent subsequence.
Since K is compact, K contains a countable dense subset, say E.
Then by the theorem 2.5.2, which states that

" If $\{f_n\}$ is pointwise bounded sequence of complex functions on a countable set E, then $\{f_n\}$ has a subsequence $\{f_{n_k}\}$ such that $\{f_{n_k}(x)\}$ converges for every $x \in E$",

$\{f_n\}$ has a subsequence $\{f_{n_k}\}$ such that $\{f_{n_k}(x)\}$ converges for every $x \in E$.

Let $f_{n_k} = g_i$.

Then $\{g_i(x)\}$ converges uniformly for every $x \in K$.
To prove that $\{g_i\}$ converges uniformly on K.
By (4.6.8.2), for every $x \in K$, there corresponds atleast one p_i with $d(x, p_i) < \delta$.
Let $U(x, \delta) = \{y \in K, d(x, y) < \delta\}$.
Since E is dense in K and K is compact, there are finitely many points $x_1, x_2, \ldots\ldots, x_m$ in E such that

$$K \subset U(x_1, \delta) \cup U(x_2, \delta) \cup \ldots\ldots\ldots \cup U(x_m, \delta).$$

Since $\{g_i(x)\}$ converges uniformly for every $x \in E$,

$\{g_i(x_s)\}$ converges for $s = 1, 2, 3, \ldots\ldots\ldots, m$.

Here, for each $s = 1, 2, \ldots\ldots, m$, and $\varepsilon > 0$, there exists such that

$$|g_i(x_s) - g_j(x_s)| < \varepsilon, \text{ whenever } i \geq N_s, j \geq N_s.$$

Let $N = \text{Max}\ (N_1, N_2, \ldots\ldots\ldots, N_m)$.

Then, $\left|g_i(x_s)-g_j(x_s)\right|<\varepsilon$, whenever $i\geq N, j\geq N$ and s = 1, 2, ……, m.

………… (4.6.8.3)

Let x be any point in K.

Then, $x\in U(x_s,\delta)$ for some s.

$$\Rightarrow d(x,x_s)<\delta.$$

Therefore by (4.6.8.2), $\left|g_i(x_s)-g_j(x_s)\right|<\varepsilon$, for every i. ………. (4.6.8.4)

From (4.6.8.3) and (4.6.8.4), for $i\geq N, j\geq N, x\in K$

$$\begin{aligned}\left|g_i(x)-g_j(x)\right| &= \left|g_i(x)-g_i(x_s)+g_i(x_s)-g_j(x_s)+g_j(x_s)-g_j(x)\right| \\ &\leq \left|g_i(x)-g_i(x_s)\right|+\left|g_i(x_s)-g_j(x_s)\right|+\left|g_j(x_s)-g_j(x)\right| \\ &< \varepsilon+\varepsilon+\varepsilon=3\varepsilon.\end{aligned}$$

Therefore, $\{g_i\}$ is uniformly convergent.

That is, $\{f_{n_k}\}$ is uniformly convergent on K.

Hence $\{f_n\}$ contains a uniformly convergent subsequence.

Hence the theorem.

4.7 Algebra of functions

Definition 4.7.1

A family A of complex functions defined on a set E is said to be an *algebra* if

i) $f+g\in$ A

ii) $fg\in$ A

iii) $cf\in$ A

for all $f\in$ A, $g\in$ A and for all complex constants c.

That is, if A is closed under addition, multiplication and scalar multiplication.

Note 4.7.2

If we consider algebras of real functions, then (iii) of the definition (4.7.1) is required to hold for all real c.

Definition 4.7.3

If A has the property that $f\in$ A whenever

i) $f_n \in \mathrm{A}$ $(n = 1, 2, 3, ...)$ and

ii) $f_n \to f$ uniformly on E,

then A is said to be uniformly closed.

Definition 4.7.4

Let B be the set of all functions which are limits of uniformly convergent sequences of members of A, then B is called uniform closure of A.

Theorem 4.7.5

Let B be the uniform closure of an algebra A of bounded functions. Then B is a uniformly closed algebra.

Proof

Let $f \in \mathrm{B}$ and $g \in \mathrm{B}$.

Then there exists uniformly convergent sequences of functions $\{f_n\}$ and $\{g_n\}$ such that

$$f_n \to f \text{ uniformly and } g_n \to g \text{ uniformly.}$$

Hence $f_n + g_n \to f + g$ uniformly,

$f_n g_n \to fg$ uniformly

and $cf_n \to cf$ uniformly, where c is a constant.

Hence, $f + g \in \mathrm{B}$, $fg \in \mathrm{B}$ and $cf \in \mathrm{B}$.

Hence, B is uniformly closed.

Definition 4.7.6

Let A be a family of functions on a set E. Then A is said to separate points on E, if, to every pair of distinct points $x_1, x_2 \in E$, there corresponds a function $f \in \mathrm{A}$ such that

$$f(x_1) \neq f(x_2).$$

Definition 4.7.7

Let A be a family of functions on a set E. Then A is said to vanish at no point of E, if, to each $x \in E$, there corresponds a function $g \in \mathrm{A}$ such that $g(x) \neq 0$.

Problems 4.7.8

1. Give an example of an algebra of functions which separates points and vanishes at no point.

Solution

The algebra of polynomials in one variable has these properties on R^1.

2. Give example of algebra of functions which does not separate points.

Solution

The set of all even polynomials on [-1, 1].

Theorem 4.7.9

Suppose A is an algebra of functions on a set E, A separates points on E, and A vanishes at no point of E. Suppose x_1, x_2 are distinct points of E and c_1, c_2 are constants (real if A is real Algebra). Then A contains a function f such that

$$f(x_1)=c_1, f(x_2)=c_2.$$

Proof

Let x_1, x_2 be distinct points of E.

Since A separates points on E, and A vanishes at no points of E, A contains functions g, h, k such that

$$g(x_1)\neq g(x_2), \qquad h(x_1)\neq 0, \qquad k(x_2)\neq 0. \qquad \text{............ (4.7.9.1)}$$

Let $u=gk-g(x_1)k$ and

$$v=gh-g(x_2)h.$$

Since A is an algebra, $g,h,k\in A$ implies $u\in A$, $v\in A$.

Also $u(x_1)=g(x_1)k(x_1)-g(x_1)k(x_1)=0$,

$$v(x_2)=g(x_2)h(x_2)-g(x_2)h(x_2)=0,$$

$$u(x_2)=g(x_2)k(x_2)-g(x_1)k(x_2)$$

$$=[g(x_2)-g(x_1)]k(x_2)=0 \qquad \text{[from (4.7.9.1)]}$$

$$v(x_1)=g(x_1)h(x_1)-g(x_2)h(x_1)$$

$$=[g(x_2)-g(x_1)]h(x_1)=0 \qquad \text{[from (4.7.9.1)]}$$

Define f on E as

$$f = \frac{c_1 v}{v(x_1)} + \frac{c_2 u}{u(x_2)}$$

Since $u, v \in \mathsf{A}$ and A is an algebra, $f \in \mathsf{A}$.

and
$$f(x_1) = \frac{c_1 v(x_1)}{v(x_1)} + \frac{c_2 u(x_1)}{u(x_2)}$$

$$= c_1 + 0 = c_1.$$

$$f(x_2) = \frac{c_1 v(x_2)}{v(x_1)} + \frac{c_2 u(x_2)}{u(x_2)}$$

$$= 0 + c_2 = c_2.$$

Hence f has the desired properties.

Hence the theorem.

4.8 The Stone - Weierstrass theorem

Theorem 4.8.1

This theorem is known as ***Weierstrass theorem***.

If f is a continuous complex function on $[a, b]$, there exists a sequence of polynomials A_n such that

$$\lim_{n\to\infty} A_n(x) = f(x)$$

uniformly on $[a, b]$. If f is real, then A_n may be taken real.

Proof

Without loss of generality, assume that $[a, b] = [0, 1]$.

Also assume that $f(0) = f(1) = 0$.

Consider $g(x) = f(x) - f(0) - x[f(1) - f(0)], \qquad 0 \le x \le 1.$

Then, $g(0) = g(1) = 0$.

Hence, if g can be obtained as the limit of polynomials, then the same is true for f also.

Hence we get the theorem.

Hence it is sufficient to prove the theorem with $f(0) = f(1) = 0$.

Let f be a continuous complex function on $[0, 1]$ with $f(0) = f(1) = 0$.

Since a continuous function on a compact space is uniformly continuous,

f is uniformly continuous on $[0, 1]$.

Define $f(x)$ to be zero for x outside [0, 1].

Then, f is uniformly continuous on the whole real line.

Define $Q_n(x) = c_n(1-x^2)^n$, $n = 1, 2, 3,$ .... (4.8.1.1)

where c_n is chosen such that

$$\int_{-1}^{1} Q_n(x)dx = 1,\ n = 1, 2, 3, \ldots\ldots\ldots\ldots \quad \text{........... (4.8.1.2)}$$

$$\Rightarrow \int_{-1}^{1} c_n(1-x^2)^n dx = 1,\ n = 1, 2, 3, \ldots\ldots\ldots\ldots \quad \text{........... (4.8.1.3)}$$

Consider $\int_{-1}^{1}(1-x^2)^n dx = 2\int_{0}^{1}(1-x^2)^n dx$

$$\geq 2\int_{0}^{1/\sqrt{n}}(1-x^2)^n dx$$

$$\geq 2\int_{0}^{1/\sqrt{n}}(1-nx^2)dx$$

$$\left[\text{Since } (1-x^2)^n = 1-nx^2+\frac{n(n-1)x^4}{2}-\ldots\ldots\ldots \geq 1-nx^2\right]$$

$$= 2\left[x-n\frac{x^3}{3}\right]_0^{1/\sqrt{n}}$$

$$= \frac{4}{3\sqrt{n}}$$

$$> \frac{1}{\sqrt{n}} \quad \text{........... (4.8.1.4)}$$

From (4.8.1.3) and (4.8.1.4),

$$1 = \int_{-1}^{1} c_n(1-x^2)^n dx > \frac{c_n}{\sqrt{n}}$$

$$c_n < \sqrt{n} \quad \text{........... (4.8.1.5)}$$

Therefore for any $\delta > 0$,

$$Q_n(x) = c_n(1-x^2)^n.$$

$$< \sqrt{n}(1-x^2)^n$$

$$< \sqrt{n}(1-\delta^2)^n, \qquad \text{for } \delta \le |x| \le 1 \qquad \text{............ (4.8.1.6)}$$

Therefore, $Q_n \to 0$ uniformly in $\delta \le |x| \le 1$.

Define $A_n(x) = \int_{-1}^{1} f(x+t)Q_n(t)dt$.

Since f is zero outside [0, 1],

$$A_n(x) = \int_{-x}^{1-x} f(x+t)Q_n(t)dt.$$

Changing the variable t as $t - x$, we get

$$A_n(x) = \int_{0}^{1} f(t)Q_n(t-x)dt, \qquad \text{............ (4.8.1.7)}$$

which is a polynomial in x.

Thus $\{A_n\}$ is a sequence of polynomials, which are real if, f is real.

Since f is uniformly continuous, for given $\varepsilon > 0$, there exists a $\delta > 0$ such that

$$|y-x| < \delta \Rightarrow |f(y) - f(x)| < \varepsilon/2 \qquad \text{............ (4.8.1.8)}$$

Let $M = \sup|f(x)|$.

For $0 \le x \le 1$,

$$|P_n(x) - f(x)| = \left|\int_{-1}^{1} f(x+t)Q_n(t)dt - f(x)\right|$$

$$= \left|\int_{-1}^{1} f(x+t)Q_n(t)dt - \int_{-1}^{1} f(x)Q_n(t)dt\right|$$

$$= \left|\int_{-1}^{1} [f(x+t) - f(x)]Q_n(t)dt\right|$$

$$\le \int_{-1}^{1} |f(x+t) - f(x)|Q_n(t)dt \qquad [\text{Since } Q_n(t) \ge 0]$$

$$= \int_{-1}^{-\delta} |f(x+t)-f(x)| Q_n(t)dt + \int_{-\delta}^{\delta} |f(x+t)-f(x)| Q_n(t)dt$$

$$+ \int_{\delta}^{1} |f(x+t)-f(x)| Q_n(t)dt$$

Now, $|f(x+t)-f(x)| \leq |f(x+t)| + |f(t)|$

$$\leq 2M$$

Hence,

$$|P_n(x)-f(x)| \leq 2M \int_{-1}^{-\delta} Q_n(t)dt + \frac{\varepsilon}{2}\int_{-\delta}^{\delta} Q_n(t)dt + 2M\int_{\delta}^{1} Q_n(t)dt$$

$$\leq 2M\sqrt{n}\,(1-\delta^2)^n \int_{-1}^{-\delta} dt + \frac{\varepsilon}{2} + 2M\sqrt{n}\,(1-\delta^2)^n \int_{\delta}^{1} dt$$

$$< 2M\sqrt{n}\;(1-\delta^2)^n(-\delta+1) + \frac{\varepsilon}{2} + 2M\sqrt{n}\;(1-\delta^2)^n(1-\delta)$$

$$< 4M\sqrt{n}\;(1-\delta^2)^n(-\delta+1) + \frac{\varepsilon}{2}$$

$< \varepsilon$, for sufficiently large n.

[since $\lim_{n\to\infty}(1-\delta^2)^n = 0$]

Hence, $\lim_{n\to\infty} A_n(x) = f(x)$.

Hence the theorem.

Theorem 4.8.2

For every interval [-a, a], there is a sequence of real polynomials A_n such that

$\lim_{n\to\infty} A_n(x) = |x|$ uniformly on [-a, a].

Proof

Since $|x|$ is a continuous function on [-a, a], by the above theorem 4.8.1, there exist a sequence $\{A_n^*\}$ of real polynomials which converges to $|x|$ uniformly on [-a, a].

That is, $\lim_{n\to\infty} A_n^*(x) = |x|$ uniformly on [-a, a]. .... (4.8.2.1)

$\Rightarrow \lim_{n\to\infty} A_n^*(0) = 0$ .... (4.8.2.2)

Let $A_n(x) = A_n^*(x) - A_n^*(0)$.

Then from (4.8.2.1) and (4.8.2.2),

$$A_n(0) = A_n^*(0) - A_n^*(0) = 0.$$

and $\lim_{n\to\infty} A_n(x) = \lim_{n\to\infty}\left[A_n^*(x) - A_n^*(0)\right]$

$$= |x| - 0$$

$$= |x| \text{ uniformly on } [-a, a].$$

Hence the theorem.

Theorem 4.8.3

This theorem is known as ***Stone – Weierstrass theorem***.

Let A be an algebra of real continuous functions on a compact set K. If A separates points on K and if A vanishes at no point of K, then the uniform closure B of A consists of all real continuous functions on K.

Proof

We shall divide the proof into four steps.

Step 1: If $f \in$ B, then $|f| \in$ B.

Proof

Let $a = \sup\{|f(x), x \in K|\}$.

Let $\varepsilon > 0$ be given.

Then by the theorem 4.8.2, which states that,

> "For every interval [-a, a], there is a sequence of real polynomials P_n such that $\lim_{n\to\infty} P_n(x) = |x|$ uniformly on [-a, a].",

there exists a sequence of real polynomials $\{P_n\}$ such that

$P_n(0) = 0$ and $\lim_{n\to\infty} P_n(y) = |y|$ uniformly on [-a, a]. .... (4.8.3.1)

Therefore, there exists real numbers c_1, c_2, ..., c_n such that

$$\left|\sum_{i=1}^{n} c_i y^i - |y|\right| < \varepsilon, \qquad -a \le y \le a \qquad \text{............ (4.8.3.2)}$$

Since B is an algebra,

the function $g = \sum_{i=1}^{n} c_i f^i$ is a member of B.

Therefore, $\quad \left|g(x) - |f(x)|\right| = \left|\sum_{i=1}^{n} c_i f^i(x) - |f(x)|\right|$

$$= \left|\sum_{i=1}^{n} c_i y^i - |y|\right|, \text{ where } y = f(x). \qquad \text{........ (4.8.3.3)}$$

Since $a = \sup\{|f(x), x \in K|\}$

$\Rightarrow -a \le f(x) \le a, \quad \forall x \in K$.

That is, $-a \le y \le a$.

Hence by (4.8.3.2),

$$\left|\sum_{i=1}^{n} c_i y^i - |y|\right| < \varepsilon. \qquad \text{............ (4.8.3.4)}$$

From (4.8.3.3) and (4.8.3.4),

$\left|g(x) - |f(x)|\right| < \varepsilon, \quad$ for all $x \in K$.

$\Rightarrow |f| \in$ B, since B is uniformly closed.

Hence step 1 is proved.

Step 2: If $f \in$ B and $g \in$ B, then max.$(f, g) \in$ B and min.$(f, g) \in$ B.

Proof

By definition,

$$\max.(f, g)(x) = \begin{cases} f(x), & \text{if } f(x) \ge g(x) \\ g(x), & \text{if } f(x) < g(x) \end{cases}$$

Therefore,

$$\max.(f, g) = \frac{f+g}{2} + \frac{|f-g|}{2}$$

Similarly, $\min.(f, g) = \dfrac{f+g}{2} - \dfrac{|f-g|}{2}$.

Since $f \in \mathsf{B}$, $g \in \mathsf{B}$ and B is closed, $f + g \in \mathsf{B}$ and $f - g \in \mathsf{B}$.

By the proof of step 1, $|f - g| \in \mathsf{B}$.

Hence,

$$\max.(f, g) = \frac{f + g}{2} + \frac{|f - g|}{2} \in \mathsf{B},$$

and $\min.(f, g) = \dfrac{f + g}{2} - \dfrac{|f - g|}{2} \in \mathsf{B}$.

Hence step 2 is proved.

By repeating this, we get that, if $f_1, f_2, \ldots\ldots, f_n \in \mathsf{B}$, then

$$\max(f_1, f_2, \ldots\ldots, f_n) \in \mathsf{B} \text{ and } \min(f_1, f_2, \ldots\ldots, f_n) \in \mathsf{B}.$$

Step 3: Given a real valued function *f*, continuous on *K*, a point $x \in K$ and $\varepsilon > 0$, there exists a function $g_x \in \mathsf{B}$ such that

$$g_x(x) = f(x)$$

And $g_x(t) > f(t) - \varepsilon, \quad \forall t \in K$. .... (4.8.3.5)

Proof

Let *x* be the given point in *K*.

Then for every y in K with $y \neq x$, by theorem 4.7.9,

taking $c_1 = f(x)$ and $c_2 = f(y)$, we get that, there exists a function $h_y \in \mathsf{A}$ satisfying

$$h_y(x) = f(x) \text{ and } h_y(y) = f(y). \qquad \text{............ (4.8.3.6)}$$

Since $\mathsf{A} \subset \mathsf{B}$, $h_y \in \mathsf{B}$.

Since h_y is continuous,

there exists an open set J_y containing *y* such that

$$h_y(t) > f(t) - \varepsilon, \quad \forall t \in J_y. \qquad \text{............ (4.8.3.7)}$$

Then $K \subset \bigcup\limits_{y \in K} J_y$.

Since *K* is compact, there is a finite set of points $y_1, y_2, \ldots\ldots, y_n$ such that

$$K \subset J_{y_1} \cup J_{y_2} \cup \ldots\ldots\ldots \cup J_{y_n}.$$

Let $g_x = \max(h_{y_1}, h_{y_2}, \ldots\ldots\ldots, h_{y_n})$.

By step 2, $g_x \in \mathsf{B}$.

Now from (4.8.3.6),

$$h_{y_i}(x) = f(x) \text{ for all } i = 1, 2, 3, \ldots., n.$$

Therefore,

$$g_x(x) = \max\left(h_{y_1}, h_{y_2}, \ldots\ldots.., h_{y_n}\right)(x) = f(x).$$

Also from (4.8.3.7),

$$h_{y_i}(t) > f(t) - \varepsilon, \quad \forall i = 1,2,\ldots\ldots,n.$$

Therefore,

$$g_x(t) = \max\left(h_{y_1}, h_{y_2}, \ldots\ldots.., h_{y_n}\right)(t)$$
$$> f(t) - \varepsilon, \quad \forall t \in K.$$

Hence g_x has the desired properties.

Hence step 3 is proved.

Step 4: Given a real valued function f continuous on K and $\varepsilon > 0$, there exists a function $h \in \mathsf{B}$ such that

$$|h(x) - f(x)| < \varepsilon, \qquad \forall x \in K.$$

That is, the uniform closure B of A consists of all real continuous functions on K.

Proof

Consider the functions g_x, for each $x \in K$ constructed in step 3.

Since each g_x is continuous on K, there exist open sets V_x containing x such that

$$g_x(t) < f(t) + \varepsilon, \qquad \forall t \in V_x. \qquad \ldots\ldots\ldots\ldots (4.8.3.8)$$

Then $K \subset \bigcup_{x \in K} V_x$.

Since K is compact, there exists a finite set of points $x_1, x_2, \ldots\ldots, x_m$ such that

$$K \subset V_{x_1} \cup V_{x_2} \cup \ldots\ldots\ldots \cup V_{x_m}.$$

Let $h = \min\left(g_{x_1}, g_{x_2}, \ldots\ldots.., g_{x_m}\right)$.

Then $h \in \mathsf{B}$, by step 2.

By (4.8.3.5), $g_{x_i}(t) > f(t) - \varepsilon, \qquad \forall t \in K, \ i = 1, 2, \ldots\ldots., n.$

Therefore,

$$h(t) > f(t) - \varepsilon, \qquad \forall t \in K. \qquad \text{............ (4.8.3.9)}$$

By (4.8.3.8), $g_{x_i}(t) < f(t) + \varepsilon, \quad \forall t \in K, \ i = 1, 2, ..., n.$

Therefore,

$$h(t) < f(t) + \varepsilon, \qquad \forall t \in K. \qquad \text{............ (4.8.3.10)}$$

From (4.8.3.9) & (4.8.3.10),

$$f(t) - \varepsilon < h(t) < f(t) + \varepsilon, \quad \forall t \in K.$$

Hence,

$$|h(t) - f(t)| < \varepsilon, \qquad \forall t \in K.$$

Hence for every real function *f* continuous on *K* and $\varepsilon > 0$, there exists a function $h \in \mathrm{B}$ such that

$$|h(t) - f(t)| < \varepsilon, \qquad \forall t \in K.$$

Hence the uniform closure B of A consists of all real continuous functions on *K*.

Hence the theorem.

4.9 Problems

1. If $\{f_n\}$ and $\{g_n\}$ converge uinformly on a set E, prove that $\{f_n + g_n\}$ converge uniformly on *E*. If in addition, $\{f_n\}$ and $\{g_n\}$ are bounded, prove that $\{f_n g_n\}$ converges uniformly on *E*.
2. Show that the sequence $\{f_n\}_{n-1}^{\infty}$ where $f_n(x) = \sin \dfrac{nx}{\sqrt{n}}$ is uniformly convergent on $[0, \pi]$.
3. Prove that every uniformly convergent sequence of bounded functions is uniformly bounded.
4. Set $f_n(x) = \dfrac{x^2}{x^2 + (1 - nx)^2}$, $0 \le x \le 1$, $n = 1, 2, ...$ Verify if the sequence of functions $\{f_n\}_{n-1}^{\infty}$ is equicontinuous.

Unit	
5	**DIFFERENTIATION OF FUNCTIONS**

5.1 Introduction

Differentiation is a process that associates to a real-valued function f another function f' called the derivative of f. The value of f' at a point c depends only on the values of f in a small interval around c. The concept of *derivative* is the main theme of differential calculus, one of the major discoveries in mathematics, and in science in general. Differentiation is the process of finding the best local linear approximation of a function.

The concept of differentiation originated from two classical problems:

1. The geometric problem of determining a tangent at a point to a curve in the plane.
2. The physical problem of determining the speed or the velocity of an object, such as a particle or a vehicle or a planet.

The notion of a derivative, which we shall study in this chapter, and the fact that it can often be computed effectively, turns out to be a key to solving the above two problems. Furthermore, the notion of a derivative has an enormous number of applications both within and outside mathematics.

Today, both Leibnitz and Newton are considered to have discovered calculus independently.

The concept of *derivative* has been applied in various domains and we mention only some of them:

- velocity and acceleration of a movement (Galileo 1638, Newton 1686);
- astronomy, verification of the law of gravitation (Newton, Kepler);
- calculation of the angles under which two curves intersect (Descartes);

- construction of telescopes (Galileo) and clocks (Huygens 1673);
- search for the maxima and minima of a function (Fermat 1638).

5.2 Derivatives of functions

Definition 5.2.1

Let f be a real-valued function defined on an open interval containing a point x. We say f is *differentiable at x*, or *f has a derivative at x*, if the limit

$$\lim_{y\to x}\frac{f(y)-f(x)}{y-x} \text{ exists.}$$

If f is differentiable at every point of a subset E of the open interval, then we say that f is differentiable on E.

Note 5.2.2

Let $f:[a,b]\to\mathbb{R}$ be any function. Let $x\in[a,b)$. We define

$$f'(x+)=\lim_{\substack{y\to x\\ x<y\le b}}\frac{f(y)-f(x)}{y-x}, \text{ if this limit exists.}$$

Let $x\in(a,b]$. We define

$$f'(x-)=\lim_{\substack{y\to x\\ a\le y<x}}\frac{f(y)-f(x)}{y-x}, \text{ if this limit exists.}$$

For $x\in(a,b)$, f is differentiable at x, if $f'(x+)=f'(x-)$.

f is differentiable at a, if $f'(a+)$ exists.

f is differentiable at b, if $f'(b-)$ exists.

Note 5.2.3

1. $f'(x+)=\lim\limits_{\substack{h\to 0\\ h>0}}\frac{f(x+h)-f(x)}{h}$ and $f'(x-)=\lim\limits_{\substack{h\to 0\\ h<0}}\frac{f(x+h)-f(x)}{h}$.

2. Suppose f is differentiable at x. Then,

   $f(y)-f(x)=[f'(x)-v(y)](y-x)$, where $v(y)\to 0$ as $y\to x$.

Proof

Suppose f is differentiable at x.

Then, $f'(x)=\lim\limits_{y\to x}\frac{f(y)-f(x)}{y-x}$.

That implies, $\lim_{y\to x}\frac{f(y)-f(x)}{y-x}-f'(x)=0$.

Hence, $\frac{f(y)-f(x)}{y-x}-f'(x)=v(y)$, where $v(y)\to 0$ as $y\to x$.

That implies,

$$f(y)-f(x)=[f'(x)+v(y)](y-x),\ \text{where}\ v(y)\to 0\ \text{as}\ y\to x.$$

Theorem 5.2.4

If f is differentiable at a point $x\in[a,b]$, then f is continuous.

Proof

Given that f is differentiable at $x\in[a,b]$.

Therefore, $\lim_{y\to x}\frac{f(y)-f(x)}{y-x}$ exists and its value is $f'(x)$.

That is, $f'(x)=\lim_{y\to x}\frac{f(y)-f(x)}{y-x}$.

We have to prove that, f is continuous at x.

That is, $\lim_{y\to x}f(y)=f(x)$.

Now, since $y\neq x$, we can write, $f(y)-f(x)=\frac{f(y)-f(x)}{y-x}(y-x)$.

That implies, $\lim_{y\to x}\left(f(y)-f(x)\right)=\lim_{y\to x}\frac{f(y)-f(x)}{y-x}\lim_{y\to x}(y-x)=f'(x).0=0.$

Hence, $\lim_{y\to x}f(y)=f(x)$.

Therefore, f is continuous at $x\in[a,b]$.

Hence proved.

Note 5.2.5

The converse of the above theorem 5.2.4 need not be true.

That is, continuity does not imply differentiability.

Let us consider an example,

Define $f:[-1,1]\to\mathbb{R}$ by $f(x)=|x|$.

Now, f is continuous on [-1, 1].

We claim that, f is continuous at $x = 0$.

$$f(0+)=\lim_{\substack{h\to 0\\ h>0}} f(0+h)=\lim_{\substack{h\to 0\\ h>0}} f(h)=\lim_{\substack{h\to 0\\ h>0}}|h|=\lim_{\substack{h\to 0\\ h>0}} h=0.$$

$$f(0-)=\lim_{\substack{h\to 0\\ h<0}} f(0+h)=\lim_{\substack{h\to 0\\ h<0}}|h|=\lim_{h\to 0}(-h)=0.$$

$$f(0)=|0|=0.$$

Therefore, f is continuous at 0.

We claim that, f is not differentiable at $x = 0$.

Now, $$f'(0+)=\lim_{\substack{h\to 0\\ h>0}}\frac{f(0+h)-f(0)}{h}=\lim_{h\to 0}\frac{h-0}{h}=1.$$

$$f'(0-)=\lim_{\substack{h\to 0\\ h<0}}\frac{f(0+h)-f(0)}{h}=\lim_{h\to 0}\frac{-h-0}{h}=-1.$$

Therefore, $f'(0+)\neq f'(0-)$.

Hence, f is not differentiable at 0.

Theorem 5.2.6

Suppose f and g are defined on $[a, b]$ and differentiable at $x\in[a,b]$, then $f+g, fg$ and f/g are differentiable at x, and

i) $(f+g)'(x)=f'(x)+g'(x)$

ii) $(fg)'(x)=f'(x)g(x)+f(x)g'(x)$

iii) $\left(\dfrac{f}{g}\right)'(x)=\dfrac{g(x)f'(x)-f(x)g'(x)}{[g(x)]^2}$, $g(x)\neq 0$.

Proof

Given that f and g are differentiable at $x\in[a,b]$.

Therefore, $f'(x)=\lim_{y\to x}\dfrac{f(y)-f(x)}{y-x}$ and .... (5.2.6.1)

$g'(x)=\lim_{y\to x}\dfrac{g(y)-g(x)}{y-x}$ exists .... (5.2.6.2)

i) Now, $\lim_{y\to x}\frac{(f+g)(y)-(f+g)(x)}{y-x}=\lim_{y\to x}\frac{f(y)+g(y)-f(x)-g(x)}{y-x}$.

$$=\lim_{y\to x}\frac{f(y)-f(x)}{y-x}+\lim_{y\to x}\frac{g(y)-g(x)}{y-x}=f'(x)+g'(x).$$

$$=f'(x)+g'(x). \qquad \text{............ (5.2.6.3)}$$

From equations (5.2.6.1) and (5.2.6.2), we say that $f'(x)$ and $g'(x)$ exists.

Therefore, $\lim_{y\to x}\frac{(f+g)(y)-(f+g)(x)}{y-x}$ exists.

That implies, $f+g$ is differentiable at x.

Hence, from equation (5.2.6.3), $(f+g)'(x)=f'(x)+g'(x)$.

This is also true for, $(f-g)'(x)=f'(x)-g'(x)$.

ii) Now, $\lim_{y\to x}\frac{(fg)(y)-(fg)(x)}{y-x}=\lim_{y\to x}\frac{f(y)g(y)-f(x)g(x)}{y-x}$.

$$=\lim_{y\to x}\frac{f(y)g(y)-f(y)g(x)+f(y)g(x)-f(x)g(x)}{y-x}.$$

$$=\lim_{y\to x}f(y)\left(\frac{g(y)-g(x)}{y-x}\right)+\lim_{y\to x}g(x)\left(\frac{f(y)-f(x)}{y-x}\right).$$

$$=\lim_{y\to x}f(y).\lim_{y\to x}\frac{g(y)-g(x)}{y-x}+g(x).\lim_{y\to x}\frac{f(y)-f(x)}{y-x}.$$

Since *f* is differentiable at x, *f* is continuous at x, and hence $\lim_{y\to x}f(y)=f(x)$

and also from equations (5.2.6.1) and (5.2.6.2), we get

$$\lim_{y\to x}\frac{(fg)(y)-(fg)(x)}{y-x}=f(x)g'(x)+g(x)f'(x). \qquad \text{............ (5.2.6.4)}$$

From equations (5.2.6.1) and (5.2.6.2), $f(x)g'(x)+g(x)f'(x)$ exists.

Hence from equation (5.2.6.4), $\lim_{y\to x}\frac{(fg)(y)-(fg)(x)}{y-x}$ exists.

Therefore, *fg* is differentiable at x.

Again, from equation (5.2.6.4), we get

$$(fg)'(x) = f(x)g'(x) + g(x)f'(x).$$

iii) $$\lim_{y\to x}\frac{(f/g)(y)-(f/g)(x)}{y-x} = \lim_{y\to x}\frac{\frac{f(y)}{g(y)}-\frac{f(x)}{g(x)}}{y-x}$$

$$= \lim_{y\to x}\frac{f(y)g(x)-f(x)g(y)}{g(y).g(x)(y-x)}$$

$$= \lim_{y\to x}\frac{f(y)g(x)-f(x)g(x)+f(x)g(x)-g(y)f(x)}{[g(y).g(x)](y-x)}.$$

$$= \frac{1}{g(x)\lim_{y\to x} g(y)}\left[\lim_{y\to x} g(x)\frac{f(y)-f(x)}{y-x} - \lim_{y\to x} f(x)\frac{g(y)-g(x)}{y-x}\right].$$

Since g is differentiable at x, g is continuous at x, and hence $\lim_{y\to x} g(y) = g(x)$ and from equations (5.2.6.1) and (5.2.6.2), we get

$$\lim_{y\to x}\frac{(f/g)(y)-(f/g)(x)}{y-x} = \frac{1}{g(x)g(x)}[g(x)f'(x)-f(x)g'(x)].$$

Therefore, $\left(\dfrac{f}{g}\right)(x) = \dfrac{g(x)f'(x)-f(x)g'(x)}{[g(x)]^2}$.

Hence the theorem.

Theorem 5.2.7

This theorem is known as **chain rule or function of function rule for differentiation.**

Suppose f is continuous on $[a, b]$, $f'(x)$ exists at some point $x \in [a,b]$. g is defined on an interval I which contains the range of f and g is differentiable at the point $f(x)$. If $h(y) = g(f(y)), a \le y \le b$, then h is differentiable at x and $h'(x) = g'(f(x)).f'(x)$.

Proof

Since f is differentiable at $x \in [a,b]$, we have

$$f(y) - f(x) = (y-x)[f'(x)+u(y)], \qquad \text{............ (5.2.7.1)}$$

where $u(y) \to 0$ as $y \to x$ and $y \in [a,b]$.

Since g is differentiable at $f(x)$, we have

$$g(s)-g(t)=(s-t)\left[g'(t)+v(s)\right], \qquad \text{............ (5.2.7.2)}$$

where $v(s)\to 0$ as $s\to t$ and $s\in I$.

For $y\in[a,b]$, $s=f(t)\in I$.

Now, $h(y)-h(x)=g(f(y))-g(f(x))=g(s)-g(t)$.

By using equation (5.2.7.2), we get

$$h(y)-h(x)=(s-t)\left[g'(t)+v(s)\right]=\left(f(y)-f(x)\right)\left[g'(t)+v(s)\right].$$

Again by using equation (5.2.7.1), we get

$$h(y)-h(x)=(y-x)\left[f'(x)+v(y)\right]\left[g'(t)+v(s)\right].$$

When $y\neq x$, we have

$$\frac{h(y)-h(x)}{y-x}=\left[f'(x)+v(y)\right]\left[g'(t)+v(s)\right]. \qquad \text{............ (5.2.7.3)}$$

Since f is continuous, when $y\to x$, $f(y)\to f(x)$. That is, $s\to t$.

Taking limit as $y\to x$ in equation (5.2.7.3), we have

$$\lim_{y\to x}\frac{h(y)-h(x)}{y-x}=\lim_{y\to x}\left\{\left[f'(x)+v(y)\right]\left[g'(t)+v(s)\right]\right\}=\left[f'(x)+0\right]\left[g'(t)+0\right].$$

$$\Rightarrow \lim_{y\to x}\frac{h(y)-h(x)}{y-x}=f'(x)\left[g'(f(x))\right]. \qquad \text{............ (5.2.7.4)}$$

Since $f'(x)$ and $g'(f(x))$ exists, $\lim\limits_{y\to x}\dfrac{h(y)-h(x)}{y-x}$ exists.

Therefore, h is differentiable at x.

Hence, equation (5.2.7.4) becomes, $h'(x)=f'(x)\left[g'(f(x))\right]$.

Hence the theorem.

Definition 5.2.8

Let f be a real function defined on a metric space X. We say f has a *local maximum at* $p\in X$, if there exists $\delta>0$ such that $f(q)\le f(p)$, for all $q\in X$, with $d(p,q)<\delta$.

Theorem 5.2.9

Suppose *f* is defined on $[a, b]$. If *f* has a local maximum at $x \in (a,b)$ and if $f'(x)$ exist, then $f'(x) = 0$.

Proof

Given that *f* has a local maximum at $x \in (a,b)$.

Therefore, there exists $\delta > 0$ such that $a < x - \delta < x < x + \delta < b$ and for all $t \in (x - \delta, x + \delta)$, we have

$$f(t) \le f(x). \qquad \text{............ (5.2.9.1)}$$

Now, for $x - \delta < t < x$, we have $t - x < 0$.

$$\text{Therefore, } \frac{f(t) - f(x)}{t - x} \ge 0. \qquad \text{............ (5.2.9.2)}$$

Now, taking limit as $t \to x$, we get $f'(x-) \ge 0$. .... (5.2.9.3)

Again, for $x < t < x + \delta$, we have $t - x > 0$.

$$\text{Therefore, } \frac{f(t) - f(x)}{t - x} \le 0.$$

Now, taking limit as $t \to x$,

we get $f'(x+) \le 0$. .... (5.2.9.4)

Since $f'(x)$ exists, $f'(x+) = f'(x-) = f'(x)$.

Therefore, equations (5.2.9.3) and (5.2.9.4) becomes

$f'(x) \le 0$ and $f'(x) \ge 0$, which implies $f'(x) = 0$.

Hence the theorem.

Theorem 5.2.10

Suppose *f* is defined on (a, b). If *f* has a local minimum at $x \in (a,b)$ and if $f'(x)$ exists, then $f'(x) = 0$.

Proof

Given *f* has a local minimum at $x \in (a,b)$.

Therefore, there exists $\delta > 0$ such that $a < x - \delta < x < x + \delta < b$ and for all $t \in (x - \delta, x + \delta)$,

we have $f(t) \ge f(x)$. .... (5.2.10.1)

Now, for $x-\delta<t<x$, we have $t-x<0$.

Therefore, $\dfrac{f(t)-f(x)}{t-x}\leq 0$. .... (5.2.10.2)

Taking limit as $t\to x$, we have $f'(x-)\leq 0$. .... (5.2.10.3)

Again for $x<t<x+\delta$, we have $t-x>0$.

Therefore, $\dfrac{f(t)-f(x)}{t-x}\geq 0$.

Taking limit as $t\to x$, we have $f'(x+)\geq 0$. .... (5.2.10.4)

Since $f'(x)$ exists, $f'(x+)=f'(x-)=f'(x)$.

Hence, from equations (5.2.10.3) and (5.2.10.4), we get

$f'(x)\leq 0$ and $f'(x)\geq 0$.

Therefore, $f'(x)=0$.

Hence proved.

Theorem 5.2.11

This theorem is known as **Generalised Mean Value theorem.**

If f and g are continuous real functions on [a, b] which are differentiable on (a, b), then there is a point $x\in(a,b)$ such that

$$(f(b)-f(a))g'(x)=(g(b)-g(a))f'(x).$$

Proof

For $a\leq x\leq b$, let $h(x)=[f(b)-f(a)]g(x)-[g(b)-g(a)]f(x)$.

.... (5.2.11.1)

Since f and g are continuous on [a, b], h is continuous on [a, b].

Since f and g are differentiable on (a, b), h is differentiable on (a, b).

Now, $h(a)=[f(b)-f(a)]g(a)-[g(b)-g(a)]f(a)=f(b)g(a)-g(b)f(a)$

.... (5.2.11.2)

Again,

$$h(b)=[f(b)-f(a)]g(b)-[g(b)-g(a)]f(b)=g(a)f(b)-g(b)f(a).$$

.... (5.2.11.3)

Therefore, from equations (5.2.11.2) and (5.2.11.3), we get

$$h(a)=h(b).$$

Now, for $a < x < b$, $h'(x) = [f(b) - f(a)]g'(x) - [g(b) - g(a)]f'(x)$.

.... (5.2.11.4)

To prove the theorem, we have to prove that,

$h'(x) = 0$, for some $x \in (a,b)$.

Case i)

Suppose h is a constant function, then

$h'(x) = 0$, for all $x \in (a,b)$.

Case ii)

Suppose $h(t) > h(a)$, for some $t \in (a,b)$.

Since h is continuous, h is bounded.

If $M = \sup_{a \leq y \leq b} h(y)$, then $M = h(x)$, for some $x \in [a,b]$.

Since $h(t) > h(a) = h(b)$, we have $x \neq a, b$.

Hence, $x \in (a,b)$.

Therefore, h has a maximum at $x \in (a,b)$ and $h'(x)$ exists.

Hence, by theorem 5.2.9, $h'(x) = 0$.

Case iii)

Suppose $h(t) < h(a)$, for some $t \in (a,b)$.

Since h is continuous, h is bounded.

If $m = \inf_{a \leq y \leq b} h(y)$, then $m = h(x)$, for some $x \in [a,b]$.

Since $h(t) < h(a) = h(b)$, we have $x \neq a, b$.

Hence, $x \in (a,b)$.

Therefore, h has a minimum at $x \in (a,b)$ and $h'(x)$ exists.

Hence, by theorem 5.2.10, $h'(x) = 0$.

Therefore, from equation 5.2.11.4, we get

$$[f(b) - f(a)]g'(x) - [g(b) - g(a)]f'(x) = 0.$$

Hence, $[f(b) - f(a)]g'(x) = [g(b) - g(a)]f'(x)$.

Hence the theorem.

Theorem 5.2.12

This theorem is known as ***Mean Value theorem***.

If f is a continuous real function on $[a, b]$ which is differentiable on (a, b), then there exists $x \in (a,b)$ such that $f(b) - f(a) = (b-a)f'(x)$.

Proof

Define $g(t) = t, \quad a \le t \le b, \quad g'(t) = 1$.

Therefore, g is continuous on $[a, b]$ and differentiable on (a, b).

Hence, by Generalized Mean Value theorem, there exists $x \in (a,b)$ such that

$$[f(b) - f(a)]g'(x) = [g(b) - g(a)]f'(x).$$

That implies, $f(b) - f(a) = (b-a)f'(x)$.

Hence the theorem.

Theorem 5.2.13

Suppose f is differentiable on (a, b), then

i) If $f'(x) \ge 0$, for all $x \in (a,b)$, then f is monotonically increasing.

ii) If $f'(x) = 0$, for all $x \in (a,b)$, then f is a constant function.

iii) If $f'(x) \le 0$ for all $x \in (a,b)$, then f is monotonically decreasing.

Proof

Let $x_1, x_2 \in (a,b)$ with $x_1 < x_2$.

Therefore, $[x_1, x_2] \subset (a,b)$.

Since f is differentiable on (a, b), f is continuous on $[x_1, x_2]$ and differentiable on (x_1, x_2).

Therefore, by Mean Value Theorem, there exists $t \in (x_1, x_2)$ such that

$$f(x_2) - f(x_1) = (x_2 - x_1)f'(t). \qquad \text{........... (5.2.13.1)}$$

i) Given that $f'(x) \ge 0$, for all $x \in (a,b)$.

Therefore, $f'(t) \ge 0$.

Also, we have $x_2 - x_1 > 0$.

Hence, $(x_2 - x_1)f'(t) \ge 0$.

Now, from equation (5.2.13.1), we have $f(x_2)-f(x_1)\geq 0$.

Therefore, $f(x_1)\leq f(x_2)$.

Hence, *f* is monotonically increasing.

ii) Given that $f'(x)=0$, for all $x\in(a,b)$.

Therefore, $f'(t)=0$.

Hence, $(x_2-x_1)f'(t)=0$.

Therefore, from equation (5.2.13.1), we get $f(x_2)-f(x_1)=0$.

That implies, $f(x_2)=f(x_1)$.

Hence, *f* is a constant function.

iii) Given that $f'(x)\leq 0$, for all $x\in(a,b)$.

Therefore, $f'(t)\leq 0$.

Also, $x_2-x_1>0$.

Hence, $(x_2-x_1)f'(t)\leq 0$.

Again, from equation (5.2.13.1), we have $f(x_2)-f(x_1)\leq 0$.

That implies, $f(x_1)\geq f(x_2)$.

Hence, *f* is monotonically decreasing.

Hence the theorem.

Theorem 5.2.14

Suppose *f* is a real differentiable function on [*a*, *b*]. If $f'(a)<\lambda<f'(b)$, then there exists $x\in(a,b)$ such that $\lambda=f'(x)$.

Proof

Define $g(t)=f(t)-\lambda t,\ \ a\leq t\leq b$.

Since *f* is differentiable on [*a*, *b*], and λt is differentiable on [*a*, *b*], *g* is differentiable on [*a*, *b*].

Therefore, $g'(t)=f'(t)-\lambda,\ \ a\leq t\leq b$. .... (5.2.14.1)

Since $f'(a)<\lambda,\quad g'(a)=f'(a)-\lambda<0$.

Let $\varepsilon=g'(a)$. Hence $\varepsilon<0$ implies that $(-\varepsilon)>0$.

Now, $\varepsilon = g'(a) = \lim_{t \to a} \frac{g(t) - g(a)}{t - a}$.

Therefore, there exists $c \in (a, b)$ such that for all $t \in (a, c)$,

$$\frac{g(t) - g(a)}{t - a} < g'(a) + (-\varepsilon) = 0.$$

Hence, $\frac{g(t) - g(a)}{t - a} < 0$, for all $t \in (a, c)$.

That is, $g(t) < g(a)$, for all $t \in (a, c)$.

Therefore, there exists $t_1 \in (a, b)$, such that $g(t_1) < g(a)$. .... (5.2.14.2)

Since $\lambda < f'(b)$, $g'(b) = f'(b) - \lambda > 0$.

Let $\varepsilon_1 = g'(b)$. Hence, $\varepsilon_1 > 0$.

Now, $\varepsilon_1 = g'(b) = \lim_{t \to b} \frac{g(t) - g(b)}{t - b}$.

Therefore, there exists $c_1 \in (a, b)$ such that for all $t \in (a, c_1)$,

$$\frac{g(t) - g(b)}{t - b} > g'(b) - \varepsilon_1 = 0.$$

Hence, $\frac{g(t) - g(b)}{t - b} > 0$, for all $t \in (a, c_1)$.

That is, $g(t) > g(b)$, for all $t \in (a, c_1)$.

Therefore, there exists $t_2 \in (a, b)$ such that $g(t_2) > g(b)$. .... (5.2.14.3)

Since g is differentiable on $[a, b]$, g is continuous on $[a, b]$.

Therefore, g is bounded.

If $m = \inf_{a \le y \le b} g(y)$, then $m = g(x)$, for some $x \in [a, b]$.

From equations (5.2.14.2) and (5.2.14.3), we get that $x \neq a, b$.

Therefore, g has a minimum at $x \in (a, b)$ and $g'(x)$ exists.

Hence, by local minimum theorem, $g'(x) = 0$.

Therefore, from equation (5.2.14.1), we have $f'(x) - \lambda = 0$.

That is, $f'(x) = \lambda$.

Hence the theorem.

Problems 5.2.15

1. Let *f* be defined by $f(x)=\begin{cases} x\sin(1/x) & x\neq 0 \\ 0 & x=0 \end{cases}$

Find $f'(x)$.

Solution

When $x\neq 0$, $f'(x)=-\frac{1}{x}\cos(1/x)+\sin(1/x)$.

When $x=0$, $\lim_{t\to 0}\sin(1/t)$ oscillates between 1 and (-1) and hence

for $t\neq 0, \lim_{t\to 0}\frac{f(t)-f(0)}{t-0}=\lim_{t\to 0}\sin(1/t)$ doesn't exists.

Hence, $f'(0)$ doesn't exist.

2. Let *f* be defined by $f(x)=\begin{cases} x^2\sin(1/x) & x\neq 0 \\ 0 & x=0 \end{cases}$

Find $f'(x)$.

Solution

When $x\neq 0$, $f'(x)=-\cos(1/x)+2x\sin(1/x)$.

When $x=0$, for $t\neq 0$, $\lim_{t\to 0}\frac{f(t)-f(0)}{t-0}=\lim_{t\to 0}\frac{t^2\sin(1/t)}{t}$ = 0 times a finite quantity = 0.

Therefore, $f'(0)=0$.

Theorem 5.2.16

This theorem is known as ***L'Hospital's rule***.

Suppose f and g are real and differentiable in (a, b) and $g'(x)\neq 0$, for all $x\in(a,b)$, where $-\infty\leq a\leq b\leq +\infty$. Suppose $\frac{f'(x)}{g'(x)}\to A$ as $x\to a$. If $f(x)\to 0$ and $g(x)\to 0$ as $x\to a$ or if $g(x)\to +\infty$ as $x\to a$, then $\frac{f(x)}{g(x)}\to A$ as $x\to a$.

Proof

Case i)

Let $-\infty \leq A \leq \infty$.

Let us choose real numbers q and r, such that $A < r < q$.

Given $\lim_{x \to a} \frac{f'(x)}{g'(x)} = A$ and $r - A > 0$.

Therefore, there exists $c \in (a,b)$ such that, for all $t \in (a,c)$,

$$\frac{f'(t)}{g'(t)} < A + r - A < r.$$

That is, $\frac{f'(t)}{g'(t)} < r$, for all $t \in (a,c)$. .... (5.2.16.1)

Let $a < x < y < c$.

Hence, $[x, y] \subset (a,c)$.

Since f and g are differentiable in (a, b), f and g are continuous on $[x, y]$ and differentiable on (x, y).

Therefore, by Generalized Mean Value theorem, there exists $t \in (x, y)$ such that

$$\frac{f(x) - f(y)}{g(x) - g(y)} = \frac{f'(t)}{g'(t)}.$$

By equation (5.2.16.1), if $a < x < y < c$, then $\frac{f(x) - f(y)}{g(x) - g(y)} < r$. .. (5.2.16.2)

If $f(x) \to 0$ and $g(x) \to 0$ as $x \to a$, then equation (5.2.16.2) becomes,

$$\frac{-f(y)}{-g(y)} < r, \quad \text{if } a < y < c.$$

That is, $\frac{f(y)}{g(y)} < q$, if $a < y < c$. .... (5.2.16.3)

Fix $a < y < c$.

Suppose $g(x) \to \infty$ as $x \to a$. Then, there exists $c_1 \in (a, y)$ such that, for all $x \in (a, c_1)$

$$g(x) > g(y) \text{ and } g(x) > 0.$$

Hence, $\frac{g(x)-g(y)}{g(x)}>0$.

Now, multiplying equation (5.2.16.2), $\frac{g(x)-g(y)}{g(x)}$, we get

$$\frac{f(x)}{g(x)}-\frac{f(y)}{g(x)}<r\left(1-\frac{g(y)}{g(x)}\right).$$

$$\frac{f(x)}{g(x)}<r+\frac{f(y)}{g(x)}-r\frac{g(y)}{g(x)}.$$

That is, for $a<x<c_1$, we have $\frac{f(x)}{g(x)}<r+\frac{1}{g(x)}[f(y)-r.g(y)]$.

.... (5.2.16.4)

Since $g(x)\to\infty$ as $x\to a$, $\frac{1}{g(x)}\to 0$ as $x\to a$.

That is, $\lim\limits_{x\to a}\frac{1}{g(x)}=0$.

Now, $q-r>0$.

Therefore, there exists $c_2\in(a,c_1)$ such that, for all $x\in(a,c_2)$, we have

$$[f(y)-r.g(y)]\frac{1}{g(x)}<q-r.$$

Hence, equation (5.2.16.4) becomes $\frac{f(x)}{g(x)}<q$, for all $x\in(a,c_2)$.

.... (5.2.16.5)

From equations (5.2.16.3) and (5.2.16.5), we get both the cases,

there exists $c_3\in(a,b)$ such that, for all $x\in(a,c_3)$, $\frac{f(x)}{g(x)}<q$. (5.2.16.6)

Case *ii*)

Let $-\infty<A\leq\infty$.

Let us choose a real number p, such that $p<A$.

Therefore, by case i), there exists $c_4\in(a,b)$, such that for all $x\in(a,c_4)$, $p<\frac{f(x)}{g(x)}$. .... (5.2.16.7)

We have to prove that, $\lim\limits_{x \to a} \frac{f(x)}{g(x)} = A$.

i) Suppose $A = +\infty$,

We can choose a real number p, such that $p < A$.

By case ii), there exists $c_4 \in (a,b)$, such that for all

$x \in (a, c_4)$, $\frac{f(x)}{g(x)} > p$.

Therefore, $\lim\limits_{x \to a} \frac{f(x)}{g(x)} = +\infty = A$.

ii) Suppose $A = -\infty$,

We can choose a real number q, such that $A < q$.

By case i), there exists $c_3 \in (a,b)$ such that for all

$x \in (a, c_3)$, $\frac{f(x)}{g(x)} < q$.

Therefore, $\lim\limits_{x \to a} \frac{f(x)}{g(x)} = -\infty = A$.

iii) Suppose A is finite.

Let $\varepsilon > 0$ be given.

Let us take $p = A - \varepsilon$, $q = A + \varepsilon$.

Hence, $p < A < q$.

Therefore, by case i), there exists $c_3 \in (a,b)$, such that for all

$$x \in (a, c_3), \quad \frac{f(x)}{g(x)} < q.$$

By case ii), there esists $c_4 \in (a,b)$, such that for all $x \in (a, c_4)$, $\frac{f(x)}{g(x)} > p$.

Let us take $c_5 = Mini.(c_3, c_4)$.

Then, $c_5 \in (a,b)$ and for all $x \in (a, c_5)$,

$$A - \varepsilon = p < \frac{f(x)}{g(x)} < q = A + \varepsilon.$$

$$-\varepsilon < \frac{f(x)}{g(x)} - A < \varepsilon .$$

Therefore, $\lim_{x \to a} \frac{f(x)}{g(x)} = A$.

Hence the theorem.

5.3 Derivatives of Higher order

Definition 5.3.1

If f has a derivative f' on an interval and if f' is itself differentiable, we denote the derivative of f' by f'' and call f'' the second derivative of f. Continuing in this manner, we obtain the functions

$$f, f', f'', f^{(3)}, \ldots\ldots, f^{(n)}, \ldots\ldots$$

each of which is the derivative of its preceding one. Here, $f^{(n)}$ is called the nth *derivative of f.*

Theorem 5.3.2

This theorem is known as ***Taylor's theorem***.

Suppose f is a real function on $[a, b]$, n is a positive integer, $f^{(n-1)}$ is continuous on $[a, b]$, $f^{(n)}(t)$ exists for every $t \in (a,b)$. Let α, β be distinct points of $[a, b]$ and define

$$p(t) = \sum_{k=0}^{n-1} \frac{f^{(k)}(\alpha)}{k!}(t-\alpha)^k ,$$

then, there exist a point x between α and β, such that

$$f(\beta) = p(\beta) + \frac{f^{(n)}(x)}{n!}(\beta - \alpha)^n .$$

Proof

Let $\alpha < \beta$.

Let us choose a number M such that $f(\beta) = p(\beta) + M(\beta - \alpha)^n$.

$$\ldots\ldots\ldots\ldots (5.3.2.1)$$

To prove the theorem, we have to prove that, $M = \frac{f^{(n)}(x)}{n!}$.

Define $g(t) = f(t) - p(t) - M(t-\alpha)^n, \quad a \le t \le b$. .... (5.3.2.2)

Hence, from equation (5.3.2.1), we have, $g(\beta)=f(\beta)-p(\beta)-M(\beta-\alpha)^n=0$.

Now, $p(t)$ is a polynomial of degree atmost $(n-1)$.

Hence, the nth derivative of $p(t)=0$. That is, $p^{(n)}(t)=0$.

Differentiating equation (5.3.2.2) n times, we get

$$g^{(n)}(t)=f^{(n)}(t)-p^{(n)}(t)-Mn!=f^{(n)}(t)-Mn!.$$

To prove $M=\dfrac{f^{(n)}(x)}{n!}$, for some $x\in(\alpha,\beta)$, it is enough, if we prove that, $g^{(n)}(x)=0$, for some $x\in(\alpha,\beta)$.

Let $0\le k\le n-1$.

Now,

$$p(t)=f(\alpha)+\frac{f'(\alpha)}{1!}(t-\alpha)+\frac{f''(\alpha)}{2!}(t-\alpha)^2+....$$

$$....+\frac{f^{(k)}(\alpha)}{k!}(t-\alpha)^k+\frac{f^{(k+1)}(\alpha)}{(k+1)!}(t-\alpha)^{k+1}+....+\frac{f^{(n-1)}(\alpha)}{(n-1)!}(t-\alpha)^{n-1}$$

Differentiating k times, we get

$$p^{(k)}(t)=\frac{f^{(k)}(t)}{k!}k!+\frac{f^{(k+1)}(t)}{(k+1)!}(k+1)!(t-\alpha)+.....$$

$$+\frac{f^{(n-1)}(\alpha)}{(n-1)!}n-1)(n-2).....\big((n-1)-(k-1)\big)(t-\alpha)^{(n-1-k)}.$$

Hence, $p^{(k)}(\alpha)=f^{(k)}(\alpha)$.

Now, differentiating equation (5.3.2.2) k times, we get

$$g^{(k)}(t)=f^{(k)}(t)-p^{(k)}(t)-M.n(n-1)(n-2).....(n-(k-1))(t-\alpha)^{n-k}.$$

Hence, $g^{(k)}(\alpha)=f^{(k)}(\alpha)-p^{(k)}(\alpha)=0$.

Therefore, $g(\alpha)=0,\ g'(\alpha)=0,\ g''(\alpha)=0,......,\ g^{(n-1)}(\alpha)=0$.

Now, $[\alpha,\beta]\subset[a,b]$.

Since g is continuous in $[\alpha,\beta]$ and differentiable in (α,β), by Mean Value theorem, there exists $x_1\in(\alpha,\beta)$ such that

$$g(\beta)-g(\alpha)=(\beta-\alpha)g'(x_1).$$

That is, $0=(\beta-\alpha)g'(x_1)$.

Therefore, $g'(x_1)=0$, for some $x_1 \in (\alpha,\beta)$.

Now, $g'(\alpha)=0$ and $g'(x_1)=0$.

Let us apply Mean Value theorem for the function g' in (α,x_1).

Therefore, there exists $x_2 \in (\alpha,x_1)$ such that $g'(x_1)-g'(\alpha)=(x_1-\alpha)g''(x_2)$.

That is, $0=(x_1-\alpha)g''(x_2)$.

Therefore, $g''(x_2)=0$, for some $x_2 \in (\alpha,x_1)$.

Repeating the above steps *n* times, we get

$$g^{(n)}(x_n)=0, \text{ for some } x_n \in (\alpha,x_{n-1}).$$

Let us take $x_n = x$.

Therefore, there exists some $x \in (\alpha,\beta)$ such that $g^{(n)}(x)=0$.

Hence, $M=\dfrac{f^{(n)}(x)}{n!}$, for some $x \in (\alpha,\beta)$.

Therefore, $f(\beta)=p(\beta)+\dfrac{f^{(n)}(x)}{n!}(\beta-\alpha)^n$.

Hence the theorem.

5.4 Differentiation of Vector Valued Function

Definition 5.4.1

Let $f(x)=f_1(x)+if_2(x)$, $a \le x \le b$, where f_1 and f_2 are real functions. We define *f* is differentiable at *x*, if and only if f_1 and f_2 are differentiable at *x*.

Also, $f'(x)=f_1'(x)+if_2'(x)$.

Let $\overline{f}:[a,b]\to R^k$ and $\overline{f}=(f_1,f_2,\ldots\ldots,f_k)$.

We define $\overline{f}$ is differentiable at *x*, if and only if $f_1,f_2,\ldots\ldots,f_k$ are differentiable at *x*.

Also, $\overline{f}'=(f_1',f_2',\ldots\ldots,f_k')$.

Note 5.4.2

1. Mean Value theorem is not true for complex valued functions.

Let us prove this statement by an example;

Define $f:[0,2\pi]\to C$ by $f(x)=e^{ix}=\cos x+i\sin x$.

Now, f is continuous in $[0,2\pi]$ and f is differentiable in $(0,2\pi)$.

We know that, $f(2\pi)=1$ and $f(0)=1$.

Hence, $f(2\pi)-f(0)=0$. .... (5.4.2.1)

Again, $f'(x)=i.e^{ix}$.

Hence, $|f'(x)|=|i.e^{ix}|=|e^{ix}||i|=1$.

Therefore, $f'(x)\neq 0$, for all $x\in(0,2\pi)$.

That implies, $(2\pi-0)f'(x)\neq 0$, for all $x\in(0,2\pi)$. .... (5.4.2.2)

From equations (5.4.2.1) and (5.4.2.2), we can conclude that, there exists no $x\in(0,2\pi)$ such that

$$f(2\pi)-f(0)=(2\pi-0)f'(x).$$

Therefore, Mean Value theorem is not true for complex valued functions.

2. L'Hospital's rule is not true for complex valued functions.

Let us prove this statement by an example;

Define $f:(0,1)\to C$ by $f(x)=x$ and $g:(0,1)\to C$ by $g(x)=x+x^2e^{i/x^2}$.

Here, f and g are differentiable in (0,1).

Now, $f'(x)=1$ and $g'(x)=2e^{i/x^2}\left(x-\dfrac{i}{x}\right)+1$.

Hence, $|g'(x)|=\left|2e^{i/x^2}\left(x-\dfrac{i}{x}\right)+1\right|$.

Since $|z_1-z_2|\geq|z_1|-|z_2|$, $|g'(x)|\geq\left|2e^{i/x^2}\left(x-\dfrac{i}{x}\right)\right|-1=2\left|x-\dfrac{i}{x}\right|-1$.

Again, since $|x+iy|\geq|y|$, $|g'(x)|\geq 2\left|\dfrac{-1}{x}\right|-1=\dfrac{2}{x}-1>0$.

Therefore, $g'(x)\neq 0$.

Now, $\left|\dfrac{1}{g'(x)}\right| \le \dfrac{x}{2-x}$.

Hence, $\left|\dfrac{f'(x)}{g'(x)}\right| \le \dfrac{x}{2-x}$.

Therefore, $\lim\limits_{x\to 0}\left|\dfrac{f'(x)}{g'(x)}\right| \le 0$, which implies $\lim\limits_{x\to 0}\dfrac{f'(x)}{g'(x)} = 0$. .... (5.4.2.3)

Now, $\lim\limits_{x\to 0}\dfrac{f(x)}{g(x)} = \lim\limits_{x\to 0}\dfrac{x}{x^2 e^{i/x^2} + x} = 1$. .... (5.4.2.4)

Hence, from equations (5.4.2.3) and (5.4.2.4), we get that

$$\lim_{x\to 0}\frac{f(x)}{g(x)} \ne \lim_{x\to 0}\frac{f'(x)}{g'(x)}.$$

Therefore, L'Hospital's rule is not true for complex valued functions.

Theorem 5.4.3

This theorem is known as ***Mean Value theorem for vector valued functions.***

Suppose $\overline{f}$ is a continuous mapping of $[a, b]$ into R^k and $\overline{f}$ is differentiable in (a, b). Then, there exists

$x \in (a,b)$ such that $\left|\overline{f}(b) - \overline{f}(a)\right| \le (b-a)\left|f'(x)\right|$.

Proof

Let $\overline{f} = (f_1, f_2, \ldots\ldots, f_k)$, $\overline{z} = \overline{f}(b) - \overline{f}(a)$ and $\overline{z} = (z_1, z_2, \ldots\ldots, z_k)$.

Define $\varphi : [a,b] \to \mathbb{R}$ as $\varphi(t) = \overline{z}.\overline{f}(t)$, where $a \le t \le b$.

That is, $\varphi(t) = z_1 f_1 + z_2 f_2 + \ldots.. + z_k f_k$.

Since $\overline{f}$ is continuous on $[a, b]$, $f_1, f_2, \ldots\ldots, f_k$ are continuous on $[a, b]$.

Therefore, φ is continuous on $[a, b]$.

Since $\overline{f}$ is differentiable on (a, b), $f_1, f_2, \ldots\ldots, f_k$ are differentiable on (a, b).

Therefore, φ is differentiable on (a, b).

Hence, by Mean Value theorem, there exists $x \in (a,b)$, such that

$\varphi(b) - \varphi(a) = (b-a)\varphi'(x)$. .... (5.4.3.1)

Now, $\varphi(b)-\varphi(a)=\bar{z}.\bar{f}(b)-\bar{z}.\bar{f}(a)=\left|\bar{z}\right|^{2}$.

Therefore, equation (5.4.3.1) becomes,

$$\left|\bar{z}\right|^{2}=(b-a)\bar{z}.\bar{f}'(x).$$

Taking modulus on both sides of the above equation and using Schwartz inequality, we get

$$\left|\bar{z}\right|^{2}\leq(b-a)\left|\bar{z}\right|\left|\bar{f}'(x)\right|. \qquad \text{............ (5.4.3.2)}$$

If $\left|\bar{z}\right|\neq 0$, then $\left|\bar{z}\right|\leq(b-a)\left|\bar{f}'(x)\right|$.

That is, $\left|\bar{f}(b)-\bar{f}(a)\right|\leq(b-a)\left|\bar{f}'(x)\right|$.

If $\left|\bar{z}\right|=0$, then $\left|\bar{f}(b)-\bar{f}(a)\right|=\left|\bar{z}\right|=0\leq(b-a)\left|f'(x)\right|$.

That is, $\left|\bar{f}(b)-\bar{f}(a)\right|\leq(b-a)\left|f'(x)\right|$.

Hence the theorem.

Unit	
6	**INTEGRATION – RS INTEGRALS**

6.1 Introduction

This chapter gives an introduction to Riemann-Steiltjes integral. Most applications in elementary probability theory are covered by Riemann-Steiltjes integral. The reason for introducing Steiltjes integral is to get a more unified approach to the theory of random variables.

6.2 Definition and Existence of the Integral

Definition 6.2.1

Let $[a, b]$ be a given interval. A partition P of $[a, b]$ is a finite set of points $x_0, x_1, x_2, \ldots\ldots, x_n$ such that $a = x_0 \leq x_1 \leq x_2 \leq \ldots\ldots \leq x_n = b$

Definition 6.2.2

Let α be a monotonically increasing function on $[a, b]$. Corresponding to any partition P of $[a, b]$,

$$\alpha_i = \alpha(x_i) - \alpha(x_{i-1}),\ i = 1, 2, 3, \ldots\ldots\ldots, n.$$

Then, $\alpha_i \geq 0$.

Let f be a bounded real valued function on $[a, b]$.

Let $U(p, f, \alpha) = \sum_{i=1}^{n} M_i \Delta\alpha_i$

$$L(p, f, \alpha) = \sum_{i=1}^{n} m_i \Delta\alpha_i$$

where $M_i = \text{Sup}\{f(x) / x \in [x_{i-1}, x_i]\}$ and

$$m_i = \text{Inf}\{f(x) / x \in [x_{i-1}, x_i]\}$$

We define the upper Riemann – Steiltjes integral of *f* as

$$\overline{\int_a^b} f d\alpha = \inf U(P, f, \alpha)$$

and the lower Riemann – Steiltjes integral of *f* as

$$\underline{\int_a^b} f d\alpha = \sup L(P, f, \alpha)$$

where the infinimum and supremum are taken over all partitions P of $[a, b]$.

If $\overline{\int_a^b} f d\alpha = \underline{\int_a^b} f d\alpha$, their common value is denoted by $\int_a^b f d\alpha$ or $\int_a^b f(x) d\alpha(x)$.

This is called the Riemann – Steiltjes integral of *f* with respect to α on $[a, b]$.

If $\int_a^b f d\alpha$ exists, then *f* is said to be integrable with respect to α on $[a, b]$.

It is written as $f \in R(\alpha)$ on $[a, b]$.

Definition 6.2.3

A partition P^* is said to be a refinement of *P*, if $P^* \supseteq P$.

Note 6.2.4

Given two partitions P_1 and P_2 of $[a, b]$, their common refinement is $P^* = P_1 \cup P_2$.

Theorem 6.2.5

If P^* is the refinement of *P*, then

$$U(P^*, f, \alpha) \le U(P, f, \alpha)$$

and $L(P^*, f, \alpha) \ge L(P, f, \alpha)$.

Proof

Assume that P^* contains just one point more than *P*.

Let this be *c* and $x_{i-1} < c < x_i$.

Let $M_i' = \text{Sup}\{f(x) / x \in [x_{i-1}, c]\}$ and

$$M_i'' = \text{Sup}\{f(x) / x \in [c, x_i]\}.$$

Then $M_i' \le M_i$ and $M_i'' \le M_i$.

Consider,

$$U(P^*, f, \alpha) = \sum_{\substack{k=1 \\ k \ne i}}^{n} M_k \Delta\alpha_k + M_i'[\alpha(c) - \alpha(x_{i-1})] + M_i''[\alpha(x_i) - \alpha(c)]$$

$$\le \sum_{\substack{k=1 \\ k \ne i}}^{n} M_k \Delta\alpha_k + M_i[\alpha(c) - \alpha(x_{i-1})] + M_i[\alpha(x_i) - \alpha(c)]$$

$$\le \sum_{\substack{k=1 \\ k \ne i}}^{n} M_k \Delta\alpha_k + M_i[\alpha(x_i) - \alpha(x_{i-1})]$$

$$= U(P, f, \alpha)$$

Similarly, we can prove that

$$L(P^*, f, \alpha) \ge L(P, f, \alpha).$$

Hence the theorem.

Theorem 6.2.6

$$\underline{\int_a^b} f d\alpha \le \overline{\int_a^b} f d\alpha .$$

Proof

Let P_1 and P_2 be any partitions of $[a, b]$.

Let $P^* = P_1 \cup P_2$.

Then, P^* is the common refinement of P_1 and P_2.

Therefore, by theorem 6.2.5,

$$U(P^*, f, \alpha) \le U(P_1, f, \alpha) \qquad \text{............ (6.2.6.1)}$$

and $L(P^*, f, \alpha) \ge L(P_2, f, \alpha)$. .... (6.2.6.2)

We also know that

$$L(P^*, f, \alpha) \le U(P^*, f, \alpha) \qquad \text{............ (6.2.6.3)}$$

From (6.2.6.1), (6.2.6.2) and (6.2.6.3), we get

$$L(P_2, f, \alpha) \le L(P^*, f, \alpha) \le U(P^*, f, \alpha) \le U(P_1, f, \alpha).$$

Therefore for any two partitions P_1 and P_2 of $[a, b]$, we have

$$L(P_2, f, \alpha) \le U(P_1, f, \alpha).$$

Keeping P_2 fixed and varying P_1 over all partitions of $[a, b]$,

$$L(P_2, f, \alpha) \le \inf U(P, f, \alpha).$$

Now, this is true for all partitions P_2 of $[a, b]$.

Therefore,

$$\text{Sup}L(P, f, \alpha) \le \inf U(P, f, \alpha).$$

Therefore,

$$\underline{\int_a^b} f d\alpha \le \overline{\int_a^b} f d\alpha.$$

Hence the theorem.

Theorem 6.2.7

$f \in R(\alpha)$ on $[a, b]$ if and only if there exists a partition P of $[a, \text{b}]$ such that

$$U(P, f, \alpha) - L(P, f, \alpha) < \varepsilon.$$

Proof

Let $f \in R(\alpha)$ on $[a, b]$.

Then, $\overline{\int_a^b} f d\alpha = \underline{\int_a^b} f d\alpha$, ... (6.2.7.1)

where $\overline{\int_a^b} f d\alpha = \inf U(P, f, \alpha)$

and $\underline{\int_a^b} f d\alpha = \sup L(P, f, \alpha)$

Therefore, by the definition of infinimum and supremum,

For given $\varepsilon > 0$, there exists a partition P_1 of $[a, b]$ such that

$$U(P_1, f, \alpha) < \overline{\int_a^b} f d\alpha + \varepsilon/2 \qquad \text{............ (6.2.7.2)}$$

and a partition P_2 of [a, b] such that

$$L(P_2, f, \alpha) > \underline{\int_a^b} f d\alpha - \varepsilon/2. \qquad \text{............ (6.2.7.3)}$$

Let $P = P_1 \cup P_2$.

Then by theorem 6.2.5,

$$U(P,f,\alpha) \le U(P_1,f,\alpha) \qquad \text{............ (6.2.7.4)}$$

and $L(P,f,\alpha) \ge L(P_2,f,\alpha)$ .... (6.2.7.5)

Therefore, from (6.2.7.1), (6.2.7.2), (6.2.7.3), (6.2.7.4) and (6.2.7.5), we get

$$\begin{aligned} U(P,f,\alpha) &\le U(P_1,f,\alpha) \\ &< \overline{\int_a^b} f d\alpha + \varepsilon/2 \\ &< \underline{\int_a^b} f d\alpha \\ &< L(P_2,f,\alpha) + \varepsilon/2 + \varepsilon/2 \\ &< L(P,f,\alpha) + \varepsilon \end{aligned}$$

Therefore, there exists a partition P of $[a, b]$ such that

$$U(P,f,\alpha) - L(P,f,\alpha) < \varepsilon.$$

Conversely,

Assume that there exists a partition P of $[a, b]$ such that

$$U(P,f,\alpha) - L(P,f,\alpha) < \varepsilon. \qquad \text{............ (6.2.7.6)}$$

For every partition P of $[a, b]$, we have

$$L(P,f,\alpha) \le \underline{\int_a^b} f d\alpha \le \overline{\int_a^b} f d\alpha \le U(P,f,\alpha) \qquad \text{............ (6.2.7.7)}$$

From (6.2.7.6) and (6.2.7.7), we get that

$$0 \le \overline{\int_a^b} f d\alpha - \underline{\int_a^b} f d\alpha \le U(P,f,\alpha) - L(P,f,\alpha) < \varepsilon.$$

This is true for every $\varepsilon > 0$.

Hence, $\overline{\int_a^b} f d\alpha - \underline{\int_a^b} f d\alpha = 0.$

That implies, $\overline{\int_a^b} f d\alpha = \underline{\int_a^b} f d\alpha.$

Therefore, $f \in R(\alpha)$ on $[a, b]$.

Hence the theorem.

Theorem 6.2.8

If f is continuous on $[a, b]$, then $f \in R(\alpha)$ on $[a, b]$.

Proof

Let $\varepsilon > 0$.

Choose $\eta > 0$ such that $[\alpha(b) - \alpha(a)]\eta < \varepsilon$.

Since f is continuous on $[a, b]$ and $[a, b]$ is compact, f is uniformly continuous on $[a, b]$.

Therefore for this $\eta > 0$, there exists a $\delta > 0$ such that

$$|f(x) - f(t)| < \eta \text{ whenever } x, t \in [a, b] \text{ with } |x - t| < \delta. \quad \text{............ (6.2.8.1)}$$

If P is any partition of $[a, b]$ such that $\Delta x_i < \delta$, then

$$M_i - m_i = Sup.\{|f(x) - f(t)| / x, t \in [x_{i-1}, x_i]\} \qquad i = 1, 2, \ldots\ldots, n$$
$$\leq \eta$$

Therefore, $U(P, f, \alpha) - L(P, f, \alpha) = \sum_{i=1}^{n} M_i \Delta\alpha_i - \sum_{i=1}^{n} m_i \Delta\alpha_i$.

$$= \sum_{i=1}^{n} (M_i - m_i)\Delta\alpha_i$$
$$\leq \eta \sum_{i=1}^{n} \Delta\alpha_i$$
$$= \eta[\alpha(b) - \alpha(a)]$$
$$< \varepsilon$$

Therefore, $f \in R(\alpha)$ on $[a, b]$.

Hence the theorem.

Theorem 6.2.9

If f is monotonic on $[a, b]$ and if α is continuous on $[a, b]$, then $f \in R(\alpha)$ on $[a, b]$.

Proof

Let α be increasing on $[a, b]$.

Let $\varepsilon > 0$.

Choose n large enough such that

$$[\alpha(b) - \alpha(a)]/n < \varepsilon / [f(b) - f(a)]$$

Choose a partition P such that $\Delta\alpha_i = [\alpha(b) - \alpha(a)]/n.$

Let f be increasing on $[a, b]$.

Hence $f(x_{i-1}) \le f(x) \le f(x_i)$ whenever $x_{i-1} \le x \le x_i$.

Therefore, $M_i = f(x_i)$ and $m_i = f(x_{i-1}),\quad i = 1,2,3,\ldots\ldots,n.$

$$\begin{aligned}
\text{Therefore, } U(P,f,\alpha) - L(P,f,\alpha) &= \sum_{i=1}^{n} M_i \Delta\alpha_i - \sum_{i=1}^{n} m_i \Delta\alpha_i \\
&= \sum_{i=1}^{n} (M_i - m_i)\Delta\alpha_i \\
&= \sum_{i=1}^{n} (f(x_i) - f(x_{i-1}))[(\alpha(b) - \alpha(a))/n] \\
&= [(\alpha(b) - \alpha(a))/n][f(b) - f(a)] \\
&< \varepsilon
\end{aligned}$$

Therefore, $f \in R(\alpha)$ on $[a, b]$.

Hence the theorem.

6.3 Properties of Riemann – Steiltjes integral

Theorem 6.3.1

If $f_1 \in R(\alpha)$ and $f_2 \in R(\alpha)$ on $[a, b]$, then

$$f_1 + f_2 \in R(\alpha) \text{ and } \int_a^b (f_1 + f_2) d\alpha = \int_a^b f_1 d\alpha + \int_a^b f_2 d\alpha.$$

Proof

Let $f = f_1 + f_2$ and P be any partition of $[a, b]$.

Consider

$$U(P,f,\alpha) = \sum_{k=1}^{n} M_k \Delta\alpha_k$$

$$\begin{aligned}
\text{where } M_k &= Sup\{f(x) / x \in [x_{k-1}, x_k]\} \\
&= Sup\{f_1(x) + f_2(x) / x \in [x_{k-1}, x_k]\} \\
&= Sup\{f_1(x) / x \in [x_{k-1}, x_k]\} + Sup\{f_2(x) / x \in [x_{k-1}, x_k]\} \\
&\le M'_k + M''_k
\end{aligned}$$

Therefore, $\sum_{k=1}^{n} M_k \Delta\alpha_k \le \sum_{k=1}^{n} M'_k \Delta\alpha_k + \sum_{k=1}^{n} M''_k \Delta\alpha_k$

Hence, $U(P, f, \alpha) \le U(P, f_1, \alpha) + U(P, f_2, \alpha)$ .... (6.3.1.1)

Similarly, $L(P, f, \alpha) \ge L(P, f_1, \alpha) + L(P, f_2, \alpha)$ .... (6.3.1.2)

Since $f_1 \in R(\alpha)$ on [a, b] and $f_2 \in R(\alpha)$ on [a, b],

for given $\varepsilon > 0$, there exists partitions P_1 and P_2 such that

$$U(P_1, f_1, \alpha) - L(P_1, f_1, \alpha) < \varepsilon/2 \qquad \text{............ (6.3.1.3)}$$

and $U(P_2, f_2, \alpha) - L(P_2, f_2, \alpha) < \varepsilon/2$ .... (6.3.1.4)

Let $P = P_1 \cup P_2$

Then we know that

$$U(P, f_1, \alpha) \le U(P_1, f_1, \alpha)$$

$$U(P, f_2, \alpha) \le U(P_2, f_2, \alpha)$$

$$L(P, f_1, \alpha) \ge L(P_1, f_1, \alpha)$$

and $L(P, f_2, \alpha) \ge L(P_2, f_2, \alpha)$.

Therefore,

$$U(P, f_1, \alpha) - L(P, f_1, \alpha) \le U(P_1, f_1, \alpha) - L(P_1, f_1, \alpha)$$
$$< \varepsilon/2 \qquad \text{............ (6.3.1.5)}$$

$$U(P, f_2, \alpha) - L(P, f_2, \alpha) \le U(P_2, f_2, \alpha) - L(P_2, f_2, \alpha)$$
$$< \varepsilon/2 \qquad \text{............ (6.3.1.6)}$$

From (6.3.1.1), (6.3.1.2), (6.3.1.5) and (6.3.1.6), we get

$$\begin{aligned} U(P, f, \alpha) - L(P, f, \alpha) &\le [U(P, f_1, \alpha) + U(P, f_2, \alpha)] - [L(P, f_1, \alpha) + L(P, f_2, \alpha)] \\ &= [U(P, f_1, \alpha) - L(P, f_1, \alpha)] + [U(P, f_2, \alpha) - L(P, f_2, \alpha)] \\ &< \varepsilon/2 + \varepsilon/2 \\ &< \varepsilon \end{aligned}$$

Hence, $f = f_1 + f_2 \in R(\alpha)$ on [a, b].

For the same partition P of [a, b],

$$
\begin{aligned}
U(P, f_1, \alpha) &< L(P, f_1, \alpha) + \varepsilon / 2 \\
&< Sup\{L(P, f, \alpha)\} + \varepsilon / 2 \\
&< \underline{\int_a^b} f_1 d\alpha + \varepsilon / 2 \\
&= \int_a^b f_1 d\alpha + \varepsilon / 2
\end{aligned}
$$

$$\left\{ \text{Since } f_1 \in R(\alpha) \text{ on } [a, b], \int_a^b f_1 d\alpha = \overline{\int_a^b} f_1 d\alpha = \underline{\int_a^b} f_1 d\alpha \right\}$$

Similarly, $U(P, f_2, \alpha) < \int_a^b f_2 d\alpha + \varepsilon / 2$

Therefore,

$$
\begin{aligned}
\int_a^b f d\alpha &= \text{Inf}\{U(P, f, \alpha)\} \\
&< U(P, f, \alpha) \\
&< U(P, f_1, \alpha) + U(P, f_2, \alpha) \\
&< \int_a^b f_1 d\alpha + \varepsilon / 2 + \int_a^b f_2 d\alpha + \varepsilon / 2 \\
&= \int_a^b f_1 d\alpha + \int_a^b f_2 d\alpha + \varepsilon
\end{aligned}
$$

Hence, $\int_a^b f d\alpha \leq \int_a^b f_1 d\alpha + \int_a^b f_2 d\alpha$ .... (6.3.1.7)

Replacing f_1 by $-f_1$ and f_2 by $-f_2$, we get

$$-\int_a^b f d\alpha \leq -\int_a^b f_1 d\alpha - \int_a^b f_2 d\alpha$$

Multiplying both sides by (-1), we get

$$\int_a^b f d\alpha \geq \int_a^b f_1 d\alpha + \int_a^b f_2 d\alpha \qquad \text{............ (6.3.1.8)}$$

From (6.3.1.7) and (6.3.1.8), we get that

$$\int_a^b f d\alpha = \int_a^b f_1 d\alpha + \int_a^b f_2 d\alpha$$

Hence the theorem.

Theorem 6.3.2

If $f \in R(\alpha)$ on $[a, b]$, then $cf \in R(\alpha)$ on $[a, b]$, for any constant c and

$$\int_a^b cf d\alpha = c\int_a^b f d\alpha.$$

Proof

If $c = 0$, then the result is true.

Assume that $c > 0$.

Since $f \in R(\alpha)$ on $[a, b]$,

for given $\varepsilon > 0$, there exists a partition P of $[a, b]$ such that $U(P, f, \alpha) - L(P, f, \alpha) < \varepsilon / c$.

Consider,

$$U(P, cf, \alpha) - L(P, cf, \alpha) = \sum_{k=1}^{n} M_k' \Delta\alpha_k + \sum_{k=1}^{n} M_k'' \Delta\alpha_k$$

where $M_k' = Sup\{(cf)(x) / x \in [x_{k-1}, x_k]\}$

$$= cSup\{f(x) / x \in [x_{k-1}, x_k]\}$$

$$= cM_k$$

Similarly, $m_k' = cm_k$

Therefore,

$$U(P, cf, \alpha) = c\sum_{k=1}^{n} M_k \Delta\alpha_k = cU(P, f, \alpha)$$

and $L(P, cf, \alpha) = c\sum_{k=1}^{n} m_k \Delta\alpha_k = cL(P, f, \alpha)$

Therefore, $U(P, cf, \alpha) - L(P, cf, \alpha) = c[U(P, f, \alpha) - L(P, f, \alpha)] < \varepsilon.$

Hence, $cf \in R(\alpha)$ on [a, b].

Therefore for the same P,

$$U(P, cf, \alpha) < L(P, cf, \alpha) + \varepsilon$$

$$\leq \text{Sup}L(P, cf, \alpha) + \varepsilon$$

$$\leq c\text{Sup}L(P, f, \alpha) + \varepsilon$$

$$\leq c\int_a^b f d\alpha + \varepsilon$$

Therefore,

$$\text{Inf } U(P, cf, \alpha) \leq U(P, cf, \alpha)$$

$$\leq c\int_a^b f d\alpha + \varepsilon$$

Therefore, $\int_a^b cf d\alpha \leq c\int_a^b f d\alpha$ .... (6.3.2.1)

Replacing f by $-f$, we get

$$-\int_a^b cf d\alpha \leq c\left[-\int_a^b f d\alpha\right]$$

Multiplying both sides by (-1), we get

$$\int_a^b cf d\alpha \geq c\int_a^b f d\alpha \qquad \text{............ (6.3.2.2)}$$

From (6.3.2.1) and (6.3.2.2),

$$\int_a^b cf d\alpha = c\int_a^b f d\alpha.$$

Hence the theorem.

Theorem 6.3.3

If $f_1 \in R(\alpha)$ on $[a, b]$, $f_2 \in R(\alpha)$ on $[a, b]$ and $f_1(x) \leq f_2(x)$ on $[a, b]$, then

$$\int_a^b f_1 d\alpha \leq \int_a^b f_2 d\alpha.$$

Proof

Let P be a partition of $[a, b]$.

Since $f_1(x) \leq f_2(x)$,

$$\text{Sup}\{f_1(x) / x \in [x_{k-1}, x_k]\} \leq \text{Sup}\{f_2(x) / x \in [x_{k-1}, x_k]\}$$

Therefore, $U(P, f_1, \alpha) \le U(P, f_2, \alpha)$

$$\Rightarrow \text{Inf}\, U(P, f_1, \alpha) \le \text{Inf}\, U(P, f_2, \alpha)$$

$$\Rightarrow \overline{\int_a^b} f_1 d\alpha \le \overline{\int_a^b} f_2 d\alpha$$

$$\Rightarrow \int_a^b f_1 d\alpha \le \int_a^b f_2 d\alpha$$

$$\left\{ \text{Since } f_1 \in R(\alpha) \text{ on } [a,b], f_2 \in R(\alpha) \text{ on } [a,b], \right.$$

$$\left. \int_a^b f_1 d\alpha = \overline{\int_a^b} f_1 d\alpha \text{ and } \int_a^b f_2 d\alpha = \overline{\int_a^b} f_2 d\alpha \right\}$$

Hence the theorem.

Theorem 6.3.4

If $f \in R(\alpha)$ on [a, b] and if $a < c < b$, then $f \in R(\alpha)$ on [a, c], $f \in R(\alpha)$ on [c, b] and

$$\int_a^c f d\alpha + \int_c^b f d\alpha = \int_a^b f d\alpha.$$

Proof

Since $f \in R(\alpha)$ on [a, b], for given $\varepsilon > 0$, there exists a partition P of [a, b] such that

$$U(P, f, \alpha) - L(P, f, \alpha) < \varepsilon.$$

Let $P_1 = P \cap [a,c]$ and $P_2 = P \cap [c,b]$.

Then P_1 is a partition of [a, c] and P_2 is a partition of [c, b].

Also on [a, c],

$$U(P_1, f, \alpha) - L(P_1, f, \alpha) < U(P, f, \alpha) - L(P, f, \alpha) < \varepsilon.$$

And on [c, b],

$$U(P_2, f, \alpha) - L(P_2, f, \alpha) < U(P, f, \alpha) - L(P, f, \alpha) < \varepsilon.$$

Therefore,

$f \in R(\alpha)$ on [a, c] and [c, b].

For any partition P of [a, b], since $P = P_1 \cup P_2$,

$$\begin{aligned} U(P,f,\alpha) &= U(P_1,f,\alpha)+U(P_2,f,\alpha) \\ &\geq \inf U(P_1,f,\alpha)+\inf U(P_2,f,\alpha) \\ &= \overline{\int_a^c} fd\alpha+\overline{\int_c^b} fd\alpha \end{aligned}$$

Therefore,

$$\inf U(P,f,\alpha) \geq \overline{\int_a^c} fd\alpha+\overline{\int_c^b} fd\alpha$$

$$\Rightarrow \overline{\int_a^b} \mathrm{fd}\alpha \geq \overline{\int_a^c} fd\alpha+\overline{\int_c^b} fd\alpha$$

$$\Rightarrow \int_a^b fd\alpha \geq \int_a^c fd\alpha+\int_c^b fd\alpha \qquad \text{............ (6.3.4.1)}$$

Similarly, using lower sums,

$$\begin{aligned} L(P,f,\alpha) &= L(P_1,f,\alpha)+L(P_2,f,\alpha) \\ &\leq \sup L(P_1,f,\alpha)+\sup L(P_2,f,\alpha) \\ &= \underline{\int_a^c} fd\alpha+\underline{\int_c^b} fd\alpha \end{aligned}$$

Therefore,

$$\sup L(P,f,\alpha) \leq \underline{\int_a^c} fd\alpha+\underline{\int_c^b} fd\alpha$$

$$\Rightarrow \underline{\int_a^b} fd\alpha \leq \underline{\int_a^c} fd\alpha+\underline{\int_c^b} fd\alpha$$

$$\Rightarrow \int_a^b fd\alpha \leq \int_a^c fd\alpha+\int_c^b fd\alpha \qquad \text{............ (6.3.4.2)}$$

From (6.3.4.1) and (6.3.4.2), we get

$$\int_a^b fd\alpha = \int_a^c fd\alpha+\int_c^b fd\alpha.$$

Hence the theorem.

Theorem 6.3.5

If $f \in R(\alpha)$ on $[a, b]$, and if $|f(x)| \le M$ on $[a, b]$, then

$$\left|\int_a^b f d\alpha\right| \le M[\alpha(b) - \alpha(a)].$$

Proof

Let P be any Partition of $[a, b]$.

Since if $|f(x)| \le M$ on $[a, b]$,

$$M_k = \sup\{f(x) / x \in [x_{k-1}, x_k]\}$$
$$\le M \text{, for all } k = 1, 2, \ldots\ldots\ldots, n.$$

Since $f \in R(\alpha)$ on $[a, b]$,

$$\int_a^b f d\alpha = \overline{\int_a^b} f d\alpha = \inf U(P, f, \alpha)$$
$$\le U(P, f, \alpha)$$
$$= \sum_{k=1}^{n} M_k \Delta\alpha_k$$
$$\le M \sum_{k=1}^{n} \Delta\alpha_k$$
$$= M[\alpha(b) - \alpha(a)] \qquad \ldots\ldots\ldots\ldots (6.3.5.1)$$

Replacing f by $-f$, we get

$$-\int_a^b f d\alpha \le M[\alpha(b) - \alpha(a)] \qquad \ldots\ldots\ldots\ldots (6.3.5.2)$$

From (6.3.5.1) and (6.3.5.2), we get

$$\left|\int_a^b f d\alpha\right| \le M[\alpha(b) - \alpha(a)]$$

Hence the theorem.

Theorem 6.3.6

If $f \in R(\alpha_1)$ on $[a, b]$ and $f \in R(\alpha_2)$ on $[a, b]$ then $f \in R(\alpha_1 + \alpha_2)$ on $[a, b]$ and $\int_a^b f d(\alpha_1 + \alpha_2) = \int_a^b f d\alpha_1 + \int_a^b f d\alpha_2$

Proof

Let P be any partition of $[a, b]$.

Let $\alpha = \alpha_1 + \alpha_2$

Then, $\Delta\alpha_k = (\alpha_1 + \alpha_2)(x_k) - (\alpha_1 + \alpha_2)(x_{k-1})$

$$= \alpha_1(x_k) + \alpha_2(x_k) - \left[\alpha_1(x_{k-1}) + \alpha_2(x_{k-1})\right]$$

$$= \alpha_1(x_k) - \alpha_1(x_{k-1}) + \alpha_2(x_k) - \alpha_2(x_{k-1})$$

$$= \Delta(\alpha_1)_k - \Delta(\alpha_2)_k$$

Consider, $U(P, f, \alpha) = \sum_{k=1}^{n} M_k \Delta\alpha_k$

$$= \sum_{k=1}^{n} M_k \left[\Delta(\alpha_1)_k + \Delta(\alpha_2)_k\right]$$

$$= \sum_{k=1}^{n} M_k \Delta(\alpha_1)_k + \sum_{k=1}^{n} M_k \Delta(\alpha_2)_k$$

$$= U(P, f, \alpha_1) + U(P, f, \alpha_2)$$

$$\leq \inf U(P, f, \alpha_1) + \inf U(P, f, \alpha_2)$$

$$= \overline{\int_a^b} f d\alpha_1 + \overline{\int_a^b} f d\alpha_2$$

Therefore,

$$\inf U(P, f, \alpha) \leq \overline{\int_a^b} f d\alpha_1 + \overline{\int_a^b} f d\alpha_2$$

$$\Rightarrow \overline{\int_a^b} f d\alpha \leq \overline{\int_a^b} f d\alpha_1 + \overline{\int_a^b} f d\alpha_2$$

$$\Rightarrow \int_a^b f d\alpha \leq \int_a^b f d\alpha_1 + \int_a^b f d\alpha_2 \qquad \text{............ (6.3.6.1)}$$

Similarly, considering the lower sums, we can prove that

$$\int_a^b f d\alpha \geq \int_a^b f d\alpha_1 + \int_a^b f d\alpha_2 \qquad \text{............ (6.3.6.2)}$$

From (6.3.6.1) and (6.3.6.2), we get

$$\int_a^b f d\alpha = \int_a^b f d\alpha_1 + \int_a^b f d\alpha_2.$$

Hence the theorem.

Theorem 6.3.7

If $f \in R(\alpha)$ on $[a, b]$ and c is a positive constant, then $f \in R(c\alpha)$ on $[a, b]$ and $\int_a^b f d(c\alpha) = c\int_a^b f d\alpha$

Proof

The proof follows from theorem 6.3.6.

Theorem 6.3.8

If $f \in R(\alpha)$ on $[a, b]$ and $g \in R(\alpha)$ on $[a, b]$, then

(i) $f^2 \in R(\alpha)$ on $[a, b]$.

(ii) $fg \in R(\alpha)$ on $[a, b]$.

(iii) $|f| \in R(\alpha)$ on $[a\ b]$ and $\left|\int_a^b f d\alpha\right| \leq \int_a^b |f| d\alpha$

Proof

(i) Let P be a partition of $[a\ b]$.

Let $M_k(f)$ denotes $\sup\{f(x) / x \in [x_{k-1}, x_k]\}$ and

$m_k(f)$ denotes $\inf\{f(x) / x \in [x_{k-1}, x_k]\}$.

Then, $M_k(f^2) = \sup\{f^2(x) / x \in [x_{k-1}, x_k]\}$

$$= [M_k(|f|)]^2$$

Similarly, $m_k(f^2) = [m_k(|f|)]^2$

Since $f \in R(\alpha)$ on $[a, b]$, f is bounded on $[a, b]$.

Therefore, there exists a $M > 0$ such that $|f(x)| \leq M$.

Then, $M_k(f) \leq M$ and

$m_k(f) \leq M$, for all $k = 1, 2, 3, \ldots.., n$.

Also for given $\varepsilon > 0$, there exists a partition P of $[a, b]$ such that

$$U(P, f, \alpha) - L(P, f, \alpha) < \varepsilon.$$

Therefore,

$$\begin{aligned} M_k(f^2) - m_k(f^2) &= [M_k(|f|)]^2 - [m_k(|f|)]^2 \\ &= [M_k(|f|) - m_k(|f|)][M_k(|f|) + m_k(|f|)] \\ &\leq 2M[M_k(|f|) - m_k(|f|)] \end{aligned}$$

Therefore,

$$\begin{aligned} U(P, f^2, \alpha) - L(P, f^2, \alpha) &= \sum_{k=1}^{n} [M_k(f^2) - m_k(f^2)]\Delta\alpha_k \\ &\leq 2M \sum_{k=1}^{n} [M_k(f) - m_k(f)]\Delta\alpha_k \\ &\leq 2M[U(P, f, \alpha) - L(P, f, \alpha)] \\ &< 2M(\varepsilon / 2M) \\ &< \varepsilon. \end{aligned}$$

Therefore, $f^2 \in R(\alpha)$ on $[a, b]$.

Hence part (i) is proved.

(ii) Since $f \in R(\alpha)$ on $[a, b]$ and $g \in R(\alpha)$ on $[a, b]$, by theorems 6.3.1 and 6.3.2,

$$f + g \in R(\alpha) \text{ on } [a, b] \text{ and } f - g \in R(\alpha) \text{ on } [a, b].$$

Therefore, by part (i) proved above,

$$(f + g)^2 \in R(\alpha) \text{ on } [a, b] \text{ and}$$

$$(f - g)^2 \in R(\alpha) \text{ on } [a, b].$$

Therefore, again by theorems 6.3.1 and 6.3.2,

$$(1/4)\left[(f+g)^2 - (f-g)^2\right] \in R(\alpha) \text{ on } [a, b].$$

$$\Rightarrow fg \in R(\alpha) \text{ on } [a, b].$$

Hence part (ii) is proved.

(iii) Since $f \in R(\alpha)$ on $[a, b]$, for given $\varepsilon > 0$, there exists a partition P of $[a, b]$ such that

$$U(P, f, \alpha) - L(P, f, \alpha) < \varepsilon.$$

Let P be a partition of $[a\ b]$.

Since $\big||f(x)| - |f(y)|\big| \le |f(x) - f(y)|$, we have

$$M_k(|f|) - m_k(|f|) = \sup\left\{\big||f(x)| - |f(y)|\big| / x, y \in [x_{k-1}, x_k]\right\}$$
$$\le \sup\left\{|f(x) - f(y)| / x, y \in [x_{k-1}, x_k]\right\}$$
$$\le M_k(f) - m_k(f)$$

Therefore,

$$U(p, |f|, \alpha) - L(P, |f|, \alpha) \le U(P, f, \alpha) - L(P, f, \alpha)$$
$$< \varepsilon.$$

Hence, $|f| \in R(\alpha)$ on $[a, b]$.

Therefore, by applying theorem 6.3.3, we get

$$\int_a^b f d\alpha \le \int_a^b |f| d\alpha \text{ and } -\int_a^b f d\alpha \le \int_a^b |f| d\alpha$$

Therefore, $\left|\int_a^b f d\alpha\right| \le \int_a^b |f| d\alpha.$

Hence part (iii) is proved.

Hence the theorem.

Theorem 6.3.9

Suppose φ is a strictly increasing continuous function that maps an interval $[A, B]$ onto $[a, b]$. Suppose α is monotonically increasing on $[a, b]$ and $f \in R(\alpha)$ on $[a, b]$. Define β and g on $[A, B]$ by

$$\beta(y) = \alpha(\varphi(y)), \quad \text{and } g(y) = f(\varphi(y)).$$

Then, $g \in R(\beta)$ and $\int_A^B g d\beta = \int_a^b f d\alpha.$

Proof

To each partition $P = \{x_0, x_1, \ldots\ldots\ldots, x_n\}$ of $[a, b]$, there exists a partition $Q = \{y_0, y_1, \ldots\ldots\ldots, y_n\}$ of $[A, B]$ such that $x_i = \varphi(y_i)$.

All partitions of $[A, B]$ can be obtained in this way.

Since $g(y) = f(\varphi(y))$ on $[A, B]$, the values taken by g on $[y_{i-1}, y_i]$ are the same as those taken by f on $[x_{i-1}, x_i]$.

Therefore,

$$U(Q, g, \beta) = U(P, f, \alpha) \text{ and}$$

$$L(Q, g, \beta) = L(P, f, \alpha). \qquad \text{............ (6.3.9.1)}$$

Since $f \in R(\alpha)$ on $[a, b]$, for given $\varepsilon > 0$, there exists a partition P of $[a, b]$ such that

$$U(P, f, \alpha) - L(P, f, \alpha) < \varepsilon.$$

Therefore,

$$U(Q, g, \beta) - L(Q, g, \beta) = U(P, f, \alpha) - L(P, f, \alpha) < \varepsilon.$$

Therefore, $g \in R(\beta)$ on $[A, B]$.

Moreover from (6.3.9.1), $\inf U(Q, g, \beta) = \inf U(P, f, \alpha)$.

Therefore,

$$\overline{\int_A^B} g d\beta = \overline{\int_a^b} f d\alpha.$$

Since $f \in R(\alpha)$ on $[a, b]$ and $g \in R(\beta)$ on $[A, B]$,

$$\int_A^B g d\beta = \overline{\int_A^B} g d\beta \text{ and}$$

$$\int_a^b f d\alpha = \overline{\int_a^b} f d\alpha.$$

Therefore,

$$\int_A^B g d\beta = \int_a^b f d\alpha.$$

Hence the theorem.

6.4 Integration and Differentiation

Theorem 6.4.1

Let $f \in R$ on $[a, b]$.(That is, f is Riemann integrable on $[a, b]$). For $a \le x \le b$, define $F(x) = \int_a^x f(t)dt$. Then, F is continuous on $[a, b]$. Furthermore, if f is continuous at a point x_0 of $[a, b]$, then F is differentiable at x_0 and $F'(x_0) = f(x_0)$.

Proof

Since $f \in R$ on $[a, b]$, f is bounded.

Therefore, if $a \leq x \leq y \leq b$, then

$$\begin{aligned}|F(y) - F(x)| &= \left|\int_a^y f(t)dt - \int_a^x f(t)dt\right| \\ &= \left|\int_x^y f(t)dt\right| \\ &\leq \int_x^y |f(t)dt| \\ &\leq M\int_x^y dt\end{aligned}$$

$\Rightarrow |F(x) - F(y)| \leq M(y - x)$.

Therefore, $|F(x) - F(y)| < \varepsilon$, provided that $|y - x| < \varepsilon / M$.

Hence, F is continuous on $[a, b]$.

Suppose, if f is continuous at x_0, then for given $\varepsilon > 0$, there exists a $\delta > 0$ such that $|f(t) - f(x_0)| < \varepsilon$, whenever $|t - x_0| < \delta$.

Hence, if $x_0 - \delta < s \leq x_0 \leq t < x_0 + \delta$ and $a \leq s < t \leq b$,

$$\begin{aligned}\left|\frac{F(t) - F(s)}{t - s} - f(x_0)\right| &= \left|\frac{1}{t - s}\int_s^t f(t)dt - f(x_0)\right| \\ &= \left|\frac{1}{t - s}\int_s^t [f(t) - f(x_0)]dt\right| \\ &\leq \frac{1}{t - s}\int_s^t |f(t) - f(x_0)|dt \\ &< \frac{\varepsilon}{t - s}|t - s| \\ &= \varepsilon\end{aligned}$$

Therefore, $\left|\frac{F(t)-F(s)}{t-s}-f(x_0)\right|<\varepsilon$ whenever $x_0-\delta<s\le x_0\le t<x_0+\delta$.

$$\Rightarrow F'(x_0)=f(x_0).$$

Hence the theorem.

Theorem 6.4.2

This theorem is known as ***Fundamental theorem of calculus.***

If $f\in R$ on $[a, b]$ and if there exists a differentiable function F on $[a, b]$ such that $F'=f$, then $\int_a^b f(x)dx=F(b)-F(a)$.

Proof

Let $\varepsilon>0$ be given.

Since $f\in R$ on $[a, b]$, there exists a partition $P=\{x_0,x_1,\dots\dots,x_n\}$ of $[a, b]$ such that $U(P,f)-L(P,f)<\varepsilon$.

By the Mean Value theorem, there exist points t_i such that

$$F(x_i)-F(x_{i-1})=f(t_i)[x_i-x_{i-1}]$$
$$=f(t_i)\Delta x_i \text{ for } i=1, 2, \dots, n.$$

Therefore, $\sum_{i=1}^{n} f(t_i)\Delta x_i=\sum_{i=1}^{n}[F(x_i)-F(x_{i-1})]$

$$=F(x_n)-F(x_0)$$
$$=F(b)-F(a).$$

Also, $L(P,f)\le\sum_{i=1}^{n} f(t_i)\Delta x_i\le U(P,f)$

And $L(P,f)\le\int_a^b f(x)dx\le U(P,f)$

Therefore, $\left|\int_a^b f(x)dx-[F(b)-F(a)]\right|=\left|\int_a^b f(x)dx-\sum_{i=1}^{n} f(t_i)\Delta x_i\right|$

$$\le U(P,f)-L(P,f)$$
$$<\varepsilon.$$

Since ε is arbitrary, $\int_a^b f(x)dx = F(b) - F(a)$.

Hence the theorem.

Theorem 6.4.3

This theorem is known as ***Integration by parts***.

Suppose F and G are differentiable functions on $[a, b]$, $F' = f \in R$ and $G' = g \in R$. Then, $\int_a^b F(x)g(x)dx = F(b)G(b) - F(a)G(a) - \int_a^b f(x)G(x)dx$.

Proof

Let $\quad H(x) = F(x)G(x)$.

Then

$$\begin{aligned} H' &= (FG)' \\ &= FG' + F'G \\ &= Fg + fG \in R. \end{aligned}$$

Applying theorem 6.4.2 to the differentiable function H,

$$\int_a^b [F(x)g(x) + f(x)G(x)]dx = H(b) - H(a).$$

$$\Rightarrow \int_a^b F(x)g(x)dx + \int_a^b f(x)G(x)dx = F(b)G(b) - F(a)G(a).$$

$$\Rightarrow \int_a^b F(x)g(x)dx = F(b)G(b) - F(a)G(a) - \int_a^b f(x)G(x)dx.$$

Hence the theorem.

6.5 Integration of vector valued functions

Definition 6.5.1

Let $f_1, f_2, \ldots\ldots, f_k$ be the real valued functions on $[a, b]$. Let $f = (f_1, f_2, \ldots\ldots, f_k)$ be the corresponding mapping of $[a, b]$ onto R^k. If α increases monotonically on $[a, b]$ and if $f_j \in R(\alpha)$ for $j = 1, 2, \ldots\ldots, k$, then we say that $f \in R(\alpha)$ on $[a, b]$ and define $\int_a^b f d\alpha$ as

$$\int_a^b f d\alpha = \left(\int_a^b f_1 d\alpha, \int_a^b f_2 d\alpha, \dots\dots\dots, \int_a^b f_k d\alpha \right)$$

That is, $\int_a^b f d\alpha$ is the point in R^k whose j^{th} coordinate is $\int_a^b f_j d\alpha$.

Let $f = (f_1, f_2, \dots\dots\dots, f_k)$ and $g = (g_1, g_2, \dots\dots\dots, g_k)$ be vector valued functions on $[a, b]$. Then by the way we have defined $\int_a^b f d\alpha$ and using the results obtained in previous sections, we get the following results.

Theorem 6.5.2

If $f \in R(\alpha)$ and $g \in R(\alpha)$ on $[a, b]$, then $f + g \in R(\alpha)$ and $\int_a^b (f+g) d\alpha = \int_a^b f d\alpha + \int_a^b g d\alpha$.

Proof

Since $f \in R(\alpha)$ and $g \in R(\alpha)$ on $[a, b]$,

$f_j, g_j \in R(\alpha)$ on $[a, b]$, for $j = 1, 2, \dots\dots, k$.

Hence $f_j + g_j \in R(\alpha)$ on $[a, b]$, for $j = 1, 2, \dots\dots, k$.

Therefore, $f + g \in R(\alpha)$ on $[a, b]$

and $\int_a^b (f+g) d\alpha = \left(\int_a^b (f_1 + g_1) d\alpha, \int_a^b (f_2 + g_2) d\alpha, \dots\dots\dots\dots, \int_a^b (f_k + g_k) d\alpha \right)$

$$= \left(\int_a^b f_1 d\alpha + \int_a^b g_1 d\alpha, \int_a^b f_2 d\alpha + \int_a^b g_2 d\alpha, \dots\dots\dots\dots, \int_a^b f_k d\alpha + \int_a^b g_k d\alpha \right)$$

$$= \left(\int_a^b f_1 d\alpha, \int_a^b f_2 d\alpha, \dots\dots\dots\dots, \int_a^b f_k d\alpha \right) + \left(\int_a^b g_1 d\alpha, \int_a^b g_2 d\alpha, \dots\dots\dots\dots, \int_a^b g_k d\alpha \right)$$

$$= \int_a^b f d\alpha + \int_a^b g d\alpha.$$

Hence the theorem.

Similarly, we can prove the following results.

Theorem 6.5.3

If $f \in R(\alpha)$ on $[a, b]$, then $cf \in R(\alpha)$ on $[a, b]$, for any constant c and

$$\int_a^b cf d\alpha = c\int_a^b f d\alpha.$$

Theorem 6.5.4

If $f \in R(\alpha)$ on $[a, b]$ and if $a < c < b$, then $f \in R(\alpha)$ on $[a, c]$, $f \in R(\alpha)$ on $[c, b]$ and

$$\int_a^c f d\alpha + \int_c^b f d\alpha = \int_a^b f d\alpha.$$

Theorem 6.5.5

If $f \in R(\alpha_1)$ on $[a, b]$ and $f \in R(\alpha_2)$ on $[a, b]$ then $f \in R(\alpha_1 + \alpha_2)$ on $[a, b]$ and $\int_a^b f d(\alpha_1 + \alpha_2) = \int_a^b f d\alpha_1 + \int_a^b f d\alpha_2$

Theorem 6.5.6

If $f \in R(\alpha)$ on $[a, b]$ and c is a positive constant, then $f \in R(c\alpha)$ on $[a, b]$ and $\int_a^b f d(c\alpha) = c\int_a^b f d\alpha$

Theorem 6.5.7

Let $f \in R$ on $[a, b]$.(That is, f is Riemann integrable on $[a, b]$). For $a \le x \le b$, define $F(x) = \int_a^x f(t)dt$. Then, F is continuous on $[a, b]$. Furthermore, if f is continuous at a point x_0 of $[a, b]$, then F is differentiable at x_0 and $F'(x_0) = f(x_0)$.

Theorem 6.5.8

If f and F maps $[a, b]$ onto R^k, if $f \in R$ on $[a, b]$ and if $F' = f$, then

$$\int_a^b f(x)dx = F(b) - F(a).$$

Theorem 6.5.9

If f maps $[a, b]$ onto R^k, if $f \in R(\alpha)$ for some monotonically increasing function α on $[a, b]$, then $|f| \in R(\alpha)$ on $[a, b]$ and $\left|\int_a^b f d\alpha\right| \le \int_a^b |f| d\alpha$

Proof

If $f_1, f_2, \ldots\ldots, f_k$ are the components of f, then $|f| = \left(f_1^2 + f_2^2 + \ldots\ldots + f_k^2\right)^{1/2}$.

Since $f \in R(\alpha)$ on $[a, b]$, by definition, each $f_j \in R(\alpha)$ for $j = 1, 2, \ldots\ldots, k$.

By theorem 6.3.8,

$$f_j^2 \in R(\alpha) \text{ for } j = 1, 2, \ldots\ldots, k.$$

Hence by theorem 6.3.1,

$$f_1^2 + f_2^2 + \ldots\ldots + f_k^2 \in R(\alpha).$$

Hence, $|f| = \left(f_1^2 + f_2^2 + \ldots\ldots + f_k^2\right)^{1/2} \in R(\alpha)$.

{Since square root of a continuous function is continuous on $[0, M]$, for every real M}.

To prove that: $\left|\int_a^b f d\alpha\right| \le \int_a^b |f| d\alpha$

Let $y = (y_1, y_2, \ldots\ldots, y_k)$, where $y_j = \int_a^b f_j d\alpha$.

Then $y = \int_a^b f d\alpha$.

And $|y|^2 = \sum_{j=1}^{k} y_j^2$

$$= \sum_{j=1}^{k} y_j \int_a^b f_j d\alpha$$

$$= \int_a^b \left(\sum_{j=1}^{k} y_j f_j\right) d\alpha$$

From Schwartz inequality,

$$\sum_{j=1}^{k} y_j f_j \le |y||f(t)|, \qquad a \le t \le b$$

Therefore,

$$|y|^2 \le |y| \int_a^b |f| d\alpha .$$

Therefore, if $y \ne 0$, dividing this inequality by $|y|$, we get

$$|y| \le \int_a^b |f| d\alpha .$$

Hence, $\left| \int_a^b f d\alpha \right| \le \int_a^b |f| d\alpha .$

Hence the theorem.

6.6 Rectifiable curves

Definition 6.6.1

A continuous mapping γ of an interval $[a, b]$ into R^k is called a *curve* in R^k or γ is a curve on $[a, b]$.

If γ is one-to-one, γ is called an *arc*.

If $\gamma(a) = \gamma(b)$, γ is called a *closed curve.*

Definition 6.6.2

To each partition $P = \{x_0, x_1, ..., x_n\}$ of $[a, b]$ and to each curve γ on $[a, b]$, we associate a number $\Lambda(P,\gamma) = \sum_{i=1}^{n} |\gamma(x_i) - \gamma(x_{i-1})|$, where $|\gamma(x_i) - \gamma(x_{i-1})|$ = distance between the points $\gamma(x_{i-1})$ and $\gamma(x_i)$.

$\Lambda(P,\gamma)$ = Length of a polygonal path with vertices at $\gamma(x_0), \gamma(x_1), ..., \gamma(x_n)$.

As the partition P becomes finer and finer, the polygon approaches γ, more and more closely.

The length of Λ is defined as

$$\Lambda(\gamma) = \sup \Lambda(P, \gamma)$$

where the supremum is taken over all partitions of $[a, b]$.
If $\Lambda(\gamma) < \infty$, then γ is rectifiable.

Theorem 6.6.3

If γ' is continuous on $[a, b]$, then γ is rectifiable and $\Lambda(\gamma) = \int_a^b |\gamma'(t)|dt$.

Proof

If $a \le x_{i-1} < x_i \le b$, then

$$|\gamma(x_i) - \gamma(x_{i-1})| = \left|\int_{x_{i-1}}^{x_i} \gamma'(t)dt\right|$$

$$\le \int_{x_{i-1}}^{x_i} |\gamma'(t)|dt$$

Hence, $\Lambda(P, \gamma) = \sum_{i=1}^{n} |\gamma(x_i) - \gamma(x_{i-1})|$

$$\le \sum_{i=1}^{n} \int_{x_{i-1}}^{x_i} |\gamma'(t)|dt$$

$\le \int_a^b |\gamma'(t)|dt$, for every partition P of $[a, b]$.

Therefore, $\Lambda(\gamma) \le \int_a^b |\gamma'(t)|dt$ .... (6.6.3.1)

To prove the opposite inequality,
Let $\varepsilon > 0$ be given.
Since γ' is uniformly continuous on $[a, b]$, there exists a $\delta > 0$ such that $|\gamma'(s) - \gamma'(t)| < \varepsilon$, whenever $|s - t| < \delta$.
Let $P = \{x_0, x_1, \ldots\ldots, x_n\}$ be a partition of $[a, b]$, with $\Delta x_i < \delta$ for all i.
Therefore, if $x_{i-1} \le t \le x_i$, then $|\gamma'(t) - \gamma'(x_i)| < \varepsilon$.
Therefore, $|\gamma'(t)| - |\gamma'| \le |\gamma'(t) - \gamma'(x_i)| < \varepsilon$

Therefore, $|\gamma'(t)| \le |\gamma'(x_i)| + \varepsilon$

Hence,

$$\int_{x_{i-1}}^{x_i} |\gamma'(t)|dt \le \int_{x_{i-1}}^{x_i} \left(|\gamma'(x_i)| + \varepsilon\right)dt$$

$$\le \left|\int_{x_{i-1}}^{x_i} \gamma'(x_i)dt\right| + \varepsilon\Delta x_i$$

$$\le \left|\int_{x_{i-1}}^{x_i} [\gamma'(t) + \gamma'(x_i) - \gamma'(t)]dt\right| + \varepsilon\Delta x_i$$

$$\le \left|\int_{x_{i-1}}^{x_i} \gamma'(t)dt\right| + \left|\int_{x_{i-1}}^{x_i} [\gamma'(x_i) - \gamma'(t)]dt\right| + \varepsilon\Delta x_i$$

$$\le |\gamma(x_i) - \gamma(x_{i-1})| + \varepsilon\Delta x_i + \varepsilon\Delta x_i$$

$$= |\gamma(x_i) - \gamma(x_{i-1})| + 2\varepsilon\Delta x_i$$

Therefore, $\displaystyle\int_a^b |\gamma'(t)|dt = \sum_{i=1}^{n} \int_{x_{i-1}}^{x_i} |\gamma'(t)|dt$

$$\le \sum_{i=1}^{n} |\gamma(x_i) - \gamma(x_{i-1})| + 2\varepsilon \sum_{i=1}^{n} \Delta x_i$$

$$\le \Lambda(P,\gamma) + 2\varepsilon(b-a)$$

$$\le \Lambda(\gamma) + 2\varepsilon(b-a)$$

Since $\varepsilon > 0$ is arbitrary,

$$\int_a^b |\gamma'(t)|\, dt \le \Lambda(\gamma) \qquad \text{............ (6.6.3.2)}$$

From (6.6.3.1) & (6.6.3.2), we get

$$\Lambda(\gamma) = \int_a^b |\gamma'(t)|\, dt$$

Hence the theorem.

6.7 Problems

1. Suppose $P_1 = \{0,1/4,1/2,3/4,1\}$ and $P_2 = \{0,1/3,2/3,1\}$. Is P_1 a refinement of P_2?
2. Can you find a refinement that is finer than both P_1 and P_2?
3. Can you prove that $\int_a^b d\alpha(x) = \alpha(b) - \alpha(a)$, directly from the definition?
4. If $f(x) = 0$ for all irrational x, $f(x) = 1$ for all rational x, is $f \in R$ on $[a, b]$ for any a < b.
5. Suppose $f \geq 0$, f is continuous on $[a, b]$ and $\int_a^b f(x)dx = 0$. Prove that $f(x) = 0 \; \forall x \in [a,b]$.
6. Suppose α is an increasing function on $[a, b]$ and α is continuous at $x_0 \in [a,b]$ and $f:[a,b] \to R$ is defined by $f(x_0) = 1$, $f(x) = 0$ and $f_x = x_0$. Prove that $f \in R(\alpha)$ and $\int f dx = 0$.
7. Give an example of a function in which $\int f dx < \overline{\int f dx}$.
8. If $a < b$, $\int_a^b f d\alpha = ?$
9. Suppose f is a bounded valued function on $[a, b]$ and $f^2 \in R(\alpha)$ on $[a, b]$. Does it follow that and $f \in R(\alpha)$ on $[a, b]$?
10. Show that $\int_0^1 x^2 dx = 3/5$ where $\alpha(x) = x^3$.
11. Show that $\int_0^1 [x]dx = 3/5$ where $\alpha(x) = x^2 = 3$.
12. Derive the identity $\sum_{k=1}^{n} \frac{1}{k} = \log n - \int_1^n \frac{x-[x]}{x^2} dx + 1$.
13. If it is a real, continuously differentiable function on $[a, b]$ with $f(a) = f(b) = 0$ and $\int_a^b f^2(x)dx = 1$, prove that $\int_a^b x f(x) f'(x) dx = -1/2$.

14. If $f:[0,1]\to R^3$ is defined by $f(t)=\left(t,t^2,t^3\right)$ and if $\alpha(t)=t,\ \forall t\in[0,1]$, find $\int_0^1 f d\alpha$.

15. If $f:[0,1]\to R^3$ is defined by $f(t)=\left(t,t^2,t^3\right)$ and if $\alpha(t)=t,\ \forall t\in[0,1]$, find $\int_0^1 f d\alpha$.

16. Let $\gamma_1,\gamma_2,\gamma_3$ be curves in the complex plane, defined on $[0,2\pi]$ by $\gamma_1(t)=e^{it},\gamma_2(t)=e^{2it},\gamma_3(t)=e^{2\pi it\sin(1/t)}$. Show that
   a) These three curves have the same range
   b) γ_1 and γ_2 are rectifiable
   c) Length of γ_1 is 2π
   d) Length of γ_2 is 4π and
   e) γ_3 is not rectifiable.

17. If γ_1 and γ_2 are curves in the complex plane defined on [0, 1] & [0, 2] respectively by $\gamma_1(t)=e^{2\pi it},\gamma_2(t)=e^{2\pi it};t\in[0,1]$, prove that length of γ_2 is twice that of γ_1.

Unit 7	DIFFERENT SPECIAL FUNCTIONS

7.1 Introduction

In this lesson we will study about power series and some properties of functions that are represented by power series.

7.2 Power series

Definition 7.2.1

A series of the form $c_0 + c_1 x + c_2 x^2 + ... = \sum_{n=0}^{\infty} c_n x^n$ is called a power series in *x*. Here, we take *x* as real. A function which can be expressed in the form

$$f(x) = \sum_{n=0}^{\infty} c_n x^n \qquad .... (7.2.1.1)$$

$$\text{or } f(x) = \sum_{n=0}^{\infty} c_n (x-a)^n \qquad .... (7.2.1.2)$$

is called an analytic function.

If (7.2.1.1) converges for all *x* in (-*R*, *R*), for some *R* > 0, we say that *f* is *expanded in a power series about the point x* = 0. Similarly, if (7.2.1.2) converges for $|x-a| < R$, *f* is said to be *expanded in a power series about a point x* = *a*.

Theorem 7.2.2

Suppose the series $\sum_{n=0}^{\infty} c_n x^n$ converges for $|x| < R$, and define $f(x) = \sum_{n=0}^{\infty} c_n x^n$, $|x| < R$. Then, $\sum_{n=0}^{\infty} c_n x^n$ converges uniformly on $[-R+\varepsilon, R-\varepsilon]$, for every $\varepsilon > 0$. Also, *f* is continuous and differentiable in

(-R, R) and $f'(x)=\sum_{n=1}^{\infty} nc_n x^{n-1}, \quad |x|<R$.

Proof

Let $\varepsilon > 0$.

$$|x|<R \quad \Rightarrow |x|<R-\varepsilon \qquad \text{...........} (7.2.2.1)$$

Therefore, we have

$$|c_n x^n| \le |c_n (R-\varepsilon)^n|.$$

Since every power series converges absolutely in the interior of the interval of convergence, $\sum c_n (R-\varepsilon)^n$ converges absolutely in the interior of the interval of convergence.

Therefore,

$$\sum_{n=0}^{\infty} |c_n (R-\varepsilon)^n| \text{ is convergent.}$$

By M-test, $\sum_{n=0}^{\infty} c_n x^n$ is uniformly convergent in $[-R+\varepsilon, R-\varepsilon]$.

Now,

$$\limsup_{n\to\infty} \sqrt[n]{n|c_n|} = \limsup_{n\to\infty}\ n^{1/n} \sqrt[n]{|c_n|}$$

$$= \limsup_{n\to\infty} \sqrt[n]{|c_n|}$$

$$[\text{Since} \quad \sqrt[n]{n} \to 1 \text{ as } n \to \infty].$$

Therefore, the series $\sum_{n=0}^{\infty} c_n x^n$ and $\sum_{n=1}^{\infty} nc_n x^{n-1}$ have the same interval of convergence.

Therefore, $\sum_{n=0}^{\infty} c_n x^n$ is convergent for $|x|<R$, implies

$$\sum_{n=1}^{\infty} nc_n x^{n-1} \text{ is convergent for } |x|<R.$$

By first part, $\forall \varepsilon > 0$, $\sum_{n=1}^{\infty} nc_n x^{n-1}$ converges uniformly in $[-R+\varepsilon, R-\varepsilon]$.

We know that, "If $\{f_n\}$ is a sequence of differentiable functions on E, and if

$\sum f_n(x_0)$ converges for some point x_0 and if $\sum f_n'$ converges uniformly, then $\sum f_n$ converges uniformly to a function f on E and $f'(x) = \sum f_n'(x)$".

Therefore,

$$f'(x) = \sum_{n=1}^{\infty} nc_n x^{n-1}, \qquad x \in [-R+\varepsilon, R-\varepsilon].$$

$$f'(x) = \sum_{n=1}^{\infty} nc_n x^{n-1}, \qquad |x| < R \text{ [By (7.2.2.1)]}.$$

Therefore, f is differentiable in $(-R, R)$ and hence f is continuous in $(-R, R)$. Hence the theorem.

Note 7.2.3

Suppose $\sum_{n=0}^{\infty} c_n x^n$ converges for $|x| < R$ and define $f(x) = \sum_{n=0}^{\infty} c_n x^n$, $|x| < R$. Then, f has derivatives of all orders in $(-R, R)$ and $f^{(k)}(x) = \sum_{n=k}^{\infty} n(n-1)\text{.......}(n-k+1)c_n x^{n-k}$. In particular, $f^{(k)}(0) = k!c_k$, $(k = 0, 1, 2, \ldots\ldots)$.

Theorem 7.2.4

This theorem is known as ***Abel's theorem***.

Suppose $\sum_{n=0}^{\infty} c_n$ converges and let $f(x) = \sum_{n=0}^{\infty} c_n x^n$, $-1 < x < 1$. Then

$$\lim_{x \to 1} f(x) = \sum_{n=0}^{\infty} c_n .$$

Proof

Let $s_n = c_0 + \text{.......} + c_n$, $s_{-1} = 0$.

Since $\sum c_n$ is convergent, $\{s_n\}$ is convergent.

Let $(s_n) \to s$.

Therefore, $\sum_{n=0}^{\infty} c_n = s$. .... (7.2.4.1)

Now, $f(x)=\sum_{n=0}^{\infty}c_n x^n$

$$
\begin{aligned}
&=\lim_{m\to\infty}\sum_{n=0}^{m}c_n x^n\\
&=\lim_{m\to\infty}\sum_{n=0}^{m}(s_n-s_{n-1})x^n\\
&=\lim_{m\to\infty}\left[\sum_{n=0}^{m}s_n x^n-\sum_{n=1}^{m}s_{n-1}x^n\right]\\
&=\lim_{m\to\infty}\left\{s_m x^m+\sum_{n=0}^{m-1}s_n x^n-\sum_{n=0}^{m-1}s_n x^{n+1}\right\}\\
&=\lim_{m\to\infty}s_m x^m+\lim_{m\to\infty}\left\{\sum_{n=0}^{m-1}s_n x^n-\sum_{n=0}^{m-1}s_n x^{n+1}\right\}\\
&=0+\lim_{m\to\infty}(1-x)\sum_{n=0}^{m-1}s_n x^n, \qquad [\because x^m\to 0]
\end{aligned}
$$

Therefore,

$$f(x)=(1-x)\sum_{n=0}^{\infty}s_n x^n. \qquad \text{............ (7.2.4.2)}$$

But, $f(x)=(1-x)\sum_{n=0}^{\infty}x^n=(1-x)(1+x+x^2+\text{............})$.

$$=(1-x)\frac{1}{(1-x)}=1. \qquad \text{............ (7.2.4.3)}$$

Since $\lim_{n\to\infty}s_n=s$, given $\varepsilon>0$, there exists $N\in J$ such that

$$|s_n-s|<\varepsilon/2,\ \ \forall n>N \qquad \text{............ (7.2.4.4)}$$

$$
\begin{aligned}
|f(x)-s|&=\left|(1-x)\sum_{n=0}^{\infty}s_n x^n-s(1-x)\sum_{n=0}^{\infty}x^n\right| \qquad \text{[since by (7.2.4.3)]}\\
&=\left|(1-x)\sum_{n=0}^{\infty}(s_n-s)x^n\right|\\
&=\left|(1-x)\sum_{n=0}^{N}(s_n-s)x^n+(1-x)\sum_{n=N+1}^{\infty}(s_n-s)x^n\right|
\end{aligned}
$$

$$\leq (1-x)\sum_{n=0}^{N}|s_n - s|x^n + (1-x)\sum_{n=N+1}^{\infty}|s_n - s|x^n .$$

$$< (1-x)\sum_{n=0}^{N}|s_n - s|x^n + (1-x)\sum_{n=N+1}^{\infty}\frac{\varepsilon}{2}x^n , \qquad \text{[by (7.2.4.4)]}$$

$$< (1-x)\sum_{n=0}^{N}|s_n - s|x^n + \frac{\varepsilon}{2} \text{ [by (7.2.4.3)]}$$

Choosing x close enough to 1, we can make $(1-x)\sum_{n=0}^{N}|s_n - s|x^n < \frac{\varepsilon}{2}$, since N is finite.

Therefore, $|f(x) - s| < \varepsilon$, if $x \to 1$.

Therefore, $\lim_{x\to 1} f(x) = s = \sum_{n=0}^{\infty} c_n$, [by (7.2.4.1)].

Hence the theorem.

Theorem 7.2.5

Given a double sequence $\{a_{ij}\}$, $i = 1, 2, 3, \ldots\ldots, j = 1, 2, 3, \ldots\ldots..$ and suppose that $\sum_{j=1}^{\infty}|a_{ij}| = b_i$ and $\sum b_i$ converges, then $\sum_{i=1}^{\infty}\sum_{j=1}^{\infty} a_{ij} = \sum_{j=1}^{\infty}\sum_{i=1}^{\infty} a_{ij}$.

Proof

Let E be a countable set consisting of points $x_0, x_1, \ldots\ldots..$ and let $(x_n) \to x_0$ as $n \to \infty$.

Define

$$f_i(x_0) = \sum_{j=1}^{\infty} a_{ij} , \; i = 1, 2, 3, \ldots\ldots\ldots$$

$$f_i(x_n) = \sum_{j=1}^{n} a_{ij} , \; i = 1, 2, 3, \ldots\ldots\ldots$$

$$g(x) = \sum_{i=1}^{\infty} f_i(x), \qquad x \in E .$$

Let $(x_n) \to x_0$.

Now, $\lim_{n\to\infty} f_i(x_n) = \lim_{n\to\infty}\sum_{j=1}^{n} a_{ij}$.

$$= \sum_{j=1}^{\infty} a_{ij} = f_i(x_0).$$

Therefore, f_i is continuous at x_0.

For $n = 1, 2, 3, \ldots\ldots$

$$\left|f_i(x_n)\right| = \left|\sum_{j=1}^{n} a_{ij}\right|$$

$$\leq \sum_{j=1}^{n} \left|a_{ij}\right|$$

$$< \sum_{j=1}^{\infty} \left|a_{ij}\right| = b_i \text{ [given]}$$

Also, $\left|f_i(x_0)\right| = \left|\sum_{j=1}^{\infty} a_{ij}\right|$

$$\leq \sum_{j=1}^{\infty} \left|a_{ij}\right|$$

$$\leq b_i$$

Therefore,

$$\left|f_i(x)\right| \leq b_i, \quad \text{for all } x \in E.$$

Given that $\sum b_i$ is convergent.

Therefore, by Weierstrass M-test, $\sum f_i$ is uniformly convergent.

Also, each f_i is continuous at x_0.

Therefore, g is continuous at x_0.

Hence, $\lim_{n\to\infty} g(x_n) = g(x_0)$.

Therefore,

$$\lim_{n\to\infty} \sum_{i=1}^{n} f_i(x_n) = \sum_{i=1}^{\infty} f_i(x_0)$$

$$\Rightarrow \lim_{n\to\infty} \sum_{i=1}^{\infty} \sum_{j=1}^{n} a_{ij} = \sum_{i=1}^{\infty} \sum_{j=1}^{\infty} a_{ij}$$

$$\Rightarrow \lim_{n\to\infty} \sum_{j=1}^{n} \sum_{i=1}^{\infty} a_{ij} = \sum_{i=1}^{\infty} \sum_{j=1}^{\infty} a_{ij}$$

$$\Rightarrow \sum_{j=1}^{\infty}\sum_{i=1}^{\infty} a_{ij} = \sum_{i=1}^{\infty}\sum_{j=1}^{\infty} a_{ij}.$$

Hence the theorem.

Theorem 7.2.6

This theorem is known as ***Taylor's theorem***.

Suppose $f(x) = \sum_{n=0}^{\infty} c_n x_n$, the series converging in $|x| < R$. If $-R < a < R$, then f can be expanded in a power series about the point $x = a$, which is convergent in $|x-a| < R-|a|$. Also,

$$f(x) = \sum_{n=0}^{\infty} \frac{f^{(n)}(a)}{n!}(x-a)^n, \qquad |x-a| < R-|a|.$$

Proof

$$f(x) = \sum_{n=0}^{\infty} c_n x^n$$

$$= \sum_{n=0}^{\infty} c_n [(x-a)+a]^n$$

$$= \sum_{n=0}^{\infty} c_n \sum_{m=0}^{n} \binom{n}{m} (x-a)^m a^{n-m}$$

$$\left[\because (x+a)^n = x^n + {}^nc_1\, x^{n-1}a + \ldots\ldots. + {}^nc_n\, a^n = \sum_{m=0}^{n}\binom{n}{m} x^{n-m} a^m = \sum_{m=0}^{n}\binom{n}{m} x^m a^{n-m}\right].$$

$$= \sum_{n=0}^{\infty}\sum_{m=0}^{n} c_n \binom{n}{m}(x-a)^m a^{n-m}. \qquad \ldots\ldots\ldots\ldots (7.2.6.1)$$

Consider,

$$\sum_{n=0}^{\infty}\sum_{m=0}^{n} \left| c_n \binom{n}{m}(x-a)^m a^{n-m}\right| = \sum_{n=0}^{\infty} |c_n| \sum_{m=0}^{n}\binom{n}{m} |(x-a)|^m |a^{n-m}|$$

$$= \sum_{n=0}^{\infty} |c_n| \left(|x-a| + |a|\right)^n.$$

This series is convergent, if $|x-a| + |a| < R$, that is, if $|x-a| < R-|a|$.

Therefore,

$$f(x)=\sum_{m=0}^{\infty}\sum_{n=m}^{\infty}c_n\binom{n}{m}(x-a)^m a^{n-m}.$$

Put $$d_m=\sum_{n=m}^{\infty}c_n\binom{n}{m}a^{n-m}$$

Therefore,

$$f(x)=\sum_{m=0}^{\infty}d_m(x-a)^m$$

$$f(x)=\sum_{n=0}^{\infty}d_n(x-a)^n, \qquad |x-a|<R-|a|.$$

Therefore, f is expanded in a power series about the point $x=a$ and this power series is convergent in $|x-a|<R-|a|$.

To prove: $f(x)=\sum_{n=0}^{\infty}\frac{f^n(a)}{n!}(x-a)^n, \qquad |x-a|<R-|a|.$

We have, $f(x)=\sum_{n=0}^{\infty}d_n(x-a)^n, \qquad |x-a|<R-|a|.$

To find d_n:

$$f^m(x)=\sum_{n=0}^{\infty}d_n n(n-1)(n-2).......(n-m+1)(x-a)^{n-m}$$

$$\Rightarrow f^m(a)=m!d_m.$$

$$\Rightarrow d_n=\frac{f^n(a)}{n!}$$

Therefore, $f(x)=\sum_{n=0}^{\infty}\frac{f^n(a)}{n!}(x-a)^n, \qquad |x-a|<R-|a|.$

Hence the theorem.

Theorem 7.2.7

Suppose the series $\sum a_n x^n$ and $\sum b_n x^n$ converge in the segment S = (-R, R).

Let E be the set of all $x \in S$ at which $\sum_{n=0}^{\infty}a_n x^n=\sum_{n=0}^{\infty}b_n x^n$. If E has the limit point in S, then $a_n = b_n$, for n = 0, 1, 2, ….. and $E = S$.

Proof

Put $c_n = a_n - b_n$ and $f(x) = \sum_{n=0}^{\infty} c_n x^n$,$x \in S$.

Therefore, $f(x) = 0$ in E.
Let A be the set of all limit points of E in S.
By hypothesis, E has a limit point in S.
Therefore, $A \neq \phi$.
Now, $A = E'$ is closed in S.
$B = S - A$ is open in S.
Claim: A is open in S.
Let $x_0 \in A$.

Therefore, $x_0 \in S$.

Let $|x_0| < R$.

By Taylor's theorem,

$$f(x) = \sum_{n=0}^{\infty} d_n (x - x_0)^n, \quad |x - x_0| < R - |x_0| \qquad \text{............ (7.2.7.1)}$$

Subclaim: $d_n = 0, \quad \forall n$.
Suppose not,
Let k be the smallest positive integer such that $d_k \neq 0$.
Therefore,

$$f(x) = d_k (x - x_0)^k + d_{k+1} (x - x_0)^{k+1} + ... \text{[Since } d_1 = d_2 = ...= d_k = 0\text{]}$$

$$= (x - x_0)^k [d_k + d_{k+1}(x - x_0) + ...]$$

$$= (x - x_0)^k g(x) \qquad \text{............ (7.2.7.2)}$$

where $g(x) = d_k + d_{k+1}(x - x_0) + ... = \sum_{m=0}^{\infty} d_{k+m} (x - x_0)^m$.

Now, g is continuous at x_0, $g(x_0) = d_k \neq 0$ and $\left|\frac{d_k}{2}\right| > 0 \Rightarrow \frac{|d_k|}{2} > 0$.

Therefore, there exists $\delta > 0$, such that $|x - x_0| < \delta \Rightarrow |g(x) - g(x_0)| < \frac{|d_k|}{2}$.

Now, $|g(x_o)| - |g(x)| \leq |g(x) - g(x_0)|$

$$\Rightarrow |d_k| - |g(x)| < \frac{|d_k|}{2}$$

$$\Rightarrow |g(x)| > \frac{|d_k|}{2}$$

By (7.2.7.2), $f(x) \neq 0$ in $0 < |x - x_0| < \delta$.

Therefore, $N_\delta(x_0) \cap E - \{x_0\} = \phi$.

Therefore, x_0 is not a limit point of *E*.

$$\Rightarrow x_0 \notin A.$$

This is a contradiction to $x_0 \in A$.
Hence, $d_n = 0$, for all *n*.
Therefore, our subclaim is proved.
Therefore,

$$f(x) = \sum_{n=0}^{\infty} d_n (x - x_0)^n = 0, \quad |x - x_0| < R - |x_0|.$$

$$\Rightarrow f(x) = 0 \quad \text{in } N_{R-|x_0|}(x_0).$$

Subclaim: $N_{R-|x_0|}(x_0) \subset A$.

Let $y \in N_{R-|x_0|}(x_0) \subset A$.

Since $N_{R-|x_0|}(x_0)$ is open, there exists $\delta_1 > 0$, such that $N_{\delta_1}(y) \subset N_{R-|x_0|}(x_0)$.

Let $N_{\delta_2}(y)$ be any neighbourhood of *y*.

Put $\delta_3 = \min.(\delta_1, \delta_2)$.

$$N_{\delta_3}(y) \subset N_{\delta_1}(y) \subset N_{R-|x_0|}(x_0) \subset E.$$

$$\Rightarrow N_{\delta_3}(y) \cap E \neq \phi.$$

Therefore, *y* is a limit point of *E*.

That implies, $y \in A$.

Therefore, $N_{R-|x_0|}(x_0) \subset A$.

Hence, our subclaim is proved.

Therefore, A is open.

Hence our claim is proved.

Now, $S = A \cup B$ and *S* is connected.

$A \neq 0$ and A and B are open in S.

Therefore, $B = \phi$.

Hence $S = A$.

Claim: $A \subset E$.

Let $y \in A$.

Therefore, y is a limit point of E in S.

Therefore, there is a sequence $\{y_n\}$ in E, such that $y_n \to y$.

Since $y_n \in E$, $f(y_n) = 0$, for all n.

Since f is continuous,

$$y_n \to y \Rightarrow f(y_n) \to f(y).$$

Therefore, $f(y) = 0$.

Hence, $y \in E$.

Therefore, $A \subset E$.

$\Rightarrow S \subset E$. [Since $A = S$]

Therefore, by definition $E \subset S$.

Hence, $S = E$.

Therefore, $\sum a_n x^n = \sum b_n x^n, \quad \forall x \in S$.

$$\Rightarrow a_n = b_n, \quad \forall n = 1,2,\ldots\ldots..$$

Hence the theorem.

7.3 Exponential and Logarithmic Functions

Definition 7.3.1

We define $E(z) = \sum_{n=0}^{\infty} \frac{z^n}{n!}$ for every complex z.

Here, $a_n = \frac{z^n}{n!}$

$$a_{n+1} = \frac{z^{n+1}}{(n+1)!}$$

$$\frac{a_{n+1}}{a_n} = \frac{z^{n+1}}{z^n} \frac{n!}{(n+1)!} = \frac{z}{n+1}.$$

$$\lim_{n\to\infty}\left|\frac{a_{n+1}}{a_n}\right| = 0 < 1.$$

Therefore, by ratio test $\sum a_n$ is convergent.

That is, $\sum_{n=0}^{\infty}\frac{z^n}{n!}$ is convergent.

Results 7.3.2

1. $E(z)E(w) = E(z + w)$

Proof

$$E(z)E(w) = \sum_{n=0}^{\infty}\frac{z^n}{n!}\sum_{m=0}^{\infty}\frac{w^m}{m!}$$

$$= \sum_{n=0}^{\infty}\sum_{k=0}^{n}\frac{z^k}{k!}\frac{w^{n-k}}{(n-k)!} \qquad \text{[By Metern's theorem].}$$

$$= \sum_{n=0}^{\infty}\frac{1}{n!}\sum_{k=0}^{n}\frac{n!}{k!(n-k)!}z^k w^{n-k}$$

$$= \sum_{n=0}^{\infty}\frac{1}{n!}\sum_{k=0}^{n}\binom{n}{k}z^k w^{n-k}$$

$$= \sum_{n=0}^{\infty}\frac{1}{n!}(z+w)^n = E(z+w).$$

Therefore, $E(z)E(w) = E(z + w)$.

2. To find $E(-z)$:

$$E(z)E(-z) = E(z - z)$$

$$= E(0) = 1$$

Therefore, $E(-z) = \frac{1}{E(Z)}$.

3. When x is real, $E(x) > 0$.

If $x > 0$, then $E(x) > 0$. .... (7.3.2.1)

If $x < 0$, then $-x > 0$

$\Rightarrow E(-x) > 0$ [By (7.3.2.1)]

Also, $E(x)\,E(-x) = 1$.

Therefore, $E(x) > 0$, for real x.

4. $E(x) \to \infty$ as $x \to \infty$.

$$E(-x) = \frac{1}{E(x)} \to 0 \text{ as } x \to \infty.$$

$$E(x) \to 0 \text{ as } x \to -\infty.$$

5. E is strictly increasing on R.

Let $0 < x < y$.

Therefore, $x^n < y^n$, for all n.

$$\Rightarrow E(x) < E(y). \quad \text{............ (7.3.2.2)}$$

Let $x < y < 0$.

$$0 < -y < -x.$$

Therefore, by (7.3.2.2),

$$E(-y) < E(-x).$$

$$\Rightarrow \frac{1}{E(y)} < \frac{1}{E(x)}.$$

Therefore, E is strictly increasing.

6. $E'(z) = E(z)$

Now, $E'(z) = \lim\limits_{h\to 0} \dfrac{E(z+h) - E(z)}{h}$

$$= \lim_{h\to 0} \frac{E(z)E(h) - E(z)}{h} \quad \text{[By result 7.3.2(1)]}$$

$$= E(z)\lim_{h\to 0} \frac{1}{h}\left[1 + \frac{h}{1!} + \frac{h^2}{2!} + \dots\dots\dots - 1\right]$$

$$= E(z)\lim_{h\to 0} \frac{1}{h}\left[h\left(1 + \frac{h}{2!} + \frac{h^2}{3!} + \dots\dots\dots\right)\right]$$

$$= E(z).$$

Therefore, $E'(z) = E(z)$.

Note 7.3.3

We define $E(x) = e^x, \quad \forall x \in R$.

Theorem 7.3.4

a) e^x is continuous and differentiable for all x.

b) $(e^x)' = e^x$

c) e^x is strictly increasing function of x and $e^x > 0$

d) $e^{x+y} = e^x + e^y$

e) $e^x \to x$ as $x \to \infty, \quad e^x \to 0$ as $x \to -\infty$

f) $\lim_{x\to\infty} x^n e^{-x} = 0$, for every n.

Proof

a) To prove: e^x is continuous and differentiable for all x.

$$\lim_{h\to 0}\frac{E(x+h)-E(x)}{h} = \lim_{h\to 0}\frac{E(x)E(h)-E(x)}{h}$$

$$= E(x)\lim_{h\to 0}\frac{1}{h}\left(1+\frac{h}{1!}+\ldots\ldots\ldots -1\right)$$

$$= E(x)\lim_{h\to 0}\left(1+\frac{h}{2!}+\ldots\ldots\ldots\right) = E(x)$$

Therefore,

$$\lim_{h\to 0}\frac{E(x+h)-E(x)}{h} \quad \text{exists.}$$

Hence, $E'(x)$ exists.

Therefore, $E(x) = e^x$ is differentiable.

Since differentiability implies continuity, $E(x) = e^x$ is continuous.

b) To prove: $(e^x)' = e^x$

By (a), $E'(x) = E(x) = e^x$.

Therefore, $(e^x)' = e^x$.

c) To prove: e^x is strictly increasing function of x and $e^x > 0$.

The proof follows from results 7.3.2 (5) and 7.3.2 (3).

d) To prove: $e^{x+y} = e^x + e^y$

The proof follows from results 7.3.2 (1).

e) To prove: $e^x \to x$ as $x \to \infty$, $e^x \to 0$ as $x \to -\infty$

$$e^x = E(x) \to +\infty \text{ as } x \to \infty.$$

$$e^x = E(-x) = \frac{1}{E(x)} \to 0 \text{ as } x \to \infty.$$

Therefore, $e^x \to 0$ as $x \to -\infty$.

f) To prove: $\lim\limits_{x\to\infty} x^n e^{-x} = 0$, for every n.

Now, $e^x = \sum\limits_{n=0}^{\infty} \frac{x^n}{n!} > \frac{x^{n+1}}{(n+1)!}$.

Therefore, $x^n e^{-x} < \frac{(n+1)!}{x}$.

Therefore,

$$\lim_{x\to\infty} x^n e^{-x} \le \lim_{x\to\infty} \frac{(n+1)!}{x} = 0.$$

Therefore, $\lim\limits_{x\to\infty} x^n e^{-x} = 0$, for every n.

Definition 7.3.5

E is a strictly increasing function on R. Hence, E is one-to-one. Also, $E(R) = (0, \infty)$.

Therefore, $E : R \to R^+$ is a bijection. We denote the inverse of E as L and L is called the *logarithmic function*. We define $L(E(x)) = x$, x is real or $E(L(y)) = y$, $y > 0$.

Note 7.3.6

1. If $L(E(x)) = x$, then $L'(y) = \frac{1}{y}$, where x is real and $y = E(x) > 0$.

Proof

Consider, $L(E(x)) = x$.

Differentiating both sides we get,

$$L'(E(x))\, E(x) = 1$$

$$\Rightarrow L'(E(x)) = \frac{1}{E(x)}.$$

Therefore, $L'(y) = \frac{1}{y}$, where $y = E(x) > 0$.

2. If $L(E(x)) = x$, then $L(1) = 0$, where x is real.

Proof

Put $x = 0$ in $L(E(x)) = x$, we get $L(E(0)) = 0$.

$$\Rightarrow L(1) = 0.$$

3. $L(E(x) + E(y)) = L(E(x)) + L(E(y))$

Proof

Let $u = E(x)$ and $v = E(y)$.

Therefore,

$$\begin{aligned} L(uv) &= L(E(x)E(y)) \\ &= L(E(x + y)) \\ &= x + y \\ &= L(E(x)) + L(E(y)) \\ &= L(u) + L(v). \end{aligned}$$

Therefore,

$$L(E(x) + E(y)) = L(E(x)) + L(E(y)).$$

4. We denote $L(x)$ by $\log x$.

Result 7.3.7

$\lim_{x\to\infty} x^{-\alpha} \log x = 0$.

Proof

Choose ε such that $0 < \varepsilon < \alpha$.

$$\text{Now, } x^{-\alpha} \log x = x^{-\alpha} \int_1^x \frac{1}{t}\,dt$$

$$< x^{-\alpha} \int_1^x \frac{t^{\varepsilon}}{t}\,dt$$

$$= \frac{x^{-\alpha}(x^{\varepsilon} - 1)}{\varepsilon}$$

$$< \frac{x^{\varepsilon - \alpha}}{\varepsilon}$$

Therefore,

$$\lim_{x \to \infty} x^{-\alpha} \log x \le \lim_{x \to \infty} \frac{x^{\varepsilon - \alpha}}{\varepsilon} \text{ [Since } \varepsilon - \alpha < 0\text{]}.$$

Therefore,

$$\lim_{x \to \infty} x^{-\alpha} \log x = 0.$$

7.4 Trigonometric Functions

Definition 7.4.1

We define $C(x) = \frac{1}{2}[E(ix) + E(-ix)]$ .... (7.4.1.1)

$$S(x) = \frac{1}{2i}[E(ix) - E(-ix)] \qquad \text{............ (7.4.1.2)}$$

$$\overline{E(ix)} = E(\overline{ix}) = E(-ix)$$

$$\overline{C(x)} = \frac{1}{2}[E(-ix) + E(ix)] = C(x)$$

$$\overline{S(x)} = \frac{1}{-2i}[E(-ix) - E(ix)] = \frac{1}{2i}[E(ix) - E(-ix)] = S(x)$$

Therefore, $C(x)$ and $S(x)$ are real.

$$C(x) + iS(x) = E(ix) \qquad \text{............ (7.4.1.3)}$$

Now, $C(0) = 1$ and $S(0) = 0$.

We will find $C'(x)$ and $S'(x)$:

$$C'(x) = \frac{i}{2}[E(ix) - E(-ix)]$$

$$= -\frac{1}{2i}[E(ix) - E(-ix)] = -S(x).$$

Similarly, $S'(x) = C(x)$.

Again, $|E(ix)|^2 = E(ix)\overline{E(ix)} = E(ix)E(-ix) = E(ix - ix) = 1$.

From equation (7.4.1.3), we get

$$(C(x))^2 + (S(x))^2 = 1 \qquad \text{............ (7.4.1.4)}$$

Claim:

There exist $x > 0$, such that $C(x) = 0$.

Suppose that, $C(x) \neq 0, \ \forall x > 0$.

Now, $C(0) = 1$ and C is continuous,

$$C(x) > 0, \ \forall x > 0.$$

That is, $S'(x) > 0, \ \forall x > 0$.

Therefore, S is strictly increasing.

Since $S(0) = 0, \ S(x) > 0, \ \forall x > 0$.

Let $0 < x < y$.

For all $t \in (x, y)$, $t > x$, that is, $x < t$.

Therefore, $S(x) < S(t)$ [since S is increasing].

$$\int_x^y S(x)dt < \int_x^y S(t)dt$$

$$S(x)(y - x) < \int_x^y -C'(t)dt$$

$$= -(C(y) - C(x))$$

$$\leq 2$$

Since $S(x) > 0$, by fixing x and choosing y large, we get

$$y - x > \frac{2}{s - x} \text{ which is a contradiction.}$$

Therefore, there exist $x > 0$, such that $C(x) = 0$.

Let $A = \{x > 0 / C(x) = 0\}$

$$= C^{-1}\{0\} \cap R^+.$$

Since C is continuous and $\{0\}$ is closed, $C^{-1}\{0\}$ is closed.

$$C(0) = 1 \neq 0.$$

$$\Rightarrow 0 \notin C^{-1}\{0\}.$$

$\Rightarrow \{0\} \cap C^{-1}\{0\} = \phi$.

Therefore, $A = C^{-1}\{0\} \cap \{R^{+} \cup \{0\}\}$.

Therefore, A is closed.

Also, x_0 = inf. A exists and $x_0 \in A$.

Therefore, x_0 is the smallest positive real number such that $C(x_0) = 0$.

Define, $\pi = 2x_0$.

$$\Rightarrow x_0 = \frac{\pi}{2}.$$

Therefore, $C\left(\frac{\pi}{2}\right) = 0$. .... (7.4.1.5)

To find: $S\left(\frac{\pi}{2}\right)$.

From equation (7.4.1.4), we get $\left(S\left(\frac{\pi}{2}\right)\right)^2 = 1$.

$$\Rightarrow S\left(\frac{\pi}{2}\right) = \pm 1.$$

Now, $C(x) > 0$, in $\left(0, \frac{\pi}{2}\right)$.

Therefore, $S'(x) > 0$ in $\left(0, \frac{\pi}{2}\right)$.

Therefore, S is strictly increasing in $\left(0, \frac{\pi}{2}\right)$, and $S(0) = 0$.

Therefore, $S\left(\frac{\pi}{2}\right) = 1$.

Hence, $E\left(\frac{i\pi}{2}\right) = C\left(\frac{\pi}{2}\right) + iS\left(\frac{\pi}{2}\right) = i$.

Again, $E(i\pi) = E\left(2\frac{i\pi}{2}\right) = \left[E\left(\frac{i\pi}{2}\right)\right]^2 = i^2 = -1$.

$$E(i2\pi) = E\left(4i\frac{\pi}{2}\right) = E\left(\frac{i\pi}{2}\right)^4 = i^4 = 1.$$

Also, $E(z+i2\pi) = E(z)E(i2\pi) = E(z)$.

Theorem 7.4.2

i) E is periodic with period $2\pi i$.

ii) The functions C and S are periodic with period 2π.

iii) If $0 < t < 2\pi$, then $E(it) \neq 1$.

iv) If z is a complex number such that $|z| = 1$, then there exists unique $t \in [0, 2\pi]$, such that $x = E(it)$.

Proof

i) To prove: E is periodic with period $2\pi i$.

$E(z + 2\pi i) = E(z)E(2\pi i) = E(z)$.

Therefore, E is periodic with period $2\pi i$.

ii) To prove: The functions C and S are periodic with period 2π.

$$C(x+2\pi) = \frac{1}{2}[E(i(x+2\pi)) + E(-i(x+2\pi))]$$

$$= \frac{1}{2}[E(ix+2\pi i) + E(-ix-2\pi i)]$$

$$= \frac{1}{2}[E(ix) + E(-ix)] = C(S).$$

Therefore, C is periodic with period 2π.

Similarly, S is periodic with period 2π.

iii) To prove: If $0 < t < 2\pi$, then $E(it) \neq 1$.

Suppose $0 < t < \frac{\pi}{2}$.

Put $E(it) = x + iy$, where x and y are real.

Also, $E(it) = C(t) + iS(t)$.

Therefore, $x = C(t)$and $y = S(t)$.

Since $0 < t < \frac{\pi}{2}$, we have $0 < x < 1; 0 < y < 1$.

Now, $E(4it) = [E(it)]^4 = (x + iy)^4$.

$$= x^4 - 6x^2y^2 + y^4 + 4ixy(x^2 - y^2).$$

If $E(4it)$ is complex, $E(4it) \neq 1$.

If $E(4it)$ is real, then $x^2 - y^2 = 0 \Rightarrow x^2 = y^2$.

Since $|E(it)| = 1$, we have $x^2 + y^2 = 1$.

Therefore, $x^2 = 1/2$ and $y^2 = 1/2$.

Therefore, $E(it) = -4\left(\frac{1}{2}\right)\left(\frac{1}{2}\right) = -1 \neq 1$.

Hence, if $0 < t < \pi/2$, then $E(4it) \neq 1$.

That is, if $0 < 4t < 2\pi$, then $E(4it) \neq 1$.

iv) To Prove; If z is a complex number such that $|z| = 1$, then there exists unique $t \in [0, 2\pi]$, such that $x = E(it)$.

Let z be a complex number such that $|z| = 1$.

Put $z = x + iy$; x and y are real.

Since $|z| = 1$, $x^2 + y^2 = 1$; $-1 \leq x, \quad y \leq 1$.

Case i):

Let $x, y \geq 0$.

Since C is a decreasing function from 1 to 0 in $[0, \pi/2]$, $C(t) = x$ for some $t \in [0, \pi/2]$.

Now, $E(it) = C(t) + iS(t)$.

Also, $C^2(t) + S^2(t) = 1$ and $x^2 + y^2 = 1$.

Therefore, $S^2(t) = y^2$.

On $[0, \pi/2]$, $S(t) \geq 0$.

Also, $y \geq 0$.

Therefore, $S(t) = y$.

Therefore, $E(it) = C(t) + iS(t) = x + iy = z$.

Therefore, $z = E(it)$, for some $t \in [0, \pi/2]$.

Case ii):

Let $x \leq 0, \quad y \leq 0$.

$-iz = -i(x+iy) = y + i(-x)$.

Here, $-x \geq 0$, and $y \leq 0$.

Therefore, by case i), there exists $t \in [0, \pi/2]$, such that $E(it) = -iz$.

Now, $z = i(-iz) = E(i\pi/2)E(it) = E\left(i\left(t+\frac{\pi}{2}\right)\right)$, where $t+\frac{\pi}{2} \in [0, \pi/2]$.

Case iii):

Let $y \leq 0$.

Now, $-z = -(x+iy) = (-x) + i(-y)$.

Now, $-y \geq 0$.

By cases (i) and (ii), there exists $t \in [0, \pi/2]$ such that $E(it) = -z$, that is,

$$z = -(-z) = E(i\pi)E(it) = E(i(\pi + t)), \text{where } \pi + t \in [0, 2\pi].$$

Therefore, for each z with $|z| = 1$, there exists $t \in [0, 2\pi]$ such that $z = E(it)$.

To prove: Uniqueness

Suppose, there exists t_1, t_2 with $0 \leq t_1 \leq t_2 \leq 2\pi$, such that

$$E(it_1) = z = E(it_2).$$

Now, $E(it_2)E(-it_1) = E(it_2 - it_1) = E(i(t_2 - t_1)) \neq 1$ [By (c)].

Therefore, $E(it_2) \neq E(it_1)$, which is a contradiction.

Hence, there exists a unique $t \in [0, 2\pi]$, such that E(it) = 2.

Hence the theorem.

7.5 The Algebraic Completeness of the Complex Field

Theorem 7.5.1

Suppose $a_0, a_1, \ldots\ldots, a_n$ are complex, $n \geq 1$ and $a_n \neq 0$. $P(z) = \sum_{k=0}^{n} a_k z^k$.

Then $P(z) = 0$, for some complex number z.

Proof

Without loss of generality, let us assume $a_n = 1$.

Put $\mu = \inf_{z\in C} .|P(z)|$ on $|z| = R$.

Then,

$$P(z) = \left|z^n + a_n z^{n-1} + ... + a_0\right|$$

$$= |z|^n \left|1 + a_{n-1} z^{-1} + ... + a_0 z^{-n}\right|$$

$$= R^n \left|1 + a_{n-1} z^{-1} + ... + a_0 z^{-n}\right|$$

$$\geq R^n \left\{1 - |a_{n-1}||z|^{-1} - |a_{n-2}||z|^{-2} - .... - |a_0||z|^{-n}\right\}$$

$$\geq R^n \left\{1 - |a_{n-1}|R^{-1} - |a_{n-2}|R^{-2} - .... - |a_0|R^{-n}\right\} \quad .... (7.5.1.1)$$

If $|z| = R \to \infty$, then the R.H.S. of (7.5.1.1) $\to \infty$, that is,

$$P(z) \to \infty \text{ as } |z| = R \to \infty.$$

Therefore, there exists $R_0 > 0$ such that $|P(z)| > \mu$, if $|z| > R_0$.

Put $D = \{z \in C / |z| \leq R_0\}$.

Since D is a closed and bounded subset of C, D is compact.

Also, $\mu = \inf_{z\in C} .|P(z)|$.

Since $|P|$ is continuous on D and D is compact, $\mu = |P(z_0)|$, for some $z_0 \in D$.

Claim: $\mu = 0$.

Suppose $\mu \neq 0$.

Put $Q(z) = \dfrac{P(z + z_0)}{P(z_0)}$.

Then, Q is a non-constant polynomial.

Therefore, $Q(0) = 1$.

Now, $|Q(z)| = \dfrac{|P(z + z_0)|}{|P(z_0)|} = \dfrac{|P(z + z_0)|}{\mu} \geq \dfrac{\mu}{\mu} = 1$.

That is, $|Q(z)| \geq 1$, for all complex number z. .... (7.5.1.2)

Let k be the smallest positive integer, with $1 \le k \le n$, such that

$$Q(z) = 1 + b_k z^k + ... + b_n z^n, b_k \neq 0.$$

[Since $Q(0) = 1$, $Q(z) = 1 + b_1 z + ... + b_n z^n$, for some b_i. Let k be the smallest positive integer, such that $b_k \neq 0$. Therefore, $b_1, b_2, ..., b_{k-1} = 0$].

By the part (d) of the previous theorem, there exists a real θ, such that

$$e^{ik\theta} = \frac{-|b_k|}{b_k} \Rightarrow b_k e^{ik\theta} = -|b_k|.$$

$$\left|1 + b_k r^k e^{ik\theta}\right| = 1 - |b_k| r^k.$$

Therefore,

$$\left|Q(re^{i\theta})\right| = \left|1 + b_k \left(re^{i\theta}\right)^k + ... + b_n \left(re^{i\theta}\right)^n\right|, \quad b_k \neq 0.$$

$$\le \left|1 + b_k r^k\right| + |b_{k+1}| r^{k+1} + ... + |b_n| r^n.$$

$$\le 1 + |b_k| r^k + |b_{k+1}| r^{k+1} + ... + |b_n| r^n.$$

$$\le 1 - r^k \left[|b_k| - |b_{k+1}| r - ..... - |b_n| r^{n-k}\right]. \qquad \text{............ (7.5.1.3)}$$

For sufficiently small r, the expression in the bracket is positive.

Therefore,

$\left|Q(re^{i\theta})\right| < 1$,is a contradiction to equation (7.5.1.2).

Therefore, $\mu = 0$.

$$\Rightarrow P(z_0) = 0.$$

$\Rightarrow P(z) = 0$, for some complex number z.

Hence the theorem.

7.6 Fourier Series

Definition 7.6.1

A trigonometric polynomial is the finite sum of the form $f(x) = a_0 + \sum_{n=1}^{N} (a_n \cos nx + b_n \sin nx)$, N is real, $a_0, a_1, ..., a_N$ and $b_0, b_1, ..., b_N$ are complex numbers. That is,

$$f(x) = a_0 + \sum_{n=1}^{N} a_n \left(\frac{e^{inx} + e^{-inx}}{2} \right) + \sum_{n=1}^{N} b_n \left(\frac{e^{inx} - e^{-inx}}{2i} \right).$$

Therefore,

$$f(x) = \sum_{n=-N}^{N} c_n e^{inx}, \quad \forall c_n \in C.$$

It is clear that every trigonometric polynomial is periodic with period 2π.

Note 7.6.2

1. To prove: $\frac{1}{2\pi} \int_{-\pi}^{\pi} e^{inx} dx = \begin{cases} 1, & \text{if } n = 0 \\ 0, & \text{if } n = \pm 1, \pm 2, \ldots\ldots \end{cases}$

   If $n = 0$, then

$$\frac{1}{2\pi} \int_{-\pi}^{\pi} e^{inx} dx = \frac{1}{2\pi} \int_{-\pi}^{\pi} dx = \frac{1}{2\pi}(\pi + \pi) = 1.$$

   If $n = \pm 1, \pm 2, \ldots\ldots$, then

$$\frac{1}{2\pi} \int_{-\pi}^{\pi} e^{inx} dx = \frac{1}{2\pi} \int_{-\pi}^{\pi} (\cos nx + i \sin nx) dx.$$

$$= \frac{1}{2\pi} \left[\left(\frac{\sin nx}{n} \right)_{-\pi}^{\pi} + i \left(\frac{-\cos nx}{n} \right)_{-\pi}^{\pi} \right].$$

$$= 0.$$

   Therefore, $\frac{1}{2\pi} \int_{-\pi}^{\pi} e^{inx} dx = \begin{cases} 1, & \text{if } n = 0 \\ 0, & \text{if } n = \pm 1, \pm 2, \ldots\ldots \end{cases}$

   Hence proved.

2. To prove: $c_{-n} = \overline{c_n}, \quad \forall n = 0, 1, 2, \ldots\ldots$

   Now, $f(x) = \sum_{n=-N}^{N} c_n e^{inx}, \quad \forall c_n \in C.$

   $f(x)$ is real, if and only if $f(x) = \overline{f(x)}$.

   if and only if $\sum_{n=-N}^{N} c_n e^{inx} = \sum_{n=-N}^{N} \overline{c_n} e^{-inx}$.

if and only if $\sum_{n=-N}^{N} c_{-n} e^{-inx} = \sum_{n=-N}^{N} \overline{c_n} e^{-inx}$.

if and only if $c_{-n} = \overline{c_n}, \quad \forall n = 0,1,2,......$

Hence proved.

Definition 7.6.3

The *trigonometric series* is a series of the form $\sum_{n=-\infty}^{\infty} c_n e^{inx}$ where x is real.

The nth partial sum of $\sum_{n=-\infty}^{\infty} c_n e^{inx}$ is $\sum_{n=-N}^{N} c_n e^{inx}$.

Definition 7.6.4

Let f be an integrable function in [-π, π]. Then, $c_n = \frac{1}{2\pi} \int_{-\pi}^{\pi} f(x) e^{-inx} dx$, where n is an integer, are called the *Fourier coefficients of* f and the series $\sum_{n=-\infty}^{\infty} c_n e^{inx}$ is called the *Fourier series of f*.

Definition 7.6.5

Let $\{\varphi_n\}$ be a sequence of complex functions defined on [a, b], such that $\int \varphi_n(x) \overline{\varphi_m(x)}\, dx = 0$, if $n \neq m$. Then, $\{\varphi_n\}$ is called the *sequence of orthonormal functions* on [a, b].

If, in addition, $\int_a^b |\varphi_n(x)|^2 dx = 1$, if n = 1, 2, 3, ..., then we say $\{\varphi_n\}$ is an *orthonormal system* on [a, b].

Example 7.6.6

Define $\varphi_n(x) = \frac{1}{\sqrt{2\pi}} e^{-inx}, x \in [-\pi, \pi]$.

When $n \neq m$,

$$\int_{-\pi}^{\pi} \varphi_n(x) \overline{\varphi_m(x)}\, dx = \int_{-\pi}^{\pi} \frac{1}{\sqrt{2\pi}} e^{-inx} \frac{1}{\sqrt{2\pi}} e^{imx} dx$$

$$= \frac{1}{2\pi}\int_{-\pi}^{\pi} e^{i(m-n)x}dx = 0 \text{ .[Since } n \neq m \text{]}$$

Also, $\int_{-\pi}^{\pi} |\varphi_n(x)|^2 dx = \int_{-\pi}^{\pi} \varphi_n(x)\,\overline{\varphi_n(x)}\,dx$.

$$= \int_{-\pi}^{\pi} \frac{1}{\sqrt{2\pi}} e^{-inx} \frac{1}{\sqrt{2\pi}} e^{inx} dx = \frac{1}{2\pi}\int_{-\pi}^{\pi} dx = 1.$$

Therefore, $\{\varphi_n\}$ is orthonormal on [-π, π].

Definition 7.6.7

Let $\{\varphi_n\}$ be orthonormal on $[a, b]$, and f be an integrable function on $[a, b]$. Then, $c_n = \int_a^b f(x)\,\overline{\varphi_n(x)}\,dx$ is called the *Fourier coefficient of f relative to* $\{\varphi_n\}$. We write $f(x) \approx \sum_{n=1}^{\infty} c_n \varphi_n(x)$. The series $\sum_{n=1}^{\infty} c_n \varphi_n(x)$ is called the *Fourier series relative to* $\{\varphi_n\}$.

Theorem 7.6.8

Let $\{\varphi_n\}$ be orthonormal on $[a, b]$. Let $s_n(x) = \sum_{m=1}^{n} c_m \varphi_m(x)$ be the nth partial sum of the series f and suppose that $t_n(x) = \sum_{m=1}^{n} \gamma_m \varphi_m(x)$. Then, $\int_a^b |f - s_n|^2 dx \leq \int_a^b |f - t_n|^2 dx$ and equality holds if and only if $\gamma_m = c_m$, m = 1, 2, …, n, …. That is, among all functions of t_n, s_n gives the best possible mean square approximation to f.

Proof

$$\int_a^b |f - t_n|^2 dx \leq \int_a^b (f(x) - t_n(x))\left(\overline{f(x)} - \overline{t_n(x)}\right)dx$$

$$= \int_a^b \left[f(x)\overline{f(x)} - f(x)\overline{t_n(x)} - \overline{f(x)}\,t_n(x) + t_n(x)\,\overline{t_n(x)}\right]dx$$

$$= \int_a^b |f(x)|^2 \, dx - \int_a^b f(x)\, \overline{t_n(x)}\, dx - \int_a^b \overline{f(x)}\, t_n(x)\, dx + \int_a^b |t_n(x)|^2 \, dx$$

.... (7.6.8.1)

Now, $\int_a^b f(x)\, \overline{t_n(x)}\, dx = \int_a^b f(x) \sum_{m=1}^{n} \overline{\gamma_m}\, \overline{\varphi_m}\, dx$

$$= \sum_{m=1}^{n} \overline{\gamma_m} \int_a^b f(x)\ \overline{\varphi_m(x)}\, dx = \sum_{m=1}^{n} \overline{\gamma_m}\, c_m \qquad \text{............ (7.6.8.2)}$$

Similarly,

$$\int_a^b \overline{f(x)}\, t_n(x)\, dx = \sum_{m=1}^{n} \gamma_m \overline{c_m} \qquad \text{............ (7.6.8.3)}$$

$$\int_a^b |t_n(x)|^2 \, dx = \int_a^b t_n(x)\, \overline{t_n(x)}\, dx$$

$$= \int_a^b \sum_{m=1}^{n} \gamma_m \varphi_m(x) \sum_{k=1}^{n} \overline{\gamma_k}\ \overline{\varphi_k(x)}\, dx$$

$$= \sum_{m=1}^{n} \sum_{k=1}^{n} \gamma_m \overline{\gamma_k} \int_a^b \varphi_m(x) \overline{\varphi_k(x)}\, dx$$

$$= \sum_{m=1}^{n} \gamma_m \overline{\gamma_m} \quad \text{[Since } \{\varphi_n\} \text{ is orthonormal]}$$

$$= \sum_{m=1}^{n} |\gamma_m|^2 \qquad \text{............ (7.6.8.4)}$$

Using equations (7.6.8.2), (7.6.8.3) and (7.6.8.4) in (7.6.8.1), we get

$$\int_a^b |f - t_n|^2 \, dx = \int_a^b |f(x)|^2 \, dx - \sum_{m=1}^{n} \overline{\gamma_m}\, c_m - \sum_{m=1}^{n} \gamma_m \overline{c_m} + \sum_{m=1}^{n} |\gamma_m|^2 \qquad \text{....... (7.6.8.5)}$$

If $\gamma_m = c_m$, we get $c_n = s_n$.

Therefore,

$$\int_a^b |f - s_n|^2 \, dx = \int_a^b |f(x)|^2 \, dx - \sum_{m=1}^{n} \overline{c_m}\, c_m - \sum_{m=1}^{n} c_m \overline{c_m} + \sum_{m=1}^{n} |c_m|^2$$

$$= \int_a^b |f(x)|^2 \, dx + \sum_{m=1}^{n} |c_m|^2 \qquad \text{............ (7.6.8.6)}$$

Subtracting equation (7.6.8.6) from equation (7.6.8.5), we get

$$\int_a^b |f - t_n|^2 dx - \int_a^b |f - s_n|^2 dx = -\sum_{m=1}^{n} \overline{\gamma_m}\, c_m - \sum_{m=1}^{n} \gamma_m \overline{c_m} + \sum_{m=1}^{n} |\gamma_m|^2 + \sum_{m=1}^{n} |c_m|^2$$

$$= \sum_{m=1}^{n} \left[\left(\gamma_m \overline{\gamma_m} - \overline{\gamma_m}\, c_m\right) + \left(c_m \overline{c_m} - \gamma_m \overline{c_m}\right) \right]$$

$$= \sum_{m=1}^{n} \left[\overline{\gamma_m}(\gamma_m - c_m) + \overline{c_m}(c_m - \gamma_m) \right]$$

$$= \sum_{m=1}^{n} (\gamma_m - c_m)\left(\overline{\gamma_m} - \overline{c_m}\right)$$

$$= \sum_{m=1}^{n} |\gamma_m - c_m|^2 \geq 0 \qquad \text{............ (7.6.8.7)}$$

Therefore,

$$\int_a^b |f - t_n|^2 dx \geq \int_a^b |f - s_n|^2 dx$$

Second part:

$$\int_a^b |f - t_n|^2 dx = \int_a^b |f - s_n|^2 dx$$

If and only if $\int_a^b |f - t_n|^2 dx - \int_a^b |f - s_n|^2 dx = 0$

If and only if $\sum_{m=1}^{n} |\gamma_m - c_m|^2 = 0$. [By equation (7.6.8.7)]

If and only if $|\gamma_m - c_m| = 0$, $m = 1, 2, 3, \ldots\ldots, n$.

If and only if $\gamma_m = c_m$, $m = 1, 2, 3, \ldots\ldots, n$.

Therefore, among all functions t_n, s_n gives the best possible mean square approximation to f.

Hence the theorem.

Note 7.6.9

$$\int_a^b |s_n(x)|^2 dx \le \int_a^b |f(x)|^2 dx.$$

Proof

By equation (7.6.8.4), we have

$$\int_a^b |t_n(x)|^2 dx = \sum_{m=1}^{n} |\gamma_m|^2$$

put $\gamma_m = c_m$.

Then, $t_n = s_n$.

Therefore,

$$\int_a^b |s_n(x)|^2 dx = \sum_{m=1}^{n} |c_m|^2$$

$$= \int_a^b |f(x)|^2 dx - \int_a^b |f - s_n|^2 dx \qquad \text{[By equation (7.6.8.6)]}$$

$$\le \int_a^b |f(x)|^2 dx$$

Therefore, $\int_a^b |s_n(x)|^2 dx \le \int_a^b |f(x)|^2 dx$.

Hence proved.

Theorem 7.6.10

This is known as ***Bessel's inequality***.

If $\{\varphi_n\}$ is orthonormal on $[a, b]$ and if $f(x) \sim \sum_{n=1}^{\infty} c_n \varphi_n(x)$, then $\sum_{n=1}^{\infty} |c_n|^2 \le \int_a^b |f(x)|^2 dx$. In particular, $\lim_{n\to\infty} c_n = 0$.

Proof

Given $\{\varphi_n\}$ is orthonormal on $[a, b]$ and $f(x) \sim \sum_{n=1}^{\infty} c_n \varphi_n(x)$.

Put $s_m(x) \sim \sum_{m=1}^{n} c_m \varphi_m(x)$.

Then, $\int_a^b |s_m(x)|^2 dx = \sum_{m=1}^{n} |c_m|^2 \le \int_a^b |f(x)|^2$ [By note (7.6.9)]

That is, $\sum_{m=1}^{n} |c_m|^2 \le \int_a^b |f(x)|^2$.

This is true for all n.

Taking limit as $n \to \infty$, we get

$$\sum_{m=1}^{\infty} |c_m|^2 \le \int_a^b |f(x)|^2 .$$

That is, $\sum_{n=1}^{\infty} |c_n|^2 \le \int_a^b |f(x)|^2$.

Particular part:

To prove: $\lim_{n\to\infty} c_n = 0$.

We have, $\int_a^b |f(x)|^2 dx < \infty$.

Therefore, $\sum_{n=1}^{\infty} |c_n|^2 < \infty$.

$$\lim_{n\to\infty} |c_n|^2 = 0.$$

$$\Rightarrow \lim_{n\to\infty} c_n = 0.$$

Hence proved.

Definition 7.6.11

Let f be a periodic function with period 2π which is integrable in [-π, π]. Put $c_n = \frac{1}{2\pi}\int_{-\pi}^{\pi} f(x)e^{inx} dx$. Then, $\sum_{n=-\infty}^{\infty} c_n e^{inx}$ is the *Fourier series of f*. The Nth partial sum is $s_N(x) = s_N(f;x) = \sum_{n=-N}^{N} c_n e^{inx}$.

Now, $\int_{-\pi}^{\pi}|s_N(x)|^2\,dx = \sum_{n=-N}^{N}|c_n|^2 \le \int_{-\pi}^{\pi}|f(x)|^2\,dx$.

Therefore, $\frac{1}{2\pi}\int_{-\pi}^{\pi}|s_N(x)|^2\,dx \le \int_{-\pi}^{\pi}|f(x)|^2\,dx$.

Definition 7.6.12

The *Dirichlet's kernel* is defined as $D_N(x) = \sum_{n=-N}^{N} e^{inx}$.

Results 7.6.13

1) $D_N(x) = \dfrac{\sin\left(N+\frac{1}{2}\right)x}{\sin(x/2)}$.

Solution

We have, $D_N(x) = \sum_{n=-N}^{N} e^{inx}$

$$\Rightarrow (e^{ix}-1)D_N(x) = (e^{ix}-1)\sum_{n=-N}^{N} e^{inx}.$$

$$= \sum_{n=-N}^{N}\left(e^{i(n+1)x} - e^{inx}\right).$$

$$\Rightarrow e^{\frac{-ix}{2}}(e^{ix}-1)D_N(x) = e^{\frac{-ix}{2}}\left[e^{i(N+1)x} - e^{-iNx}\right].$$

$$\Rightarrow \left(e^{\frac{ix}{2}} - e^{\frac{-ix}{2}}\right)D_N(x) = e^{i\left(N+\frac{1}{2}\right)x} - e^{-i\left(N+\frac{1}{2}\right)x}.$$

$$\Rightarrow 2i\sin\left(\frac{x}{2}\right)D_N(x) = 2i\sin\left(N+\frac{1}{2}\right)x.$$

$$\Rightarrow D_N(x) = \frac{\sin\left(N+\frac{1}{2}\right)x}{\sin(x/2)}.$$

Hence proved.

2) $\frac{1}{2\pi}\int_{-\pi}^{\pi} D_N(x)dx = 1$.

Solution

$$\frac{1}{2\pi}\int D_N(x) = \frac{1}{2\pi}\int_{-\pi}^{\pi} \sum_{n=-N}^{N} e^{inx}dx$$

$$= \sum_{n=-N}^{N} \frac{1}{2\pi}\int_{-\pi}^{\pi} e^{inx}dx$$

$$= 1 \quad [\text{Since } \frac{1}{2\pi}\int_{-\pi}^{\pi} e^{inx}dx = \begin{cases} 1, & \text{if } n = 0 \\ 0, & \text{if } n = \pm 1, \ldots \end{cases}]$$

Hence proved.

3) $s_N(f;x) = \frac{1}{2\pi}\int_{-\pi}^{\pi} f(x-t)D_N(t)dt$

Proof

$$s_N(f;x) = \sum_{n=-N}^{N} c_n e^{inx}$$

$$= \sum_{n=-N}^{N} e^{inx} \frac{1}{2\pi}\int_{-\pi}^{\pi} f(t)e^{-\text{int}}dt$$

$$= \frac{1}{2\pi}\int_{-\pi}^{\pi} f(t) \sum_{n=-N}^{N} e^{in(x-t)}dt$$

$$= \frac{1}{2\pi}\int_{-\pi}^{\pi} f(t)D_N(x-t)dt$$

Put $u = x - t$.
Therefore, $du = -dt$.
When $t = -\pi, u = x + \pi$.
When $t = \pi$, $u = x - \pi$.
Therefore,

$$s_N(f;x) = \frac{1}{2\pi}\int_{x+\pi}^{x-\pi} f(x-u)D_N(u)(-du)$$

$$= \frac{1}{2\pi} \int_{x-\pi}^{x+\pi} f(x-t) D_N(t) dt .$$

$$= \frac{1}{2\pi} \int_{-\pi}^{\pi} f(x-t) D_N(t) dt$$

[Since f and D_N are periodic in the period 2π].

Hence proved.

Theorem 7.6.14

If for some x, there are constants $\delta > 0$ and $M < \infty$ such that $|f(x+t) \div f(x)| \le M(t), \ \forall t \in (-\delta, \delta)$, then $\lim_{N \to \infty} s_N(f; x) = f(x)$.

Proof

Define $g(t) = \dfrac{f(x-t) - f(x)}{\sin t/2}, \qquad 0 < |t| \le \pi .$

Therefore, $g(0) = 0$.

Also, $\dfrac{1}{2\pi} \displaystyle\int_{-\pi}^{\pi} D_N(x) dx = 1 .$

Now, $s_N(f; x) - f(x) = \dfrac{1}{2\pi} \displaystyle\int_{-\pi}^{\pi} f(x-t) D_N(t) dt - f(x) \dfrac{1}{2\pi} \int_{-\pi}^{\pi} D_N(t) dt$

[By results (7.6.13.2) and (7.6.13.3)]

$$= \frac{1}{2\pi} \int_{-\pi}^{\pi} [f(x-t) - f(x)] D_N(t) dt$$

$$= \frac{1}{2\pi} \int_{-\pi}^{\pi} g(t) \sin\frac{t}{2} \frac{\sin\left(N + \frac{1}{2}\right)t}{\sin\frac{t}{2}} dt$$

[By result (7.6.13.1)]

$$= \frac{1}{2\pi} \int_{-\pi}^{\pi} g(t) [\sin(Nt)\cos(t/2) + \cos(Nt)\sin(t/2)] dt$$

$$= \frac{1}{2\pi} \int_{-\pi}^{\pi} [g(t)\cos(t/2)] \sin(Nt) dt + \frac{1}{2\pi} \int_{-\pi}^{\pi} [g(t)\sin(t/2)] \cos(Nt) dt$$

.... (7.6.14.1)

Since f is bounded in [-π, π], there exists $k > 0$, such that $|f(x)| \le k, \ \forall x \in [-\pi, \pi]$.

$$\text{Now, } |g(t)\sin(t/2)| = |f(x-t) - f(x)|$$
$$\le |f(x-t)| + |f(x)|$$
$$\le k + k = 2k$$

Therefore, $g(t).\sin t/2$ is bounded in [-π, π].

By hypothesis, $|t| < \delta \Rightarrow |f(x-t) - f(x)| \le M|t|$ .... (7.6.14.2)

$$|g(t)\cos(t/2)| = |g(t)||\cos(t/2)|$$
$$\le |g(t)| \qquad [\text{since } |\cos t/2| \le 1]$$
$$= \frac{|f(x-t) - f(x)|}{|\sin t/2|}$$
$$\le \frac{M|t|}{\sin|t/2|}, \quad \text{if } |t| < \delta \qquad [\text{By equation (7.6.14.2)}]$$
$$< \frac{2M|t/2|}{\sin|t/2|}, \quad \left[\text{since, if } 0 < x < \pi/2, \text{ then } \frac{\sin x}{x} > \frac{2}{\pi}\right]$$
$$< \pi M .$$

Therefore,

$$|g(t)\cos(t/2)| < \pi M \qquad \text{............ (7.6.14.3)}$$

If $\delta \le |t| \le \pi$, then $\delta/2 \le |t|/2 \le \pi/2$.

Therefore, $\sin\delta/2 \le \sin|t|/2$ [Since sin is increasing in [δ/2, π/2].

$$\frac{1}{\sin\delta/2} \ge \frac{1}{\sin|t|/2}.$$

Therefore,

$$|g(t)\cos(t/2)| \le |g(t)|$$
$$= \left|\frac{f(x-t) - f(x)}{\sin t/2}\right|$$
$$\le \frac{|f(x-t)| + |f(x)|}{\sin t/2}$$

$$\leq \frac{2k}{\sin \delta / 2}$$

Also, $|g(0)\cos(0/2)| = 0$.

Therefore, $g(t).\cos(t/2)$ is bounded in $[-\pi, \pi]$.

We know that, if $\{\varphi_n\}$ is orthonormal on $[a, b]$, then $\lim\limits_{n\to\infty} \int_a^b f(x)\, \overline{\phi_n(x)}\, dx = 0$.

Define $\varphi_n(x) = \dfrac{\sin nx}{\sqrt{\pi}}$ or $\dfrac{\cos nx}{\sqrt{\pi}}$, $x \in [-\pi, \pi]$.

Then, $\{\varphi_n(x)\}$ is orthonormal in $[-\pi, \pi]$.

Therefore,

$$\lim_{N\to\infty} \int_{-\pi}^{\pi} g(t)\cos(t/2)\sin Nt\; dt = 0 \text{ and } \lim_{N\to\infty} \int_{-\pi}^{\pi} g(t)\sin(t/2)\cos Nt\; dt = 0.$$

Substituting in equation (7.6.14.1), we get

$$\lim_{n\to\infty} s_N(f;x) = f(x).$$

Hence the theorem.

Corollory 7.6.15

If $f(x) = 0$, for all x in some segment J, then $\lim\limits_{n\to\infty} s_N(f;x) = f(x)$, for all $x \in J$.

Solution

Let $x \in J$.

Since J is open, there exists $\delta > 0$, such that $(x - \delta, x + \delta) \in J$.

By data, $f(t) = 0$, for all $t \in J$.

Therefore , $f(t) = 0$, for all $t \in (x - \delta, x + \delta)$.

For all $t \in (-\delta, \delta)$, $x + t \in (x - \delta, x + \delta)$.

Therefore, $f(x + t) = 0$ and $f(x) = 0$.

Therefore, $|f(x+t) - f(x)| = 0 \leq |t|, \quad \forall t \in (-\delta, \delta)$.

Therefore, by the previous theorem,

$$\lim_{N\to\infty} s_N(f;x) = f(x) = 0, \qquad \text{for all } x \in J.$$

Note 7.6.16

Here is another formulation of this corollary.

If $f(t) = g(t)$, for all t, in some neighbourhood of x, then $s_N(f-g;x)=0$ as $N \to \infty$.

Proof

This can be proved by taking $f - g$ instead of f in the previous corollory. This is called ***localization theorem***.

Theorem 7.6.17

If f is continuous (with period 2π) and if $\varepsilon > 0$, there is a trigonometric polynomial p, such that $|p(x)-f(x)|<\varepsilon$, for all real x.

Proof

If we identify x and $x + 2i$, we may regard the 2π periodic functions on R, as functions on the circle T, by means of the mapping $x \to e^{ix}$.

The trigonometric polynomial for a self adjoint algebra A, which separates points on T and which vanishes at no point of T.

Since T is compact, A is dense in $\mathscr{C}(T)$.

Hence the theorem.

Theorem 7.6.18

This is known as ***Parseval's theorem.***

Suppose f and g are R-integrable function with period 2π and

$$f(x) \sim \sum_{-\infty}^{\infty} c_n e^{inx}, \qquad g(x) \sim \sum_{-\infty}^{\infty} \gamma_n e^{inx}, \text{ then}$$

i) $$\lim_{N\to\infty} \frac{1}{2\pi}\int_{-\pi}^{\pi} |f(x)-s_N(f;x)|^2\,dx = 0.$$

ii) $$\frac{1}{2\pi}\int_{-\pi}^{\pi} f(x)\,\overline{g(x)}\,dx = \sum_{-\infty}^{\infty} c_n \overline{\gamma_n}.$$

iii) $$\frac{1}{2\pi}\int_{-\pi}^{\pi} |f(x)|^2\,dx = \sum_{-\infty}^{\infty} |c_n|^2.$$

Proof

We use the notation,

$$\|h\|_2 = \left[\frac{1}{2\pi}\int_{-\pi}^{\pi}|h(x)|^2\,dx\right]^{1/2}$$

To prove: $\lim_{N\to\infty}\frac{1}{2\pi}\int_{-\pi}^{\pi}|f(x)-s_N(f;x)|^2\,dx = 0$.

Let ε > 0 be given.

Since f ϵ R and f has period 2π, there exists a continuous 2π-periodic function f, such that

$$\|f-g\|_2 < \varepsilon \qquad \text{........... (7.6.18.1)}$$

By the previous theorem, there exists a trigonometric polynomial P, such that $|P(x)-h(x)|<\varepsilon$, for real x.

Now, $$\|h-P\|_2 = \left[\frac{1}{2\pi}\int_{-\pi}^{\pi}|h(x)-P(x)|^2\,dx\right]^{1/2}$$

$$< \left[\frac{1}{2\pi}\int_{-\pi}^{\pi}\varepsilon^2\,dx\right]^{1/2} = \varepsilon$$

Therefore,

$$\|h-P\|_2 < \varepsilon. \qquad \text{........... (7.6.18.2)}$$

If P has degree N_0, for all $N \geq N_0$,

$$\|h-s_N(h)\|_2 = \left[\frac{1}{2\pi}\int_{-\pi}^{\pi}|h(x)-s_N(h)|^2\,dx\right]^{1/2}$$

$$\leq \left[\frac{1}{2\pi}\int_{-\pi}^{\pi}|h(x)-P(x)|^2\,dx\right]^{1/2}$$

$$= \|h-P\|_2 < \varepsilon \text{ [By equation (7.6.18.2)]}$$

Therefore, $\|h-s_N(h)\|_2 < \varepsilon$ .... (7.6.18.3)

Now, $\|s_N(h)-s_N(f)\|_2 = \|s_N(h-f)\|_2$

$$= \left[\frac{1}{2\pi} \int_{-\pi}^{\pi} |s_N(h-f)|^2 \, dx \right]^{1/2}$$

$$\leq \left[\frac{1}{2\pi} \int_{-\pi}^{\pi} |h(x) - f(x)|^2 \, dx \right]^{1/2}$$

$$= \|h - f\|_2 < \varepsilon \quad \text{[By equation (7.6.18.1)]}$$

Therefore, $\|s_N(h) - s_N(f)\|_2 < \varepsilon$ .... (7.6.18.4)

Therefore, for all $N \geq N_0$,

$$\|f - s_N(f)\|_2 \leq \|f - h\|_2 + \|h - s_N(h)\|_2 + \|s_N(h) - s_N(f)\|_2$$

$< \varepsilon + \varepsilon + \varepsilon$ [By equations (7.6.18.1), (7.6.18.3) and (7.6.18.4)]

$< 3\varepsilon$.

Therefore,

$$\lim_{N \to \infty} \|f - s_N(f)\|_2 = 0 \qquad \text{............ (7.6.18.5)}$$

Hence, $\displaystyle \lim_{N \to \infty} \left[\frac{1}{2\pi} \int_{-\pi}^{\pi} |f - s_N(f)|^2 \, dx \right]^{1/2} = 0$.

Therefore, $\displaystyle \lim_{N \to \infty} \left[\frac{1}{2\pi} \int_{-\pi}^{\pi} |f - s_N(f)|^2 \, dx \right] = 0$.

Hence part (i) is proved.

ii) $$\frac{1}{2\pi} \int_{-\pi}^{\pi} s_N(f) \, \overline{g(x)} \, dx = \frac{1}{2\pi} \int_{-\pi}^{\pi} \sum_{n=-N}^{N} c_n e^{inx} \, \overline{g(x)} \, dx$$

$$= \sum_{n=-N}^{N} c_n \frac{1}{2\pi} \int_{-\pi}^{\pi} e^{inx} \, \overline{g(x)} \, dx$$

$$= \sum_{n=-N}^{N} c_n \overline{\frac{1}{2\pi} \int_{-\pi}^{\pi} e^{-inx} \, g(x) \, dx}$$

$$= \sum_{n=-N}^{N} c_n \overline{\gamma_n}$$

Therefore,

$$\frac{1}{2\pi}\int_{-\pi}^{\pi} s_N(f)\,\overline{g(x)}\,dx = \sum_{n=-N}^{N} c_n \overline{\gamma_n} \qquad \text{............ (7.6.18.6)}$$

Now,

$$\left|\int_{-\pi}^{\pi} f(x)\ \overline{g(x)}\ dx - \int_{-\pi}^{\pi} s_N(f)\ \overline{g(x)}\ dx\right| = \left|\int_{-\pi}^{\pi} \left(f(x) - s_N(f)\right)\ \overline{g(x)}\ dx\right|$$

$$\leq \left\| f - s_N(f) \right\|_2 \left\| \overline{g} \right\|_2$$

[By Schwarz inequality]

By using equation (7.6.18.5),

$$\lim_{N\to\infty}\left|\int_{-\pi}^{\pi} f(x)\ \overline{g(x)}\ dx - \int_{-\pi}^{\pi} s_N(f)\ \overline{g(x)}\ dx\right| = 0$$

Therefore,

$$\lim_{N\to\infty}\frac{1}{2\pi}\int_{-\pi}^{\pi} f(x)\,\overline{g(x)}\,dx = \lim_{N\to\infty}\frac{1}{2\pi}\int_{-\pi}^{\pi} s_N(f)\,\overline{g(x)}\,dx\,.$$

$$\Rightarrow \frac{1}{2\pi}\int_{-\pi}^{\pi} f(x)\,\overline{g(x)}\,dx = \lim_{N\to\infty}\sum_{n=-N}^{N} c_n \overline{\gamma_n}$$

$$= \sum_{n=-\infty}^{\infty} c_n \overline{\gamma_n}$$

Therefore, $\dfrac{1}{2\pi}\displaystyle\int_{-\pi}^{\pi} f(x)\,\overline{g(x)}\,dx = \sum_{n=-\infty}^{\infty} c_n \overline{\gamma_n}$.

Hence part (ii) is proved.

iii) By part (ii) we have

$$\frac{1}{2\pi}\int_{-\pi}^{\pi} f(x)\,\overline{g(x)}\,dx = \sum_{n=-\infty}^{\infty} c_n \overline{\gamma_n}$$

Take $g = f$.

$$\frac{1}{2\pi}\int_{-\pi}^{\pi} f(x)\,\overline{f(x)}\,dx = \sum_{n=-\infty}^{\infty} c_n \overline{c_n}\,.$$

$$\Rightarrow \frac{1}{2\pi}\int_{-\pi}^{\pi}|f(x)|^2 dx = \sum_{n=-\infty}^{\infty}|c_n|^2 .$$

Hence part (iii) is proved.
Hence the theorem.

7.7 Gamma function

Definition 7.7.1

For $0 < x < \infty$, the *gamma function* is defined as

$$\Gamma(x) = \int_0^{\infty} t^{x-1}e^{-t}dt .$$

Definition 7.7.2

A real valued function *f* defined on (*a*, *b*) is said to be *convex*, if
$f(\lambda x+(1-\lambda)y) \le \lambda f(x)+(1-\lambda)f(y), \quad \forall x, y \in (a,b)$ and $0 < \lambda < 1$.

Theorem 7.7.3

a) $\Gamma(x+1) = x\Gamma(x), \quad 0 < x < \infty$.

b) $\Gamma(n+1) = n!, \qquad$ for $n = 1,2,.......$

c) $\log\Gamma$ is convex on $(0,\infty)$.

Proof

a) To prove: $\Gamma(x+1) = x\Gamma(x), \quad 0 < x < \infty$

$$\Gamma(x+1) = \int_0^{\infty} t^{x}e^{-t}dt .$$

By taking $u = t^x$ and $dv = e^{-t}dt$, we get

$du = xt^{x-1}dt$ and $v = -e^{-t}$.

By using an integration by parts, we get

$$\int_0^{\infty} t^x e^{-t}dt = \left[-e^{-t}t^x\right]_0^{\infty} + \int_0^{\infty} e^{-t}xt^{x-1}dt .$$

$$\Rightarrow \Gamma(x+1) = x\int_0^{\infty} t^{x-1}e^{-t}dt$$

$\Rightarrow \Gamma(x+1) = x\Gamma(x)$.

Hence part (a) is proved.

b) To prove: $\Gamma(n+1) = n!$

By (a), we have

$$\Gamma(n+1) = n\Gamma(n).$$
$$= n(n-1)\Gamma(n-1).$$
$$= n(n-1)(n-2)..........1.\Gamma(1) \qquad .... (7.7.3.1)$$

Now, $\Gamma(1) = \int_0^\infty t^0 e^{-t} dt = 1$.

Therefore, equation (7.7.3.1) becomes,

$$\Gamma(n+1) = n!$$

Hence part (b) is proved.

c) To prove: log Γ is convex on $(0,\infty)$.

Let $x, y \in (0,\infty), \quad 0 < \lambda < 1$.

To prove: $\log\Gamma(\lambda x + (1-\lambda)y) \le \lambda \log\Gamma(x) + (1-\lambda)\log\Gamma(y)$.

Let $p = \frac{1}{\lambda}$.

$$\Rightarrow \frac{1}{p} = \lambda \quad \text{and} \quad \frac{1}{q} = 1-\lambda.$$

Now, since $0 < \lambda < 1$, we have $\infty > \frac{1}{\lambda} > 1$.

$$1 < \frac{1}{\lambda} < \infty.$$

Therefore, $\frac{1}{p} + \frac{1}{q} = 1$.

Now, $$\Gamma\left(\frac{x}{p} + \frac{y}{q}\right) = \int_0^\infty t^{\frac{x}{p}+\frac{y}{q}-1} e^{-t} dt.$$

$$= \int_0^\infty t^{\frac{x}{p}+\frac{y}{q}-\left(\frac{1}{p}+\frac{1}{q}\right)} e^{-t\left(\frac{1}{p}+\frac{1}{q}\right)} dt .$$

$$= \int_0^\infty \left(t^{\frac{1}{p}(x-1)} e^{-\left(\frac{1}{p}\right)t} \right)\left(t^{\frac{1}{q}(y-1)} e^{-\left(\frac{1}{q}\right)t} \right) dt .$$

$$\leq \left[\int_0^\infty \left(t^{\frac{1}{p}(x-1)} e^{-\left(\frac{1}{p}\right)t} \right)^p dt \right]^{1/p} \left[\int_0^\infty \left(t^{\frac{1}{q}(x-1)} e^{-\left(\frac{1}{q}\right)t} \right)^q dt \right]^{1/q} .$$

$$\leq \left[\int_0^\infty \left(t^{(x-1)} e^{-t} \right) dt \right]^{1/p} \left[\int_0^\infty \left(t^{(x-1)} e^{-t} \right)^q dt \right]^{1/q} .$$

$$= (\Gamma x)^{1/p} (\Gamma y)^{1/q} .$$

Taking log on both sides, we get

$$\log \Gamma\left(\frac{x}{p} + \frac{y}{q} \right) \leq \frac{1}{p} \log \Gamma x + \frac{1}{q} \log \Gamma y .$$

$$\log \Gamma(\lambda x + (1-\lambda) y) \leq \lambda \log \Gamma x + (1-\lambda) \log \Gamma y .$$

Therefore, log Γ is convex on $(0,\infty)$.

Hence the theorem.

Theorem 7.7.4

Let *f* be a positive function defined on $(0,\infty)$ such that

a) $f(x+1) = x f(x)$

b) $f(1) = 1$

c) log f is convex.

Then, $f(x) = \Gamma(x)$ on $(0,\infty)$.

Proof

Since $\Gamma(x+1) = x\Gamma x$, $\Gamma(1) = 1$ and $\log \Gamma$ is convex on $(0,\infty)$, Γ satisfies (a), (b) and (c).

To prove the theorem, it is enough if we prove that, the conditions (a), (b) and (c) determines a unique function.

Let $\phi = \log f$ on $(0, \infty)$.

$$\phi(x+1) = \log f(x+1)$$

$$= \log xf(x) \qquad \text{[By (a)]}$$

$$= \log x + \log f(x)$$

$$\Rightarrow \phi(x+1) = \log x + \phi(x) \qquad \text{............ (7.7.4.1)}$$

Let n be a positive integer.

Therefore,

$$\phi(n+1) = \log n + \phi(n)$$

$$= \log n + \log(n-1) + \ldots\ldots + \log 1 + \phi(1) \quad \text{[By equation (7.7.4.1)]}$$

$$= \log n! + \log f(1)$$

$$= \log n! + \log 1. \text{ [By (b)]}$$

Therefore,

$$\phi(n+1) = \log n! \qquad \text{............ (7.7.4.2)}$$

Let n be a positive integer and $0 < x < 1$.

Claim: $\log n \le \dfrac{\phi(n+1+x) - \phi(n+1)}{x} \le \log(n+1)$.

$$\phi(n+1) = \phi\left(\frac{(1+x)(n+1)}{(1+x)}\right)$$

$$= \phi\left(\frac{1+x+n+nx}{1+x}\right)$$

$$= \phi\left(\frac{1+x+n}{1+x} + \frac{nx}{1+x}\right)$$

Put $\lambda = \dfrac{1}{1+x}$, $0 < \lambda < 1$.

$$1 - \lambda = 1 - \frac{1}{1+x} = \frac{x}{1+x}$$

Since ϕ is convex,

$$\phi(n+1) \le \frac{1}{1+x}\phi(n+1+x) + \frac{x}{1+x}\phi(n)$$

$$\Rightarrow (1+x)\phi(n+1) \le \phi(n+1+x) + x\phi(n)$$

$$\Rightarrow x(\phi(n+1) - \phi(n)) \le \phi(n+1+x) - \phi(n+1)$$

$$\Rightarrow x(\log n! - \log(n-1)!) \le \phi(n+1+x) - \phi(n+1) \qquad \text{[By equation (7.7.4.2)]}$$

$$\Rightarrow \log n \le \frac{\phi(n+1+x)-\phi(n+1)}{x}.$$

Similarly, we can prove that

$$\frac{\phi(n+1+x)-\phi(n+1)}{x} \le \log(n+1)$$

Therefore,

$$\log n \le \frac{\phi(n+1+x)-\phi(n+1)}{x} \le \log(n+1)\,. \qquad \text{............ (7.7.4.3)}$$

Our claim is proved.

From equation (7.7.4.1), we have

$$\phi(n+1+x) = \log(n+x) + \phi(n+x)$$

$$= \log(n+x) + \log((n-1)+x) + ... + \log(1+x) + \log x + \phi(x)$$

$$= \log[x(1+x).........(n+x)] + \phi(x)\,. \qquad \text{............ (7.7.4.4)}$$

From equation (7.7.4.3), we get

$$x\log n \le \phi(n+1+x) - \phi(n+1) \le x\log(n+1) \qquad \text{............ (7.7.4.5)}$$

$$\Rightarrow 0 \le \phi(n+1+x) - \phi(n+1) - x\log n$$

$$\Rightarrow 0 \le x\log(n+1) - x\log n \quad \text{[By equation (7.7.4.5)]}$$

$$\Rightarrow 0 \le x\log\left(1+\frac{1}{n}\right) \qquad \text{............ (7.7.4.6)}$$

Also, we have

$$0 \le \phi(n+1+x) - \phi(n+1) - x\log n$$

$$= \log[x(1+x)......(n+x)] + \phi(x) - \log n! - \log n^x$$

[By equations (7.7.4.2) & (7.7.4.4)]

$$= \phi(x) - \log\frac{n!n^x}{x(1+x)....(n+x)}$$

$$\le x\log\left(1+\frac{1}{n}\right) \text{[By equation (7.7.4.6)]}$$

We know that,

$$\lim_{n\to\infty} x\log\left(1+\frac{1}{n}\right)=0.$$

Therefore,

$$\lim_{n\to\infty}\left[\phi(x)-\log\frac{n!n^x}{x(1+x)....(n+x)}\right]=0.$$

$$\Rightarrow \phi(x)=\lim_{n\to\infty}\left[\log\frac{n!n^x}{x(1+x)....(n+x)}\right] \qquad .... (7.7.4.7)$$

Since the limit of a convergent sequence is unique, $\phi(x)$ is uniquely determined.

Therefore, $\log f(x)$ is uniquely determined.

Therefore, $f(x)$ is uniquely determined by conditions (a), (b) and (c).

Therefore, $f(x)=\Gamma x, \quad 0<x<\infty$.

Hence the theorem.

Note 7.7.5

By equation (7.7.4.7) of the above theorem (7.7.4), we have

$$\phi(x)=\lim_{n\to\infty}\log\left(\frac{n!n^x}{x(1+x).......(n+x)}\right)$$

$$\Rightarrow \log\Gamma(x)=\lim_{n\to\infty}\log\left(\frac{n!n^x}{x(1+x).......(n+x)}\right)$$

$$\Rightarrow \Gamma(x)=\lim_{n\to\infty}\left(\frac{n!n^x}{x(1+x).......(n+x)}\right)$$

Definition 7.7.6

For each $x > 0$ and $y > 0$, we define the *Beta function* as

$$\beta(x,y)=\int_0^1 t^{x-1}(1-t)^{y-1}\,dt.$$

Theorem 7.7.7

$$\beta(x, y) = \frac{\Gamma(x)\Gamma(y)}{\Gamma(x+y)}$$

Proof

Claim 1: $\beta(1, y) = \dfrac{1}{y}$.

$$\beta(x, y) = \int_0^1 t^{x-1}(1-t)^{y-1}\,dt\,.$$

Therefore,

$$\beta(1, y) = \int_0^1 (1-t)^{y-1}\,dt\,.$$

$$= \frac{1}{y}.$$

Therefore, $\beta(1, y) = 1/y$.

Claim 2: $\beta(x+1, y) = \dfrac{x}{x+y}\beta(x, y)$.

Now, $\beta(x+1, y) = \displaystyle\int_0^1 t^x (1-t)^{y-1}\,dt$.

$$= \int_0^1 \frac{t^x}{(1-t)^x}(1-t)^{x+y-1}\,dt\,.$$

By taking, $u = \left(\dfrac{t}{1-t}\right)^x$ and $dv = (1-t)^{x+y-1}\,dt$, we get

$$du = \frac{xt^{x-1}}{(1-t)^{x+1}}\,dt \quad \text{and}\, v = -\frac{(1-t)^{x+y}}{x+y}$$

$$\int_0^1 \left(\frac{t}{1-t}\right)^x (1-t)^{x+y-1}\,dt = \left[-\left(\frac{t}{1-t}\right)^x \frac{(1-t)^{x+y}}{x+y}\right]_0^1 - \int_0^1 \frac{(1-t)^{x+y}}{x+y}\frac{xt^{x-1}}{(1-t)^{x+1}}\,dt$$

$$= \frac{x}{x+y}\int_0^1 t^{x-1}(1-t)^{y-1}\,dt$$

$$= \frac{x}{x+y}\beta(x, y).$$

Therefore, $\beta(x+1, y) = \frac{x}{x+y}\beta(x, y)$.

Claim 3: For each y, $\log\beta(x, y)$ is a convex function.

Let $(x_1, x_2) \in (0, \infty), 0 < \lambda < 1$.

To prove: $\log\beta(\lambda x_1 + (1-\lambda)x_2, y) \le \lambda\log\beta(x_1, y) + (1-\lambda)\log\beta(x_2, y)$.

Let $\frac{1}{p} = \lambda, \quad 1 < p < \infty$ and $\frac{1}{q} = 1-\lambda$.

Therefore, $\frac{1}{p} + \frac{1}{q} = 1$.

Now, $\beta\left(\frac{x_1}{p} + \frac{x_2}{q}, y\right) = \int_0^1 t^{\frac{x_1}{p}+\frac{x_2}{q}-1}(1-t)^{y-1}\,dt$

$$= \int_0^1 t^{\frac{x_1}{p}+\frac{x_2}{q}-\left(\frac{1}{p}+\frac{1}{q}\right)}(1-t)^{y\left(\frac{1}{p}+\frac{1}{q}\right)-\left(\frac{1}{p}+\frac{1}{q}\right)}\,dt$$

$$= \int_0^1 \left[t^{\frac{1}{p}(x_1-1)}(1-t)^{\frac{1}{p}(y-1)}\right]\left[t^{\frac{1}{q}(x_2-1)}(1-t)^{\frac{1}{q}(y-1)}\right]dt$$

$$\le \left\{\int_0^1 \left[t^{\frac{1}{p}(x_1-1)}(1-t)^{\frac{1}{p}(y-1)}\right]^p dt\right\}^{1/p} \left\{\int_0^1 \left[t^{\frac{1}{q}(x_2-1)}(1-t)^{\frac{1}{q}(y-1)}\right]^q dt\right\}^{1/q}$$

[By Holder's inequality]

$$= \left\{\int_0^1 t^{x_1-1}(1-t)^{y-1}\,dt\right\}\left\{\int_0^1 t^{x_2-1}(1-t)^{y-1}\,dt\right\}$$

$$= [\beta(x_1, y)]^{\lambda}[\beta(x_2, y)]^{1-\lambda}.$$

Therefore,

$$\log\beta(\lambda x_1 + (1-\lambda)x_2, y) \le \lambda\log\beta(x_1, y) + (1-\lambda)\log\beta(x_2, y).$$

Therefore, for each y, $\log\beta$ is a convex function.

Define, $f(x) = \frac{\beta(x, y)\Gamma(x+y)}{\Gamma(y)}$.

We have to prove that, $f(x)=\Gamma(x)$ on $(0,\infty)$.

For that, it is enough to prove that,

i) $f(x + 1) = x f(x)$

ii) $f(1) = 1$

iii) $\log f$ is convex.

i) $$f(x+1)=\frac{\beta(x+1,y)\Gamma(x+y+1)}{\Gamma(y)}.$$

$$=\frac{\frac{x}{x+y}\beta(x,y).(x+y)\Gamma(x+y)}{\Gamma(y)}.$$

$$= x f(x).$$

ii) $$f(1)=\frac{\beta(1,y)\Gamma(1+y)}{\Gamma(y)}.$$

$$=\frac{\frac{1}{y}y\Gamma(y)}{\Gamma(y)}=1.$$

Therefore, $f(1) = 1$.

iii) To prove: $\log f$ is convex.

We have, $\log f=\log\beta(x,y)+\log\Gamma(x+y)-\log\Gamma(y)$.

Let $x_1,x_2\in(0,\infty)$, $0<\lambda<1$.

We have to prove that,

$$\log f(\lambda x_1+(1-\lambda)x_2\le\lambda\log f(x_1)+(1-\lambda)\log f(x_2)$$
$$=\lambda[\log\beta(x_1,y)+\log\Gamma(x_1,y)-\log\Gamma(y)]$$
$$+(1-\lambda)[\log\beta(x_2,y)+\log\Gamma(x_2,y)-\log\Gamma(y)].$$

Now, $$\log f(\lambda x_1+(1-\lambda)x_2)=\log\beta(\lambda x_1+(1-\lambda)x_2,y)$$
$$+\log\Gamma(\lambda x_1+(1-\lambda)x_2+y)-\log\Gamma(y).$$

$$\le\lambda\log\beta(x_1,y)+(1-\lambda)\log\beta(x_2,y)$$
$$+\log\Gamma(\lambda(x_1+y)+(1-\lambda)(x_2+y))-\log\Gamma(y).$$

$$\leq \lambda \log \beta(x_1, y) + (1-\lambda)\log \beta(x_2, y)$$

$$+ \lambda \log \Gamma(x_1 + y) + (1-\lambda)\log \Gamma(x_2 + y) - \log \Gamma(y).$$

[Since log β and log Γ are convex].

$$= \lambda[\log \beta(x_1, y) + \log \Gamma(x_1 + y) - \log \Gamma(y)]$$

$$+ (1-\lambda)[\log \beta(x_2, y) + \log \Gamma(x_2 + y) - \log \Gamma(y)].$$

Therefore, $\log f$ is convex.

Hence, $f(x) = \Gamma(x)$.

$$\Rightarrow \Gamma(x) = \frac{\beta(x, y)\Gamma(x, y)}{\Gamma(y)}.$$

Therefore, $\beta(x, y) = \dfrac{\Gamma(x)\Gamma(y)}{\Gamma(x, y)}$.

Hence the theorem.

Note 7.7.8

1) $\beta(x, y) = 2\int_0^{\pi/2} \sin^{(2x-1)}\theta . \cos^{(2y-1)}\theta d\theta$.

Solution

$$\beta(x, y) = \int_0^1 t^{x-1}(1-t)^{y-1} dt.$$

put $t = \sin^2\theta$ and $1-t = \cos^2\theta$.

$$\Rightarrow dt = 2\sin\theta\cos\theta d\theta.$$

When $t = 0$, sin θ = 0 $\Rightarrow \theta = 0$.

When $t = 1$, sin θ = 1 $\Rightarrow \theta = \pi/2$.

Therefore,

$$\beta(x, y) = \int_0^{\pi/2} (\sin^2\theta)^{x-1}(\cos^2\theta)^{y-1} 2\sin\theta\cos\theta d\theta.$$

$$= 2\int_0^{\pi/2} (\sin^{2x-1}\theta)(\cos^{2y-1}\theta) d\theta \qquad \text{............ (7.7.8.1)}$$

Hence proved.

2) $\Gamma(1/2) = \sqrt{\pi}$.

Solution

Put $x = y = ½$ in equation (7.7.8.1), we get

$$\beta(1/2,1/2) = 2\int_0^{\pi/2} \sin^0\theta\cos^0\theta d\theta = \pi .$$

$$\Rightarrow \frac{\Gamma(1/2)\Gamma(1/2)}{\Gamma((1/2)+(1/2))} = \pi .$$

$$\Rightarrow \Gamma(1/2) = \sqrt{\pi}.$$

Hence proved.

3) $\int_{-\infty}^{\infty} e^{-y^2} dy = \Gamma(1/2).$

Solution

We know that, $\Gamma(x) = \int_0^{\infty} t^{x-1}e^{-t}dt$.

Put $t = y^2 \Rightarrow dt = 2ydy.$

When $t = 0, y = 0.$

When $t = \infty,\ y = \infty$.

Therefore, $\Gamma(x) = \int_0^{\infty} (y^2)^{x-1}e^{-y^2} 2ydy$.

$$= 2\int_0^{\infty} y^{2x-1}e^{-y^2} dy .$$

put $x = ½$.

Therefore, $\Gamma(1/2) = 2\int_0^{\infty} e^{-y^2} dy$.

$$\Rightarrow \Gamma(1/2) = \int_{-\infty}^{\infty} e^{-y^2} dy .$$

Hence proved.

Theorem 7.7.9

This is known as ***Stirling's Formula***.

This provides a simple approximate expression for $\Gamma(x + 1)$, then x is large, (hence for $n!$ when n is large). The formula is

$$\lim_{x\to\infty}\left\{\frac{\Gamma(x+1)}{\left(\frac{x}{e}\right)^x\sqrt{2\pi x}}\right\}=1.$$

Proof

$$\Gamma(x+1)=\int_0^\infty t^x e^{-t}dt.$$

put$t = (1 + u)x. \Rightarrow u=\frac{t-x}{x}.$

Therefore, $du=\frac{1}{x}dt$.

When $t = 0$, $u = -1$.

When $t=\infty,\ u=\infty$.

Therefore, $\Gamma(x+1)=\int_{-1}^{\infty}(1+u)^x x^x e^{-(1+u)x}xdu$.

$$\Rightarrow \Gamma(x+1)=x^{x+1}e^{-x}\int_{-1}^{\infty}\left((1+u)e^{-u}\right)^x du. \qquad \text{...........} (7.7.9.1)$$

When $-1<u<\infty, u\neq 0$, define a function $h(u)$ as,

$$(1+u)e^{-u}=e^{-\frac{u^2}{2}}du. \qquad \text{...........} (7.7.9.2)$$

$h(0) = 1$.

Therefore, $[\log(1+u)]-u=-\frac{u^2}{2}h(u)$.

$$\Rightarrow h(u)=\frac{2}{u^2}[u-\log(1+u)] \qquad \text{...........} (7.7.9.3)$$

It follows that, h is continuous and that $h(u)$ decreases monotonically from $\infty$to 0, as u increases from -1 to ∞.

put $u = s\sqrt{2/x}$ in equation (7.7.9.1), we get

$$du = \sqrt{\frac{2}{x}}ds .$$

When u = -1, $s = -\sqrt{\frac{x}{2}}$.

When $u = \infty$, $s = \infty$.

Therefore, $\Gamma(x+1) = x^{x+1}e^{-x} \int_{-\sqrt{\frac{x}{2}}}^{\infty} \left[e^{-\frac{s^2(2/x)}{2}h\left(s\sqrt{\frac{2}{x}}\right)} \right]^x \sqrt{\frac{2}{x}}ds$.

$$= x^{x+1}e^{-x}\sqrt{\frac{2}{x}} \int_{-\sqrt{\frac{x}{2}}}^{\infty} e^{-s^2h\left(s\sqrt{\frac{2}{x}}\right)}ds.$$

$$= x^x e^{-x}\sqrt{2x} \int_{-\infty}^{\infty} \psi_x(s)ds \qquad \text{........... (7.7.9.4)}$$

where

$$\psi_x(s) = \begin{cases} e^{-s^2h\left(s\sqrt{2/x}\right)}, & \text{if } \sqrt{-x/2} < s < \infty \\ 0, & \text{if } s \le -\sqrt{x/2} \end{cases}$$

Note the following fact about $\psi_x(s)$:

For every s, $\psi_x(s) \to e^{-s^2}$, as $s \to \infty$.

The convergence in (7.7.9.4) is uniform on [-A, A], for every A < ∞.

When $s < 0$, $0 < \psi_x(s) < e^{-s^2}$.

When $s > 0$ and $x > 1$, then $0 < \psi_x(s) < \psi_1(s)$.

$$\int_0^{\infty} \psi_1(s)ds < \infty .$$

By convergence theorem, the integral of equation (7.7.9.4) converges to $\sqrt{\pi}$, as $x \to \infty$.

Therefore, equation (7.7.9.4) becomes,

$$\lim_{x\to\infty}\Gamma(x+1)=\lim_{x\to\infty}x^{x}e^{-x}\sqrt{2x}\sqrt{\pi}\,.$$

$$=\lim_{x\to\infty}\left(\frac{x}{e}\right)^{x}\sqrt{2\pi x}\,.$$

Therefore, $\lim_{x\to\infty}\dfrac{\Gamma(x+1)}{(x/e)^{x}\sqrt{2\pi x}}=1\,.$

Unit	
8	**FUNCTIONS OF SEVERAL VARIABLES - DIFFERENTIATION**

8.1 Introduction

This chapter begins with the discussion of set of vectors in Euclidean space $\mathbb{R}^n$. The algebraic facts presented in this chapter extend without changing to finite-dimensional vector spaces over any field of scalars. The definitions and important theorems in differentiability, partial derivatives, continuously differentiable functions are discussed in this chapter. We will study that if *f* is a differentiable function, then its partial derivatives exist and the second order partial derivatives can be extended to higher order partial derivatives.

8.2 Linear Transformations

Definition 8.2.1

A nonempty $X \subset R^n$ is said to be a *vector space* if

i) $x + y \in X$ and

ii) $cx \in X$ for all $x \in X, y \in Y$ and for all scalars c.

Definition 8.2.2

If $x_1, x_2, ..., x_k \in R^n$ and $c_1, c_2, ..., c_k$ are scalars, the vector $c_1 x_1 + c_2 x_2 + ... + c_k x_k$ is called a *linear combination* of $x_1, x_2, ..., x_k \in R^n$.

Definition 8.2.3

If $S \subset R^n$ and if E is the subset of all linear combinations of elements of S, then we say that S spans E or that E is the *span of S*.

Definition 8.2.4

A set consisting of vectors $x_1, x_2, \ldots\ldots, x_k$ is said to be *linearly independent* if the relation $c_1 x_1 + c_2 x_2 + \ldots\ldots + c_k x_k = 0$ implies that $c_1 = c_2 = \ldots\ldots = c_k = 0$ otherwise, $\{x_1, x_2, \ldots\ldots, x_k\}$ is said to be *linearly dependent.*

Definition 8.2.5

If a vector space *X* contains an independent set of *r* vectors but contains no independent set of $r+1$ vectors, then we say that *X* has dimension *r*. It is written as $\dim X = r$.

Definition 8.2.6

An independent subset of a vector space *X* which spans *X* is called a *basis of X.*

Note 8.2.7

1. The set consisting of 0 alone is a vector space and its dimension is 0.
2. If $B = \{x_1, x_2, \ldots\ldots, x_r\}$ is a basis of X, then B spans X and B is independent. Hence every $x \in X$ has a unique representation of the form $x = c_1 x_1 + c_2 x_2 + \ldots\ldots + c_r x_r$. The numbers $c_1, c_2, \ldots\ldots, c_r$ are called the coordinates of X with respect to the basis B.
3. Let r be a positive integer. If a vector space X is spanned by a set of r vectors, then $\dim X = r$.
4. $\dim R^n = n$.
5. Suppose X is a vector space, and dim. X = n.
   a) A set E of n vectors in X spans X if and only if E is independent.
   b) X has a basis and every basis consists of n vectors.
   c) If $1 \le r \le n$ and $\{y_1, y_2, \ldots\ldots, y_r\}$ is an independent set in X, then X has a basis containing $\{y_1, y_2, \ldots\ldots, y_r\}$.

Problem 8.2.8

Check whether the set $\{e_1, e_2, \ldots\ldots, e_n\}$ forms the basis of R^n.

Solution

The set $\{e_1, e_2, \ldots\ldots, e_n\}$, where e_j is the vector in R^n whose j[th] coordinate is 1 and whose other coordinates are all 0.

That is,

$$e_1 = (1,0,0,\ldots\ldots.,0)$$

$$e_2 = (0,1,0,\ldots\ldots.,0)$$

$$\ldots\ldots\ldots\ldots\ldots\ldots\ldots\ldots$$

$$\ldots\ldots\ldots\ldots\ldots\ldots\ldots\ldots$$

and $e_n = (0,0,0,\ldots\ldots.,1)$

Now, if $x = (x_1, x_2, \ldots\ldots, x_n) \in R^n$, then

$$x = x_1(1,0,0,\ldots..,0) + x_2(0,1,0,\ldots..,0) + \ldots\ldots. + x_n(0,0,0,\ldots..,1)$$

$$= x_1e_1 + x_2e_2 + \ldots\ldots + x_ne_n.$$

Hence $\{e_1, e_2, \ldots\ldots, e_n\}$ forms a basis of R^n.

The set $\{e_1, e_2, \ldots\ldots, e_n\}$ is called the standard basis of R^n.

Definition 8.2.9

A mapping A of a vector space X into a vector space Y is said to be a *linear transformation* if

i) $A(x_1 + x_2) = A(x_1) + A(x_2)$ and

ii) $A(cx) = cA(x)$ for all $x, x_1, x_2 \in X$ and all scalars c.

If A is linear, then $A(x)$ is simply written as Ax.

If A is linear, then $A0 = 0$.

Note 8.2.10

A linear transformation A of X into Y is completely determined by its action on any basis.

Proof

If $\{x_1, x_2, \ldots\ldots, x_n\}$ is a basis of X, then every $x \in X$ has a unique representation of the form

$$x = \sum_{i=1}^{n} c_i x_i \ .$$

Since A is linear,

$$Ax = A\sum_{i=1}^{n} c_i x_i$$

$$= \sum_{i=1}^{n} c_i A x_i \ .$$

Hence Ax can be computed from the vectors Ax_1, Ax_2, ..., Ax_n and the coordinates c_1, c_2, ..., c_n.

That is, $R(A)$ is the span of the set $Q = \{Ax_1, Ax_2, ..., Ax_n\}$.

Definition 8.2.11

Linear transformations of X into X are called *linear operators* on X.

Definition 8.2.12

If A is a linear operator on X such that

i) A is one – to - one

ii) A maps X onto X,

then A is said to be *invertible* and the operator A^{-1} on X is defined such that $A^{-1}(Ax) = x$ for all $x \in X$.

Theorem 8.2.13

A linear operator A on a finite dimensional vector space X is one-to-one if and only if the range of A is all X.

Proof

Let X be a vector space of dimension n and $\{x_1, x_2, ..., x_n\}$ be a basis of X.

Since A is linear, $R(A)$ is the span of the set $Q = \{Ax_1, Ax_2, ..., Ax_n\}$.

We know that, "A set E of n vectors in X spans X if and only if E is independent".

Hence $R(A) = X$ if and only if Q is independent.

Hence, it is enough to prove that Q is independent if and only if A is one-to-one.

To prove;

Q is independent if and only if A is one-to-one.

Assume that, A is one-to-one.

Suppose $\sum_{i=1}^{n} c_i A x_i = 0$.

Then, $A\left(\sum_{i=1}^{n} c_i x_i\right) = \sum_{i=1}^{n} c_i A x_i = 0$ [Since A is linear].

$\Rightarrow \sum_{i=1}^{n} c_i x_i = 0$ [Since A is one-to-one].

$\Rightarrow c_1 = c_2 = ... = c_n = 0$ [Since $\{x_1, x_2, ..., x_n\}$ is linearly independent].

Hence $Q = \{Ax_1, Ax_2, ..., Ax_n\}$ is linearly independent.

Conversely,

Assume that Q is linearly independent.

Suppose if,

$$A\left(\sum_{i=1}^{n} c_i x_i\right) = 0.$$

Then, $\sum_{i=1}^{n} c_i A x_i = 0$ [Since A in linear].

$\Rightarrow c_1 = c_2 = ... = c_n = 0$ [Since Q is independent].

$\Rightarrow x = \sum_{i=1}^{n} c_i x_i = 0$.

That is, $Ax = 0$, only if $x = 0$. ....(8.2.13.1)

Therefore, if $Ax = Ay$, then $A(x - y) = Ax - Ay = 0$.

By equation (8.2.13.1), $x - y = 0$.

That is, $x = y$.

Hence, A is one-to-one.

Therefore, A is one-to-one if and only if $R(A) = X$.

Hence the theorem.

Definition 8.2.14

$L(X, Y)$ denote the set of all linear transformations of the vector space X into the vector space Y. $L(X, X)$ can be simply written as $L(X)$. That is, $L(X)$ denotes the set of all operators on X.

Definition 8.2.15

If $A_1, A_2 \in L(X,Y)$ and if c_1, c_2 are scalars, $c_1A_1 + c_2A_2$ is defined by

$$(c_1A_1 + c_2A_2)(x) = c_1A_1x + c_2A_2x \text{ for } x \in X.$$

Since A_1 and A_2 are linear, $c_1A_1 + c_2A_2$ is also linear. Hence, $c_1A_1 + c_2A_2 \in L(X,Y)$.

Definition 8.2.16

If X, Y, Z are vector spaces and if $A \in L(X,Y), B \in L(Y,Z)$, their product BA is defined to be the composition of A and B.

That is, $(BA)(x) = B(Ax)$, for all $x \in X$. Then, $BA \in L(X,Z)$.

Definition 8.2.17

For $A \in L(R^n, R^m)$, the *norm of A* is defined as

$$\|A\| = \sup.\{|A(x)|, x \in R^n \text{ with } |x| \le 1\}.$$

Note 8.2.18

$$|Ax| \le \|A\| |x|, \text{ for all } x \in R^n.$$

Proof

Let $x \in R^n$.

Define $y = \dfrac{x}{|x|}$.

Then, $|y| = 1$.

Therefore, by definition of $\|A\|$, $|A(y)| \le \|A\|$.

$$\Rightarrow \left| A\left(\frac{x}{|x|}\right) \right| \le \|A\|.$$

$$\Rightarrow \frac{|A(x)|}{|x|} \le \|A\|.$$

$$\Rightarrow |A(x)| \le \|A\| |x|, \text{ for all } x \in R^n.$$

Hence proved.

Also, if λ is such that $|Ax| \le \lambda |x|$, for all $x \in R^n$, then $\|A\| \le \lambda$.

Theorem 8.2.19

a) If $A \in L(R^n, R^m)$, then $\|A\| < \infty$ (that is, A is finite) and A is a uniformly continuous mapping of R^n into R^m.

b) If $A, B \in L(R^n, R^m)$ and c is a scalar, then $\|A+B\| \le \|A\| + \|B\|$, $\|cA\| = |c| \|A\|$.

With the distance between A and B as $\|A-B\|$, $L(R^n, R^m)$ is a metric space.

c) If $A \in L(R^n, R^m), B \in L(R^m, R^k)$, then $\|BA\| \le \|B\| \|A\|$.

Proof

a) Let $A \in L(R^n, R^m)$.

Let $\{e_1, e_2, \ldots\ldots, e_n\}$ be a basis of R^n.

Suppose that $x = \sum_{i=1}^{n} c_i e_i$ with $\|x\| \le 1$.

Then, $|c_i| \le 1$ for all i = 1, 2, ….., n.

$$\text{Therefore, } |Ax| = \left| A\left(\sum_{i=1}^{n} c_i e_i \right) \right|$$

$$= \left| \sum_{i=1}^{n} c_i A e_i \right|$$

$$\le \sum_{i=1}^{n} c_i |Ae_i|$$

$$\le \sum_{i=1}^{n} |Ae_i|$$

Therefore, $\sup\{|A(x)|, x \in R^n \text{ with } |x| \leq 1\} \leq \sum_{i=1}^{n} Ae_i < \infty$.

$$\Rightarrow \|A\| < \infty.$$

To prove that A is uniformly continuous.

We know that, $|A(x)| \leq \|A\||x|$, for all $x \in R^n$.

Therefore, $|Ax - Ay| \leq \|A\||x - y|$, for all $x, y \in R^n$.

Therefore, for any $\varepsilon > 0$, if we choose $\delta = \dfrac{\varepsilon}{\|A\|}$ (independent of x and y), then

$$|x - y| < \delta \Rightarrow |Ax - Ay| \leq \|A\||x - y| < \varepsilon.$$

Hence, for any $\varepsilon > 0$, there exists a $\delta > 0$, such that $|Ax - Ay| < \varepsilon$, $|x - y| < \delta$, $x, y \in R^n$.

Hence, A is uniformly continuous.

b) Consider, $|(A + B)x| = |Ax + Bx| \leq |Ax| + |Bx| \leq (\|A\| + \|B\|)|x|$.

Therefore, if $|x| \leq 1$, then $|(A + B)x| \leq \|A\| + \|B\|$.

Therefore,

$$\sup\{|(A + B)(x)|,\ x \in R^n, |x| \leq 1\} \leq \|A\| + \|B\|.$$

$$\|A + B\| \leq \|A\| + \|B\|.$$

Consider, $|(cA)x| = |c(Ax)|$

$$= |c||Ax|.$$

Therefore, $\sup\{|(cA)(x), x \in R^n \text{ and } |x| \leq 1|\} = |c| \sup\{|Ax|, x \in R^n \text{ and } |x| \leq 1\}$

$$\|cA\| = |c|\|A\|.$$

To prove; $L(R^n, R^m)$ is a metric space.

Define the metric on $L(R^n, R^m)$ as follows;

If $A, B \in L(R^n, R^m)$, define the distance between A and B as $\|A - B\|$.

Then,

i) $\|A-B\| = \text{Sup}\{|(A-B)(x)|, x \in R^n \text{ and } |x| \leq 1\} > 0$.

ii) $\|A-B\| = 0 \Leftrightarrow \sup \{|(A-B)(x)|, x \in R^n \text{ and } |x| \leq 1\} = 0$

$$\Leftrightarrow |(A-B)(x)| = 0, \text{ for all } x \in R^n \text{ with } |x| \leq 1$$

$$\Leftrightarrow |Ax - Bx| = 0, \text{ for all } x \in R^n \text{ with } |x| \leq 1$$

$$\Leftrightarrow Ax = Bx, \quad \text{for all } x \in R^n \text{ with } |x| \leq 1 \qquad \text{............ (8.2.19.1)}$$

Suppose if $y \in R^n$, with $y \neq 0$, then

$x = \dfrac{y}{|y|}$ is such that $|x| = 1$.

Therefore, $Ax = Bx \Rightarrow A\left(\dfrac{y}{|y|}\right) = B\left(\dfrac{y}{|y|}\right)$

$$\Rightarrow \frac{1}{|y|} Ay = \frac{1}{|y|} By$$

$\Rightarrow Ay = By$ [Since $|y| \neq 0$, we can divide by $|y|$].

Using this in equation (8.2.19.1), we get

$$\|A-B\| = 0 \Leftrightarrow Ay = By, \qquad \text{for all } y \in R^n.$$

Hence $\|A-B\| = 0 \Leftrightarrow A = B$.

iii) $\|A-B\| = \sup\{|(A-B)(x)|, x \in R^n \text{ and } |x| \leq 1\}$

$$= \sup\{|(B-A)(x)|, x \in R^n \text{ and } |x| \leq 1\}$$

$$= \|B-A\|.$$

iv) $\|A-B\| = \sup\{|(A-B)(x)|, x \in R^n \text{ and } |x| \leq 1\}$

$$= \sup\{|Ax - Bx|, x \in R^n \text{ and } |x| \leq 1\}$$

$$= \sup\{|Ax - Cx + Cx - Bx|, x \in R^n \text{ and } |x| \leq 1\}$$

$$\leq \sup\{|Ax - Cx|, x \in R^n \text{ and } |x| \leq 1\} + \sup\{|Cx - Bx|, x \in R^n \text{ and } |x| \leq 1\}$$

$$= \|A-C\| + \|C-B\|.$$

Hence the triangle inequality is satisfied.

Hence the metric we defined on $L(R^n, R^m)$ satisfies all the required properties.

Hence (b) is proved.

c) To prove that, if $A \in L(R^n, R^m)$ and $B \in L(R^m, R^k)$, then

$$\|BA\| = \|B\|\|A\|.$$

Consider, $|(BA)(x)| = |B(Ax)|$

$$= \|B\| |Ax|$$

$$= \|B\| \|A\| |x|$$

Therefore, $\sup\{|(BA)(x)|, x \in R^n \text{ and } |x| \leq 1\} = \|B\| \ \|\text{A}\|$.

$$\Rightarrow \|BA\| = \|B\| \|A\|.$$

Hence (c) is proved.

Hence the theorem.

Problems 8.2.20

1. If $A_1, A_2 \in L(X, Y)$ and if c_1, c_2 are scalars, prove that $c_1A_1 + c_2A_2$ is linear, that is, $c_1A_1 + c_2A_2 \in L(X, Y)$.

Solution

For $x_1, x_2 \in X$ and scalars a, b, consider

$$(c_1A_1 + c_2A_2)(ax_1 + bx_2) = c_1A_1(ax_1 + bx_2) + c_2A_2(ax_1 + bx_2)$$

[By the definition of $c_1A_1 + c_2A_2$]

$$= c_1(aA_1x_1 + bA_1x_2) + c_2(aA_2x_1 + bA_2x_2)$$

$$= c_1aA_1x_1 + c_1bA_1x_2 + c_2aA_2x_1 + c_2bA_2x_2$$

$$= a(c_1A_1 + c_2A_2)(x_1) + b(c_1A_1 + c_2A_2)(x_2)$$

Hence, $c_1A_1 + c_2A_2$ is linear.

2. If $X = Y = Z$, is $BA = AB$?

Solution

BA need not be the same as AB.

Theorem 8.2.21

Let Ω be the set of all invertible linear operators on R^n.

i) If $A \in \Omega, B \in L(R^n)$ and $\|B - A\| \|A^{-1}\| < 1$, then $B \in \Omega$.

ii) Ω is an subset of $L(R^n)$ and the mapping $A \to A^{-1}$ is continuous on Ω. [This mapping is also a one-to-one mapping of Ω onto Ω, which is its own inverse.]

Proof

i) Let $\|A^{-1}\| = 1/\alpha$ and

$$\|B - A\| = \beta.$$

Then,

$$\|B - A\|\|A^{-1}\| < 1 \text{ implies } \beta/\alpha < 1, \text{ that is, } \beta < \alpha.$$

For every $x \in R^n$,

$$\begin{aligned}
&\alpha|x| = \alpha|A^{-1}Ax| \\
&\le \alpha\|A^{-1}\| \, |Ax| \\
&= |Ax| = |Ax - Bx + Bx| \\
&\le |Ax - Bx| + |Bx| \\
&= |(A - B)(x)| + |Bx| \\
&\le \|A - B\||x| + |Bx| \\
&\le \beta|x| + |Bx| \\
&(\alpha - \beta)|x| \le |Bx|, \qquad \forall x \in R^n
\end{aligned} \qquad \text{............ (8.2.21.1)}$$

Since $\alpha - \beta > 0$, from (8.2.21.1) we get that,

$$Bx \neq 0, \text{ if } x \neq 0.$$

Hence, B is 1-1.

Therefore, by theorem 8.2.13, $\mathcal{R}(B) = R^n$.

Hence B is onto.

Therefore, B is both 1-1 and onto.

Hence B is invertible.

Hence, $B \in \Omega$.

ii) To prove that the mapping $A \to A^{-1}$ is continuous on Ω.

Replacing x by $B^{-1}y$ in (8.2.21.1), we get

$$(\alpha - \beta)|B^{-1}y| \le |B(B^{-1}y)| = |y|, \qquad \forall y \in R^n.$$

Therefore,

$$|B^{-1}y| \le \frac{|y|}{\alpha - \beta}$$

$$\Rightarrow \sup\left\{|B^{-1}y| \,/\, y \in R^n \text{ and } |y| \le 1\right\} \le \frac{1}{\alpha - \beta}$$

$$\Rightarrow \|B^{-1}\| \le \frac{1}{\alpha - \beta}$$

Consider the identity,

$$\begin{aligned} B^{-1} - A^{-1} &= B^{-1}AA^{-1} - B^{-1}BA^{-1} \\ &= B^{-1}(A - B)A^{-1} \end{aligned}$$

$$\begin{aligned} \Rightarrow \|B^{-1} - A^{-1}\| &= \|B^{-1}(A - B)A^{-1}\| \\ &\le \|B^{-1}\|\,\|A - B\|\,\|A^{-1}\| \\ &\le \frac{\beta}{\alpha(\alpha - \beta)} \end{aligned}$$

$\Rightarrow \|B^{-1} - A^{-1}\| \to 0$ as $\beta > 0$, that is, as $\|B - A\| \to 0$.

Hence the mapping $A \to A^{-1}$ is continuous on Ω.

Hence the theorem.

Matrix representation of a linear transformation 8.2.22

Suppose $\{x_1, x_2, \ldots\ldots, x_n\}$ and $\{y_1, y_2, \ldots\ldots, y_m\}$ are basis of the vector spaces X and Y respectively.

Let $A \in L(X,Y)$.

Since $Ax_j \in Y$ for $j = 1, 2, \ldots\ldots, n$, and $\{y_1, y_2, \ldots\ldots, y_m\}$ is a basis of Y,

$$Ax_j = \sum_{i=1}^{n} a_{ij} y_j \text{, for } j = 1, 2, \ldots\ldots, n. \qquad \ldots\ldots\ldots\ldots (8.2.22.1)$$

These numbers can be put into a rectangular array of m rows and n columns, called m by n matrix as follows;

$$[A] = \begin{bmatrix} a_{11} & a_{12} & \ldots & \ldots & a_{1n} \\ a_{21} & a_{22} & \ldots & \ldots & a_{2n} \\ \ldots & \ldots & \ldots & \ldots & \ldots \\ \ldots & \ldots & \ldots & \ldots & \ldots \\ a_{m1} & a_{m2} & \ldots & \ldots & a_{mn} \end{bmatrix}$$

where the coordinates a_{ij} of the vectors Ax_j appear in the jth column of $[A]$. Therefore, the vectors Ax_j are called column vectors of $[A]$.

With this terminology, the range of A is said to be *spanned by the column vectors of* $[A]$.

If $x = \sum_{j=1}^{n} c_j x_j$, then

$$\begin{aligned} Ax &= A\left(\sum_{j=1}^{n} c_j x_j\right) \\ &= \sum_{j=1}^{n} c_j Ax_j \\ &= \sum_{j=1}^{n} c_j \left(\sum_{i=1}^{m} a_{ij} y_i\right) \\ &= \sum_{j=1}^{n} \left(\sum_{i=1}^{m} a_{ij} c_j\right) y_i \end{aligned}$$

Therefore,

$$Ax = \sum_{j=1}^{n} \left(\sum_{i=1}^{m} a_{ij} c_j\right) y_i. \qquad \ldots\ldots\ldots\ldots (8.2.22.2)$$

Thus the coordinates of Ax are $\sum_{j=1}^{n} a_{ij} c_j$.

Suppose a m by n matrix is given with real entries a_{ij} and if A is defined as in (8.2.22.2), then $A \in L(X,Y)$ and $[A]$ will be the given matrix.

Hence, there is a one-to-one correspondence between $L(X, Y)$ and the set of all m by n matrices.

Note 8.2.23

$[A]$ depends not only on A but also on the choice of bases in X and Y. The same A may give rise to many different matrices, if we change bases and vice versa.

8.3 Differentiation

Definition 8.3.1

Let us recall the definition of differentiation discussed in chapter 5. If f is a real function with domain $(a,b) \subset R^1$ and if $x \in (a,b)$, then $f'(x)$ is defined to be the real number

$$\lim_{h \to 0} \frac{f(x+h) - f(x)}{h} \qquad \text{............ (8.3.1.1)}$$

provided that this limit exists. Thus

$$f(x+h) - f(x) = f'(x)h + r(h) \qquad \text{............ (8.3.1.2)}$$

where the remainder $r(h)$ is small, such that

$$\lim_{h \to 0} \frac{r(h)}{h} = 0 . \qquad \text{............ (8.3.1.3)}$$

Definition 8.3.2

Suppose E is an open set in R^n, f maps E into R^m and $x \in E$. If there exists a linear transformation A of R^n into R^m such that

$$\lim_{h \to 0} \frac{|f(x+h) - f(x) - Ah|}{|h|} = 0 , \qquad \text{............ (8.3.2.1)}$$

then we say that f is *differentiable at* x and we write $f'(x) = A$.

If f is differentiable at every $x \in E$, we say that f is *differentiable in E.*

Note 8.3.3

The relation (8.3.2.1) can be rewritten in the form

$$f(x+h)-f(x)=f'(x)h+r(h)$$

where the remainder $r(h)$ satisfies $\lim\limits_{h\to 0}\dfrac{r(h)}{h}=0$.

Suppose E is an open set in R^n, f maps E into R^m and $x\in E$ and f is differentiable in E. For every $x\in E, f'(x)$ is then a function, namely, a linear transformation of R^n into R^m. But f' is also a function which maps E into $L(R^n, R^m)$.

f is continuous at any point at which f is differentiable.

Theorem 8.3.4

Suppose E is an open set in R^n, f maps E into R^m, f is differentiable at $x_0 \in E$, g maps an open set containing $f(E)$ into R^k, and g is differentiable at $f(x_0)$. Then, the mapping F of E into R^k defined by

$$F(x) = g(f(x))$$

is differentiable at x_0, and

$$F'(x_0)=g'(f(x_0))f'(x_0). \qquad \text{............ (8.3.4.1)}$$

Proof

Put $y_0=f(x_0), A=f'(x_0), B=g'(y_0)$ and

Define $u(h)=f(x_0+h)-f(x_0)-Ah$,

$$v(k)=g(y_0+h)-g(y_0)-Bk,$$

for all $h\in R^n$ and $k\in R^m$ for which $f(x_0+h)$ and $g(y_0+k)$ are defined.

Then,

$$|u(h)|=\varepsilon(h)|h|, \qquad |v(k)|=\eta(k)|k|, \qquad \text{............ (8.3.4.2)}$$

where $\varepsilon(h)\to 0$ as $h\to 0$ and $\eta(k)\to 0$ as $k\to 0$.

Given h, put $k=f(x_0+h)-f(x_0)$.

Then

$$|k|=|Ah+u(h)|\le[\|A\|+\varepsilon(h)]|h|, \qquad \text{............ (8.3.4.3)}$$

and

$$F(x_0+h)-F(x_0)-BAh = g(y_0+k)-g(y_0)-BAh$$
$$= B(k-Ah)+v(k)$$
$$= Bu(h)+v(k)\,.$$

Hence, equations (8.3.4.2) and (8.3.4.3) imply, for $h \neq 0$,

$$\frac{|F(x_0+h)-F(x_0)-BAh|}{|h|} \leq \|B\|\varepsilon(h) + [\|A\| + \varepsilon(h)]\eta(k)\,.$$

Let $h \to 0$. Then $\varepsilon(h) \to 0$.

Also, by equation (8.3.4.3), $\eta(k) \to 0$ as $k \to 0$.

$$\Rightarrow F'(x_0) = BA.$$

Hence the theorem.

8.4 Partial Derivatives

Definition 8.4.1

Consider a function f that maps an open set $E \subset R^n$ into R^m. Let $\{e_1, e_2, \ldots\ldots, e_n\}$ and $\{u_1, u_2, \ldots\ldots, u_m\}$ be the standard bases of R^n and R^m. The components of f are real functions $f_1, f_2, \ldots., f_m$ defined by

$$f(x) = \sum_{i=1}^{m} f_i(x)u_i, \qquad x \in E\,.$$

or equivalently, by $f_i(x) = f(x)u_i$, $1 \leq i \leq m$.

For $x \in E$, $1 \leq i \leq m$, $1 \leq j \leq n$, we define

$$(D_j f_i)(x) = \lim_{t \to 0} \frac{f_i(x+te_j) - f_i(x)}{t},$$

provided the limit exists. When we write $f_i(x_1, x_2, \ldots\ldots\ldots, x_n)$ in the place of $f_i(x)$, we say that $D_j f_i$ is the derivative of f_i with respect to x_j, keeping the variables fixed. $D_j f_i$ is called a *paritial derivative*.

Note 8.4.2

If f is differentiable at a point x, then its partial derivatives exist at x and they determine the linear transformation $f'(x)$ completely.

Theorem 8.4.3

Suppose f maps an open set $E \subset R^n$ into R^m, and f is differentiable at a point $x \in E$. Let $\{e_1, e_2, ..., e_n\}$ and $\{u_1, u_2, ..., u_m\}$ be the standard bases of R^n and R^m. Then the partial derivatives $(D_j f_i)(x)$ exist, and

$$f'(x)e_j = \sum_{i=1}^{m} (D_j f_i)(x) u_i, \qquad 1 \le j \le n.$$

Proof

Let us fix j. Given that f is differentiable at x. Therefore, we have

$$f(x+te_j) - f(x) = f'(x)(te_j) + r(te_j)$$

where $\frac{|r(te_j)|}{t} \to 0$ as $t \to 0$. Therefore, the linearity of $f'(x)$ shows that

$$\lim_{t \to 0} \frac{f(x+te_j) - f(x)}{t} = f'(x)e_j. \qquad .... (8.4.3.1)$$

Now, if we represent f in terms of its components, then (1) becomes

$$\lim_{t \to 0} \sum_{i=1}^{m} \frac{f(x+te_j) - f(x)}{t} u_i = f'(x)e_j.$$

It follows that as $t \to 0$, each quotient in this sum has a limit, so that each $(D_j f_i)(x)$ exists.

Therefore,

$$f'(x)e_j = \sum_{i=1}^{m} (D_j f_i)(x) u_i, \qquad 1 \le j \le n.$$

Hence the theorem.

Theorem 8.4.4

Suppose f maps a convex open set $E \subset R^n$ into R^m, f is differentiable at a point $x \in E$, and there is a real number M such that $\|f'(x)\| \le M$ for every $x \in E$.

Then

$$|f(b) - f(a)| \le M|b-a|, \qquad \forall a, b \in E.$$

Proof

Let us choose $a, b \in E$ and

Define $\gamma(t) = (1-t)a + tb, \quad \forall t \in R^1$ such that $\gamma(t) \in E$.

Given that E is convex.

Therefore, we have

$$\gamma(t) \in E \text{ if } 0 \le t \le 1.$$

Let us assume that

$$g(t) = f(\gamma(t)).$$

That implies $\quad g'(t) = f'(\gamma(t)).\gamma'(t) = f'(\gamma(t)).(b-a),$

such that $\quad |g'(t)| \le \|f'(\gamma(t))\| |b-a| \le M|b-a|$ for all $t \in [0,1]$.

Therefore, by the theorem "Suppose *f* is a continuous mapping of $[a, b]$ into R^k and f is differentiable in (a, b). Then there exists $x \in (a,b)$ such that $|f(b) - f(a)| \le (b-a)|f'(x)|$", we have

$$|g(1) - g(0)| \le M|b-a|.$$

But $g(0) = f(a)$ and $g(1) = f(b)$.

Hence the theorem.

Definition 8.4.5

A differentiable mapping f of an open set $E \subset R^n$ into R^m is said to be *continuously differentiable* in E if f' is a continuous mapping of E into $L(R^n, R^m)$.

If, for every $x \in E$ and to every $\varepsilon > 0$ corresponds a $\delta > 0$ such that

$$\|f'(y) - f'(x)\| < \varepsilon, \text{ if } y \in E \text{ and } |x - y| < \delta,$$

then, we say that *f* is a $\mathcal{C}'$–mapping, or that $f \in \mathcal{C}'(E)$.

Note 8.4.6

Suppose f maps an open set $E \subset R^n$ into R^m. Then $f \in \mathcal{C}'(E)$ if and only if the partial derivatives $D_j f_i$ exist and are continuous on E for $1 \le i \le m$, $1 \le j \le n$. If F is the composition of two differentiable functions, then F is differentiable and its derivatives exist.

8.5 Contraction Principle

Definition 8.5.1

Let X be a metric space with metric d. If φ maps X into X and if there is a number $c < 1$ such that

$$d(\varphi(x),\varphi(y)) \leq cd(x,y), \qquad \forall x, y \in X,$$

then φ is said to be a contraction of X into X.

Theorem 8.5.2

If X is a complete metric space and if φ is a contraction of X into X, then there exists one and only one point $x \in X$ such that $\varphi(x) = x$.

Proof

Choose $x_0 \in X$ arbitrarily.

Define $\{x_n\}$ recursively, by setting

$$x_{n+1} = \varphi(x_n),\ n = 1, 2, 3, \qquad \text{............ (8.5.2.1)}$$

Since φ is a contraction of X into X, there exists a number $c < 1$ such that

$$d(\varphi(x),\varphi(y)) \leq cd(x,y), \qquad \forall x, y \in X. \qquad \text{............ (8.5.2.2)}$$

Therefore,

$$d(x_{n+1},x_n) = d(\varphi(x_n),\varphi(x_{n-1}))$$
$$\leq cd(x_n,x_{n-1})$$

Again,

$$d(x_n,x_{n-1}) \leq cd(x_{n-1},x_{n-2})$$
$$\Rightarrow d(x_{n+1},x_n) \leq c^2 d(x_{n-1},x_{n-2})$$

Therefore, by induction,

$$d(x_{n+1},x_n) \leq c^n d(x_1,x_0),\ n = 0, 1, 2, \ldots..$$

Therefore, if $n < m$,

$$d(x_m,x_n) \leq d(x_m,x_{m-1}) + d(x_{m-1},x_{m-2}) + \ldots\ldots\ldots + d(x_{n+1},x_n)$$
$$= \sum_{i=n+1}^{m} d(x_i,x_{i-1})$$
$$\leq \sum_{i=n+1}^{m} c^{i-1} d(x_1,x_0)$$

$$= \left(c^n + c^{n+1} + \ldots\ldots + c^{m-1}\right)d(x_1, x_0)$$

$$= c^n(1 + c + c^2 + \ldots\ldots + c^{m-n-1})d(x_1, x_0)$$

$$= \frac{c^n}{1-c}d(x_1, x_0)$$

Since $c < 1$,

$$\lim_{n\to\infty} c^n = 0.$$

Therefore,

$$d(x_m, x_n) \to 0 \quad \text{as} \quad m, n \to \infty.$$

Hence $\{x_n\}$ is a Cauchy sequence in X.

Since X is complete, every Cauchy sequence of points of X converges to a point in X.

Hence $\{x_n\}$ converges to a point, say x in X.

That is,

$$\lim_{n\to\infty} x_n = x.$$

Since φ is a contraction, φ is continuous on X.

Hence,

$$\varphi(x) = \varphi\left(\lim_{n\to\infty} x_n\right)$$

$$= \lim_{n\to\infty} \varphi(x_n)$$

$$= \lim_{n\to\infty} x_{n+1}$$

$$= x.$$

Hence there exists a point $x \in X$ such that $\varphi(x) = x$.

Hence the theorem.

8.6 Inverse function theorem

Definition 8.6.1

Suppose E is an open subset in R^n, f maps E into R^m and $x \in E$. f is said to be *differentiable at* x, if there exists a linear transformation A of R^n into R^m such that

$$\lim_{h\to 0}\frac{|f(x+h)-f(x)-Ah|}{|h|}=0$$

and we write $f'(x)=A$. If f is differentiable at every $x\in E$, then f is said to be *differentiable in E.*

Definition 8.6.2

A differentiable mapping f of an open set $E\subset R^n$ into R^m is said to be *continuously differentiable in E* if f' is a continuous mapping of E into $L(R^n, R^m)$, that is, if for every $x\in E$ and for every $\varepsilon>0$, there exists a $\delta>0$ such that

$$\|f'(y)-f'(x)\|<\varepsilon,\ \text{if } y\in E \text{ and } |x-y|<\delta,$$

then, f is said to be $\mathcal{C}'$ *mapping* or $f\in\mathcal{C}'$.

Theorem 8.6.3

This theorem is known as ***Inverse Function theorem***.

Suppose f is a $\mathcal{C}'$ mapping of an open subset $E\subset R^n$ into R^n, $f'(a)$ is invertible for some $a\in E$ and $b=f'(a)$. Then

a) there exist open sets U and V in R^n such that $a\in U, b\in V$, f is one-to-one on U and $f(U)=V$.

b) If g is the inverse of f, defined in V by $g(f(x))=x$, $x\in U$, then $g\in\mathcal{C}'(V)$.

[Writing the equation $y=f(x)$ in component form, the conclusion of the theorem can be interpreted as:

The system of n equations $y_i=f_i(x_1,x_2,\ldots\ldots,x_n)$, $i=1, 2, \ldots\ldots, n$ can be solved for $x_1,x_2,\ldots\ldots,x_n$ in terms of $y_1,y_2,\ldots\ldots,y_n$. If x and y are restricted to be small enough neighbourhoods of a and b, the solutions are unique and continuously differentiable.]

Proof

a) Let $f'(a)=A$.

Choose λ such that $2\lambda\|A^{-1}\|=1$. .... (8.6.3.1)

Since f is a $\mathcal{C}'$ mapping on E and $a\in E$, f' is continuous at a.

Hence, there exists an open ball $U \subset E$, with center at a, such that

$$\|f'(x) - A\| < \lambda, \ \forall x \in U . \qquad \text{............ (8.6.3.2)}$$

To each $y \in R^n$, we associate a function φ on E as

$$\varphi(x) = x + A^{-1}(y - f(x)), \ x \in E, \qquad \text{............ (8.6.3.3)}$$

So that if $f(x) = y$, then $\varphi(x) = x$.

Conversely,

If $\varphi(x) = x$, then $A^{-1}(y - f(x)) = 0$.

$$\Rightarrow y - f(x) = 0.$$

$$\Rightarrow y = f(x).$$

Hence, $f(x) = y$ if and only if $\varphi(x) = x$.

That is, x is a fixed point of φ. .... (8.6.3.4)

Now, $\varphi'(x) = 1 - A^{-1} f'(x)$.

$$= A^{-1}A - A^{-1}f'(x).$$

$$= A^{-1}(A - f'(x)).$$

Therefore, for $x \in E$,

$$\|\varphi'(x)\| = \|A^{-1}(a - f'(x))\|$$

$$\leq \|A^{-1}\|\|A - f'(x)\|$$

$$< \left(\frac{1}{2\lambda}\right)\lambda \qquad \text{[From equations (8.6.3.1) and (8.6.3.2)]}$$

Therefore, $\|\varphi'(x)\| < 1/2$ for all $x \in U$.

Therefore by the result,

"Suppose f maps a convex open set $E \subset R^n$ into R^m. If f is differentiable in E, and there is a real number M such that

$$\|f'(x)\| \leq M, \text{ for every } x \in E.$$

Then,

$$|f(b) - f(a)| \leq M|b - a|, \text{ for all } a \in E, b \in E.\text{"},$$

we get that,

$$|\varphi(x_1)-\varphi(x_2)| \le (1/2)|x_1-x_2|, \text{ for all } x_1, x_2 \in U. \qquad \text{............ (8.6.3.5)}$$

Therefore,

φ is a contraction of *U* into *U*.

Therefore,

φ has atmost one fixed point in *U*. [By theorem 8.5.2]

Therefore,

f(*x*) = *y* for atmost one $x \in U$. [From equation (8.6.3.4)]

Hence, *f* is one-to-one in *U*.

Let *V* = *f*(*U*).

To prove: *V* is open.

Let $y_0 \in V$.

Then, $y_0 = f(x_0)$ for some $x_0 \in U$.

Let *B* be the open ball with center at x_0 and radius $r > 0$, so small that its closure $\overline{B}$ lies in *U*.

Claim: $y \in V$ whenever $|y-y_0| < \lambda r$.

Let *y* be such that $|y-y_0| < \lambda r$.

From equation (8.6.3.3), we have

$$\begin{aligned}|\varphi(x_0)-x_0| &= |A^{-1}(y-y_0)| \\ &\le \|A^{-1}\| |y-y_0| \\ &< \|A^{-1}\| \lambda r \\ &= (1/2)r \qquad \text{[From equation (8.6.3.1)]}\end{aligned}$$

Hence, $|\varphi(x_0)-x_0| < r/2$. .... (8.6.3.6)

If $x \in B$, then by equation (8.6.3.5), we have

$$|\varphi(x)-\varphi(x_0)| < (1/2)|x-x_0| \qquad \text{............ (8.6.3.7)}$$

From equations (8.6.3.6) and (8.6.3.7), we get that whenever $x \in E$,

$$\begin{aligned}|\varphi(x)-x_0| &\le |\varphi(x)-\varphi(x_0)| + |\varphi(x_0)-x_0| \\ &< (1/2)|x-x_0| + (r/2) \le r\end{aligned}$$

Hence, $|\varphi(x) - x_0| \le r$, whenever $x \in B$.

$\Rightarrow \varphi(x) \in B$.

Hence φ maps B into B.

Since φ is a contraction of U into U and $\overline{B} \subset U$, φ is a contraction of $\overline{B}$ into $\overline{B}$.

Since $\overline{B}$ is a closed subset of R^n and R^n is complete, $\overline{B}$ is complete.
Hence by theorem 8.5.2,

φ has a fixed point $x \in \overline{B}$.

That is, $\varphi(x) = x$.
For this x, $f(x) = y$. [From equation (8.6.3.4)]

Hence, $y \in f(\overline{B}) \subset f(U) = V$.

Hence, $y \in V$ whenever $|y - y_0| < \lambda r$.

Hence V is open.
Hence part a) is proved.

b) Let $y \in V$, $y + t \in V$.
Since $V = f(U)$, there exists a $x \in U$, $x + h \in U$ such that

$y = f(x)$ and $y + k = f(x + h)$.

From equation (8.6.3.3), we get

$$\begin{aligned}\varphi(x+h) - \varphi(x) &= \lfloor (x+h) + A^{-1}(y - f(x+h)) \rfloor - \lfloor x + A^{-1}(y - f(x)) \rfloor \\ &= h + A^{-1}\lfloor f(x) - f(x+h) \rfloor \\ &= h - A^{-1}k \qquad \text{............ (8.6.3.8)}\end{aligned}$$

By equation (8.6.3.5),

$$|\varphi(x+h) - \varphi(x)| \le (1/2)|(x+h) - x| = (1/2)|h|. \qquad \text{............ (8.6.3.9)}$$

From equations (8.6.3.8) and (8.6.3.9), we get

$$|h - A^{-1}k| = |\varphi(x+h) - \varphi(x)| \le (1/2)|h|.$$

Therefore, $|A^{-1}k| = |h - (h - A^{-1}k)|$

$$\ge |h| - |h - A^{-1}k|$$

$$\geq |h| - \frac{|h|}{2} = |h|/2$$

Hence, $\left|A^{-1}k\right| \geq |h|/2$.

Therefore, $|h| \leq 2\left|A^{-1}k\right|$

$$\leq 2\left\|A^{-1}\right\| |k|$$

$$= \lambda^{-1}|k| \qquad \text{............ (8.6.3.10)}$$

Therefore, $\frac{1}{|k|} \leq \frac{1}{\lambda|h|}$ .... (8.6.3.11)

Let g be the inverse of f, defined in V by $g(f(x)) = x$, $x \in U$.

Let $x \in U$.

From equation (8.6.3.2), $\|f'(x) - A\| < \lambda = 1/\left\|A^{-1}\right\|$, where $A = f'(a)$ is invertible.

Hence by theorem 8.2.21, $f'(x)$ is invertible.

Let the inverse of $f'(x)$ be T.

Consider, $g(y+k) - g(y) = g(f(x+h)) - g(f(x)) - Tk$

$$= x + h - x - Tk$$

$$= h - Tk$$

$$= Tf'(x)h - Tk$$

[Since T is the inverse of $f'(x)$, $T(f'(x)) = 1$]

$$= -T[k - f'(x)h]$$

$$= -T[f(x+h) - f(x) - f'(x)h]$$

Therefore,

$$|g(y+k) - g(y) - Tk| = |T[f(x+h) - f(x) - f'(x)h]|$$

$$\leq \|T\| \, |f(x+h) - f(x) - f'(x)h| \qquad \text{............ (8.6.3.12)}$$

From equations (8.6.3.11) and (8.6.3.12),

$$\frac{|g(y+k) - g(y) - Tk|}{|k|} \leq \frac{\|T\|}{\lambda} \frac{|f(x+h) - f(x) - f'(x)h|}{|h|} \qquad \text{......... (8.6.3.13)}$$

From equation (8.6.3.10), it is clear that $h \to 0$ as $k \to 0$.

Also, since f is differentiable at x,

$$\lim_{h\to 0}\frac{|f(x+h)-f(x)-f'(x)h|}{|h|}=0.$$

Therefore, from equation (8.6.3.13),

$$\lim_{k\to 0}\frac{|g(y+k)-g(y)-Tk|}{|k|}=0.$$

This implies that, $g'(y)$ exists and $g'(y)=T$.

Since T is the inverse of $f'(x)=f'(g(y))$,

$$g'(y)=\{f'(g(y))\}^{-1}, \quad \forall y\in V.$$

Hence, g is differentiable on V.

Hence, g is a continuous mapping of V onto U.

Also, f' is a continuous mapping U into the set Ω of all invertible elements of $L(R^n)$.

Also, the inversion is a continuous mapping of Ω onto Ω.

Hence, g' is a continuous mapping on V.

Hence, $g\in C'(V)$.

Hence the theorem.

Theorem 8.6.4

If f is a C' mapping of an open set $E\subset R^n$ into R^n and if f' is invertible for every $x \in E$, then $f(W)$ is an open subset of R^n for every open set $W\subset E$. In other words, f is an open mapping of E into R^n.

Proof

Let $W\subset E$ be an open set.

To prove: $f(W)$ is an open subset of R^n.

Let $b\in f(W)$.

Then there exists a $a\in W$ such that $b = f(a)$.

Since f' is invertible for every $x \in E$, and $a\in W\subset E$, $f'(a)$ is invertible.

Hence by part (a) of Inverse function theorem,

There exist open sets $U_a \subset W$ and $V_b \subset f(W)$ such that $a \in U, b \in V$, f is one-to-one on U and $f(U) = V$.

Hence $f(W)$ can be written as $f(W) = \bigcup_{b \in f(W)} V_b$.

Since union of open sets is open, $f(W)$ is open.

Hence the theorem.

8.7 Implicit function theorem

Introduction 8.7.1

If $x = (x_1, x_2, \ldots\ldots, x_n) \in R^n$ and $y = (y_1, y_2, \ldots\ldots, y_m) \in R^m$, we write,

$$(x, y) = (x_1, x_2, \ldots\ldots, x_n, y_1, y_2, \ldots\ldots, y_m) \in R^{n+m}.$$

The first entry in (x, y) will always be a vector in R^n, the second will be a vector in R^m.

Every vector $A \in L(R^{n+m}, R^n)$ can be split into two linear transformation A_x and A_y, defined by

$$A_x h = A(h, 0),$$

$$A_y k = A(0, k), \text{ for any } h \in R^n, k \in R^m.$$

Then, $A_x \in L(R^n)$, $A_y \in L(R^m)$ and

$$\begin{aligned} A(h,k) &= A[(h,0) + (0,k)] \\ &= A(h,0) + A(0,k) \quad \text{[Since } A \text{ is linear]} \\ &= A_x h + A_y k. \end{aligned}$$

Theorem 8.7.2

This theorem is known as the ***Linear version of Implicit function theorem***.

If $A \in L(R^{n+m}, R^n)$ and if A_x is invertible, then there corresponds to every $k \in R^m$, a unique $h \in R^n$ such that $A(h, k) = 0$. This h can be computed from k by the formula

$$h = -(A_x)^{-1} A_y k.$$

Proof

Now, $A(h, k) = 0$ if and only if $A_x h + A_y k = 0$.

Since A_x is invertible, $(A_x)^{-1}$ exists.

Therefore, pre multiplying $A_x h + A_y k = 0$ by $(A_x)^{-1}$, we get

$$(A_x)^{-1}(A_x h + A_y k) = 0$$

$$h + (A_x)^{-1} A_y k = 0.$$

Therefore, $h = -(A_x)^{-1} A_y k$.

Hence the theorem.

Theorem 8.7.3

This theorem is known as ***The Implicit function theorem***.

Let f be a C'- mapping of an open set $E \subset R^{n+m}$ into R^n, such that $f(a, b) = 0$ for some point $(a,b) \in E$. Put $A = f'(a,b)$ and assume that A_x is invertible. Then there exists open sets $U \subset R^{n+m}$ and $W \subset R^m$, with $(a,b) \in U$ and $b \in W$, having the following property;

To every $y \in W$ corresponds a unique x such that $(x, y) \in U$ and $f(x, y) = 0$. .... (8.7.3.1)

If this x is defined to be $g(y)$, then g is a C'-mapping of W into R^n,

$$g(b) = a,$$

$$f(g(y), y) = 0,\ y \in W, \qquad \text{........... (8.7.3.2)}$$

and $$g'(b) = -(A_x)^{-1} A_y. \qquad \text{........... (8.7.3.3)}$$

The function g is implicitly defined by the equation (8.7.3.2).
Hence the name of the theorem.

Proof

Define $F : E \to R^{n+m}$ by $F(x, y) = (f(x, y), y)$ for $(x, y) \in E$.

Since f is a C'-mapping, F is a C'-mapping of E into R^{n+m}.

Claim: $F'(a,b)$ is an invertible element of $L(R^{n+m})$.

Since $f'(a,b)$ exists and equals A,

$$f(a+h,b+k)=f(a,b)+A(h,k)+r(h,k),$$

where r is the remainder that occurs in the definition of $f'(a,b)$.

Therefore,

$$f(a+h,b+k)=A(h,k)+r(h,k). \text{ [Since } f(a,b)=0] \qquad \ldots\ldots\ldots (8.7.3.4)$$

Consider,

$$\begin{aligned} F(a+h,b+k)-F(a,b) &= (f(a+h,b+k),b+k)-(f(a,b),b) \\ &= (f(a+h,b+k),b+k)-(0,b) \\ &= (f(a+h,b+k),k) \\ &= (A(h,k)+r(h,k),k+0) \\ & \qquad \text{[From equation (8.7.3.4)]} \\ &= (A(h,k),k)+(r(h,k),0) \end{aligned}$$

Therefore, $F'(a,b)$ is the linear operator on R^{n+m} that maps (h, k) to $(A(h, k), k)$.

If this image vector is 0,

then $A(h, k) = 0$ and $k = 0$.

Hence $A(h, 0) = 0$.

This implies that, $h = 0$. [By theorem (8.2.13)]

Therefore, if the image vector of $F'(a,b)$ at (h, k) is 0, then both $h = 0$ and $k = 0$.

Hence $F'(a,b)$ is one-to-one.

Therefore, by theorem (8.2.21), $F'(a,b)$ is onto.

Therefore, $F'(a,b)$ is both one-to-one and onto.

Hence $F'(a,b)$ is invertible.

Therefore, our claim is proved.

That implies, Inversion theorem can be applied for F.

Hence by part (a) of Inversion theorem,

There exists open sets U and V in R^{n+m} with $(a,b)\in U$, $(0,b)\in V$, such that F is a one-to-one mapping of U onto V.

Let $W=\{y\in R^m /(0,y)\in V\}$.

Therefore, $b \in W$. [Since $(0,b) \in V$]

Since V is open, W is also open.

If $y \in W$, then $(0, y) \in V = F(U)$.

Hence there exists a point (x, y) in R^{n+m} such that $(0, y) = F(x, y)$.

Therefore, $(0, y) = ((f(x, y), y)$.

$\Rightarrow f(x, y) = 0$, for this x.

Suppose for the same y, there exists another x' in R^n, such that $f(x', y) = 0$.

Then, $F(x', y) = (f(x', y), y)$

$$= (0, y)$$
$$= (f(x, y), y)$$
$$= F(x, y).$$

Since F is one-to-one in U, this implies that $x' = x$.

Hence this x is unique.

Hence there exist open sets $U \subset R^{n+m}$ and $W \subset R^m$, with $(a,b) \in U$ and $b \in W$, with the property that:

To every $y \in W$ corresponds a unique x such that $(x, y) \in U$ and $f(x, y) = 0$.

Hence the first part of the theorem is proved.

To prove the second part:

As we have seen in the first part,

To every $y \in W$, there is a unique x such that $(x, y) \in U$ and $f(x, y) = 0$.

Define this x to be $g(y)$ for every $y \in W$.

Then, $g(b) = a$. [Since $f(a, b) = 0$]

Also, $(g(y), y) = (x, y) \in U$ and $f(g(y), y) = 0$. .... (8.7.3.5)

Therefore,

$F(g(y), y) = (f(g(y), y), y) = (0, y)$, for all $y \in W$ .... (8.7.3.6)

Let g be the mapping of V into U that inverts F, that is, $G = F^{-1}$.

Also $F \in C', \Rightarrow G \in C'$.

Then from equation (8.7.3.6), we get

$G[F(g(y), y)] = G(0, y)$, for all $y \in W$.

That is, $(g(y), y) = G(0, y)$, for all $y \in W$. .... (8.7.3.7)

Since $G \in C'$, equation (8.7.3.7) shows that $g \in C'$.

To compute $g'(b)$:

Let $(g(y), y) = \varphi(y)$.

Then $\varphi'(y)k = (g'(y)k, yk)$, for all $y \in W$ and $k \in R^m$ .... (8.7.3.8)

From equation (8.7.3.5), we get

$$f(\varphi(y)) = f(g(y), y) = 0.$$

Therefore by chain rule, $f'(\varphi(y))\varphi'(y) = 0$. .... (8.7.3.9)

When $y = b$, $\varphi(y) = \varphi(b)$

$$= (g(b), b)$$

$$= (a, b).$$

Therefore,

$$f'(\varphi(y)) = f'(\varphi(b)) = f'(a, b) = A.$$

Substituting in equation (8.7.3.9), we get

$$A\varphi'(b) = 0. \quad \text{............ (8.7.3.10)}$$

We know that, for all $h \in R^n$ and $k \in R^m$, $A(h, K) = A_x h + A_y k$,

where $A_x h = A(h, K)$ and $A_y k = A(0, k)$.

Therefore,

$$A_x g'(b)k + A_y k = A(g'(b)k, k)$$

$$= A\varphi'(b)k \text{ [from equation (8.7.3.8)]}$$

$$= 0, \text{ for every } k \in R^m. \qquad \text{[from equation (8.7.3.10)]}$$

Therefore, $A_x g'(b) + A_y = 0$.

Therefore, $g'(b) = -(A_x)^{-1} A_y$.

Hence the theorem.

Note 8.7.4

The equation $f(x, y) = 0$ can be written as a system of n equations in $n + m$ variables:

$$f_1(x_1, x_2, \ldots\ldots, x_n, y_1, y_2, \ldots\ldots, y_m) = 0$$

$$\ldots\ldots\ldots\ldots\ldots\ldots\ldots\ldots\ldots\ldots\ldots\ldots \qquad \ldots\ldots\ldots\ldots (8.7.4.1)$$

$$f_n(x_1, x_2, \ldots\ldots, x_n, y_1, y_2, \ldots\ldots, y_m) = 0$$

The assumption that A_x is invertible means that the n by n matrix

$$\begin{bmatrix} D_1 f_1 & .. & .. & D_n f_1 \\ .. & .. & .. & .. \\ .. & .. & .. & .. \\ D_1 f_n & .. & .. & D_n f_n \end{bmatrix}$$

evaluated at (a, b) defines an invertible linear operator in R^n.

In other words, its column vectors should be independent, that is, its determinant should not be equal to zero.

If equation (8.7.4.1) holds when $x = a$, $y = b$, then the conclusion of the theorem is that: equation (8.7.4.1) can be solved for $x_1, x_2, \ldots\ldots, x_n$ in terms of $y_1, y_2, \ldots\ldots, y_m$, for every y near b, and that these solutions are continuously differentiable functions of y.

Problem 8.7.5

Assume $n = 2$ and $m = 3$.

Consider the mapping $f = (f_1, f_2)$ of R^5 into R^2 given by

$$f_1(x_1, x_2, y_1, y_2, y_3) = 2e^{x_1} + x_2 y_1 - 4y_2 + 3.$$

$$f_2(x_1, x_2, y_1, y_2, y_3) = x_2 \cos x_1 - 6x_1 + 2y_1 - y_3.$$

Take $a = (0, 1)$ and $b = (3, 2, 7)$, then $f(a, b) = 0$.

Now, $D_1 f_1(x_1, x_2, y_1, y_2, y_3) = 2e^{x_1}$.

Therefore, $D_1 f_1(0,1,3,2,7) = 2$.

Similarly, $D_2 f_1(0,1,3,2,7) = 3$.

$D_3 f_1(0,1,3,2,7) = 1$.

$D_4 f_1(0,1,3,2,7) = -4$.

$D_5 f_1(0,1,3,2,7) = 0$.

$D_1 f_2(0,1,3,2,7) = -6$.

$$D_2 f_2(0,1,3,2,7) = 1.$$

$$D_3 f_2(0,1,3,2,7) = 2.$$

$$D_4 f_2(0,1,3,2,7) = 0.$$

$$D_5 f_2(0,1,3,2,7) = -1.$$

Therefore, the matrix of the transformation $A = f'(a,b)$ is

$$[A] = \begin{bmatrix} 2 & 3 & 1 & -4 & 0 \\ -6 & 1 & 2 & 0 & -1 \end{bmatrix}.$$

Hence,

$$[A_x] = \begin{bmatrix} 2 & 3 \\ -6 & 1 \end{bmatrix} \text{ and } [A_y] = \begin{bmatrix} 1 & -4 & 0 \\ 2 & 0 & -1 \end{bmatrix}.$$

Since the column vectors of [Ax], i.e $\begin{bmatrix} 2 \\ -6 \end{bmatrix}$ and $\begin{bmatrix} 3 \\ 1 \end{bmatrix}$ are independent, A_x is invertible.

Therefore, by implicit function theorem,

A C'-mapping g exists in a neighborhood of $b = (3, 2, 7)$ such that $g(b) = g(3, 2, 7) = (0, 1) = a$ and $f(g(y)) = 0$.

Hence, $[(A_x)^{-1}] = \dfrac{1}{20}\begin{bmatrix} 1 & -3 \\ 6 & 2 \end{bmatrix}.$

Moreover, from $g'(b) = -(A_x)^{-1} A_y$, we get

$$g'(3,2,7) = -\frac{1}{20}\begin{bmatrix} 1 & -3 \\ 6 & 2 \end{bmatrix}\begin{bmatrix} 1 & -4 & 0 \\ 2 & 0 & -1 \end{bmatrix}$$

$$= \begin{bmatrix} \frac{1}{4} & \frac{1}{5} & -\frac{3}{20} \\ -\frac{1}{2} & \frac{6}{5} & \frac{1}{10} \end{bmatrix}.$$

In terms of partial derivatives, the conclusion is

At the point (3, 2, 7),

$D_1 g_1 = 1/4$ $D_2 g_1 = 1/5$ $D_3 g_1 = -3/20$

$D_1 g_2 = -1/2$ $D_2 g_2 = 6/5$ $D_3 g_2 = 1/10$.

8.8 The Projection

Definition 8.8.1

Let X and Y be linear spaces. Let $A \in L(X,Y)$. Then, $N(A) = \{x \in X; Ax = 0\}$ is called the *Null space of A* or *the Kernel of A*. $N(A)$ is the subspace of X.

Also, $R(A) = \{Ax \in Y; x \in X\} = Ax$ is the *range of A*. $R(A)$ is a subspace of Y. The dimension of $R(A)$ is called the *Rank of A*. It is denoted by Rank A. The dimension of $N(A)$ is called the *Nullity of A*.

Results 8.8.2

1. Rank of A + Nullity of A = Dim. of X.
2. If $A \in L(X,Y)$ and $Ax = 0$, for all $x \in X$, then $R(A) = \{0\}$.

Definition 8.8.3

Let X be a linear space. An operator $P \in L(X)$ is called a *projection* in X, if $P^2x = Px$, for all $x \in X$. $P(Px) = Px$ implies that Px belongs to the range of A. Therefore, if P is a projection, then P fixes every vector in its range.

Properties of Projection 8.8.4

Property 1

Let P be a projection in X. Then, every vector x can be uniquely expressed in the form $x = x_1 + x_2$, where $x_1 \in R(P), x_2 \in N(P)$.

Proof

Let x be an arbitrary vector in X.

Choose $x_1 = Px$. Then, $x_1 \in R(P)$.

We have, $x_2 = x - x_1$. Then, $P(x_2) = P(x - x_1) = 0$.

Therefore, $x_2 \in N(P)$.

Thus, $x = x_1 + x_2$, where $x_1 \in R(P), x_2 \in N(P)$.

To prove: Uniqueness

Suppose, if possible, let there be another representation $x = y_1 + y_2$, where $x_1 \in R(P), x_2 \in N(P)$.

To prove: $y_1 = x_1$ and $y_2 = x_2$.

Now, $x_1 = Px = P(y_1 + y_2)$

$= P(y_1)$ [since $P(y_2) = 0$]

$= y_1$ [since P fixes every element in $R(P)$]

Similarly, $x_2 = y_2$.

Hence proved.

Property 2

Let X be linear space. Let X_1 be a subspace of X. Then, a projection P in X can be defined such that Range of $P = X_1$.

Proof

If Dim.$X_1 = 0$, then $X_1 = \{0\}$.

Therefore, the required projection is $Px = 0$ for all $x \in X$.

Suppose, Dim. $X_1 > 0$.

Let Dim. $X_1 = k$.

Then, there exists a basis, $\{u_1, u_2, \ldots\ldots, u_k\}$ for X_1.

This can be completed to a basis $\{u_1, u_2, \ldots\ldots, u_k, u_{k+1}, \ldots\ldots, u_n\}$ for X. [Since Dim.$X = n$].

Let $x \in X$.

Then, $x = c_1u_1 + \ldots.. + c_ku_k + c_{k+1}u_{k+1} + \ldots.. + c_nu_n$ for a unique set of scalars c_i's.

Define $Px = c_1u_1 + \ldots.. + c_ku_k \in X_1$.

Clearly, P is a function from $X \to X$ having its range X_1.

Also, $P \in L(X)$.

Claim: P is a projection.

$$P(Px) = P(c_1u_1 + \ldots.. + c_ku_k + 0.u_{k+1} + \ldots.. + 0.u_n)$$

$$= c_1u_1 + \ldots.. + c_ku_k = Px.$$

Thus, we have a projection P, whose range is X_1.

8.9 Rank theorem

Theorem 8.9.1

Suppose *m*, *n*, *r* are non – negative mapping, and $m \geq r, n \geq r$. Let *F* be a C' - mapping of an open set $E \subset R^n$ into R^m. Let $F'(x)$ has rank *r*, for every $x \in E$. Fix $a \in E$. Put $A = F'(a)$. Let Y_1 be the range of *A*. Let P be a projection in R^m, whose range is Y_1. Let Y_2 be the null space of *P*. Then, there are open sets *U* and *V* in R^n, with $a \in U \subset E$ and there is a 1-1 C' - mapping *H* of *V* onto *U*, such that $F(H(x)) = Ax + \varphi(Ax), \forall x \in U$, ... (8.9.1.1)

where φ is a C' - mapping of the open set $A(V) \subset Y_1$ into Y_2.

Proof

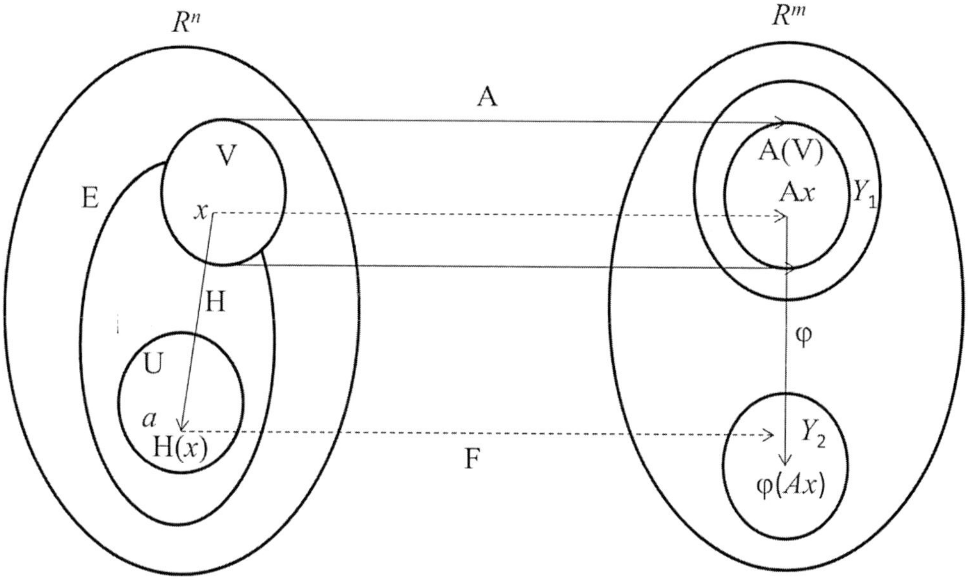

Case (i) Suppose $r = 0$.

Let $Y_1 = 0$.

$F'(x)$ has rank 0, for all $x \in E$.

Therefore, *F*(*x*) is a constant in some neighbourhood *U* of *a*.

Hence, we can take $F(x) = F(a)$, for all $x \in U$.

Choose $V = U$ and $H(x) = x$.

Since, Rank of $A = 0$ and $A = F'(a)$, Dim. $R(A) = 0$.

Therefore, $Ax = 0$, for all $x \in V$.

Therefore, $F(H(x)) = Ax + \varphi(Ax)$, holds true if we take $\varphi(0) = F(a)$.

Thus, we have proved result (1) for $r = 0$.

Case (ii) Suppose $r > 0$.

Let Dim. $Y_1 = r$.

Therefore, there exists a basis $\{y_1, y_2, \ldots.., y_r\}$ for Y_1.

Choose $z_i \in R^n$ so that $Az_i = y_i$, $1 \le i \le r$.

Using this, we define a linear map $S : Y_1 \to R^n$ by setting

$$S(c_1 y_1 + \ldots\ldots + c_r y_r) = c_1 z_1 + \ldots\ldots + c_r z_r \in R^n.$$

Then, $AS_{y_i} = Az_i = y_i$ for $1 \le i \le r$.

Therefore, we get $AS_y = y$, for all $y \in Y_1$. .... (8.9.1.2)

Define a mapping g of E into R^n, by setting $G(x) = x + SP\,[F(x) - Ax]$, for $x \in E$. .... (8.9.1.3)

Then, $G'(a) = I + SP[F'(a) - A] = I \quad [\text{Since} F'(a) = A]$.

Therefore, $G'(a)$ is simply the identity operator on R^n.

Clearly, this is invertible.

Hence, we apply Inverse Function theorem.

Therefore, there exists open sets U and V in R^n, such that $a \in U \subset E$, $G(a) \in V$. G is 1-1 on U and $G(U) = V$. Inverse of G, say $H : V \to U$ exists such that $H[G(x)] = x$, for all $x \in U$ and H is a C'-mapping on V.

Moreover, U and V can be shrunk suitably, so that V is convex and $H'(x)$ is invertible, for all $x \in V$.

Claim: ASPA = A.

Let $x \in E$ and $Ax \in Y_1$.

$$\Rightarrow P(Ax) = Ax.$$

$\Rightarrow AS(PAX) = PAx.$ [by equation (8.9.1.2)]

$\Rightarrow ASPA = A.$ .... (8.9.1.4)

Operate both sides of equation (8.9.1.3) by A.

$A[G(x)] = Ax + ASP\,[F(x) - Ax].$

$= ASPF(x).$ [by equation (8.9.1.4)]

Therefore,

$A[G(x)] = PF(x)$ [Since $PF(x) \in Y_1$ and by (8.9.1.2)] .... (8.9.1.5)

This is true for all $x \in U$.

In particular, if $x \in V$, $H(x) \in U$.

Hence, replacing x by $H(x)$ in equation (8.9.1.5), we get

$A.G(H(x)) = P.F(H(x))$.

$\Rightarrow \quad Ax = P.F(H(x))$. .... (8.9.1.6)

Define $\psi(x) = F(H(x)) - Ax, \ \forall x \in V$ .... (8.9.1.7)

Then, $P\psi(x) = P.F(H(x)) - Ax = 0$. [by equation (8.9.1.6)]

Therefore, $\psi(x) \in Y_2$.

Hence, ψ is a C' - mapping from V into Y_2.

Now, V is open.

Therefore, $A(V)$ is an open subset of Y_1.

Hence, to complete the proof, we have to prove that

There exists a C' - map $\varphi : A(V) \to Y_2$, such that $\phi(Ax) = \psi\, x, \ \forall x \in V$.

Therefore, define $\varphi(Ax) = \psi\, x, \forall x \in V$ .... (8.9.1.8)

and prove (i) φ is a well – defined function

(ii) φ is a C' - mapping.

i) *To prove*: φ is a well – defined function.

Let $x_1, x_2 \in V$ and $Ax_1 = Ax_2$.

We have to prove that, $\varphi(Ax_1) = \varphi(Ax_2)$.

Equivalently, we prove that, if $x_1, x_2 \in V$ and $Ax_1 = Ax_2$, then $\psi(x_1) = \psi(x_2)$.

Define, $\overline{\varphi}(x) = F(H(x)), \ \forall x \in V$.

Then, $\underline{\overline{\varphi}}'(x) = F(H(x)).H'(x)$.

$$\text{Rank of } F'(x) = r, \ \forall x.$$

Therefore, $\quad$ Rank of $\underline{\overline{\varphi}}'(x) = r$.

Fix $x \in V$.

Let M be the range of $\underline{\overline{\varphi}}'(x)$.

Then, $M \subset R^m$ and dim.of M is r.

Differentiating equation (8.9.1.6) with respect to x, we get

$$A = P.F'(H(x)).H'(x) = P.\underline{\overline{\varphi}}'(x) \qquad \text{............ (8.9.1.10)}$$

Therefore, P maps the range of $\underline{\overline{\varphi}}'(x)$ into the range of Y_1 or P maps M into Y_1.

Since M and Y_1 have the same dimension, we get that P restricted to M is 1-1.

Suppose, $Ah = 0$, for some $h \in R^n$.

Then, $P.\underline{\overline{\varphi}}'(x).h = Ah = 0$.

But, $\underline{\overline{\varphi}}'(x).h \in M$ and P is 1-1 on M.

Therefore, $P.\underline{\overline{\varphi}}'(x).h = 0 \Rightarrow \underline{\overline{\varphi}}'(x).h = 0$.

Thus, we have proved that, if $x \in V$ and $Ah = 0$, then $\underline{\overline{\varphi}}'(x).h = 0$.

Now, from equation (8.9.1.7), we have

$$\psi'(x) = F'(H(x)).H'(x) - A = \underline{\overline{\varphi}}'(x) - A.$$

$$\Rightarrow \psi'(x).h = \underline{\overline{\varphi}}'(x).h$$

Thus, if $x \in V$ and $Ah = 0$, we get that $\psi'(x).h = 0$. .... (8.9.1.11)

Using this, we will prove that

$$x_1, x_2 \in V \text{ and } Ax_1 = Ax_2 \Rightarrow \psi(x_1) = \psi(x_2).$$

Let $x_1, x_2 \in V$. Let $h = x_2 - x_1$.

Then, $Ah = Ax_2 - Ax_1$.

Therefore, $Ah = 0$ implies $Ax_2 = Ax_1$.

Define $g(t) = \psi(x_1 + th), \quad t \in [0,1]$.

Since V is convex, $x_1 + th \in V, \quad$ for all $t \in [0,1]$.

Here, $g'(t) = \psi'(x_1 + th).h = 0$. [by equation (8.9.1.11)]

Therefore, $g(t)$ is a constant.

Hence $g(0) = g(1)$.

But, $g(0) = \psi(x_2)$ and

$g(0) = \psi(x_1)$.

Thus, we have proved that, if $x_1, x_2 \in V$ and $Ax_1 = Ax_2$, then $\psi(x_1) = \psi(x_2)$.

Thus, φ is a well-defined function on $A(V)$.

ii) To prove: $\varphi \in C'[A(V)]$.

Fix $y_0 \in A(V)$.

Therefore, there exists an open set W in $A(V)$, such that $y_0 \in W \subseteq A(V) \subset Y_1$.

Also, there exists $x_0 \in V$, such that $Ax_0 = y_0$.

The neighbourhood W has the property that the vector $x = x_0 + S(y - y_0)$ lies in V, for all $y \in W$.

Then, $Ax = Ax_0 + AS(y - y_0) = y$. [by equation (8.9.1.2)]

$$\Rightarrow \varphi(y) = \varphi(Ax) = \psi(x) = \psi(x_0 - Sy_0 + Sy), \qquad y \in W .$$

Thus, we have expressed φ in terms of ψ.

Since ψ is a C'-mapping, the above formula gives that φ is a C'-mapping in W.

Since y_0 was chosen arbitrarily in $A(V)$, same result is true for neighbourhood of all points in $A(V)$.

Therefore, φ is a C'-mapping on $A(V)$.

Hence the theorem.

8.10 Derivatives of higher order

Definition 8.10.1

Suppose f is a real function defined in an open set $E \subset R^n$, with partial derivatives $D_1 f, D_2 f, \ldots\ldots, D_n f$. If the functions are $D_j f$ are themselves differentiable, then the second-order partial derivatives of f are defined by $D_{ij} f = D_i D_j f$, $i, j = 1, 2, 3, \ldots\ldots, n$.

If all these functions $D_{ij} f$ are continuous in E, then f is said to be of class C'' in E or that $f \in C''(E)$.

A mapping f of E into R^m is said to be of class C'', if each component of f is of class C''.

Theorem 8.10.2

This theorem is known as ***Mean Value theorem for Functions of Two Variables***.

Suppose f is defined in an open set $E \subset R^2$ and $D_1 f$ and $D_{21} f$ exist at every point of E. Suppose $Q \subset E$ is a closed rectangle with sides parallel to the coordinate axes, having (a, b) and $(a + h, b + k)$ as opposite vertices $(h \neq 0, k \neq 0)$.

Put $\Delta(f, Q) = f(a+h, b+k) - f(a+h, b) - f(a, b+k) + f(a, b)$.

Then, there is a point (x, y) in the interior of Q such that $\Delta(f, Q) = hk(D_{21} f)(x, y)$.

Proof

Let $u(t) = f(t, b + k) - f(t, b)$. .... (8.10.2.1)

Then, $\Delta(f, Q) = f(a+h, b+k) - f(a+h, b) - f(a, b+k) + f(a, b)$.

$= u(a + h) - u(a)$.

$= h\, u'(x)$, for some $x \in (a, a + h)$. .... (8.10.2.2)

[Since, by the Mean Value theorem of real functions of single variable, "If f is a real continuous function on $[a, b]$, which is differentiable in (a, b), then there is a point $x \in (a, b)$ at which $f(b) - f(a) = (b - a)f'(x)$".]

From equation (8.10.2.1), we have

$$u'(t) = D_1 f(t, b+k) - D_1 f(t, b).$$

Substituting in equation (8.10.2.2), we get

$$\Delta(f,Q) = h[D_1 f(x,b+k) - D_1 f(x,b)]. \qquad \text{..........} (8.10.2.3)$$

Again applying the mean value theorem,

There exists a point y ϵ $(b, b + k)$ such that

$$D_1f(x, b + k) - D_1f(x, b) = kD_{21}f(x, y).$$

Substituting in equation (8.10.2.3), we get that

$$\Delta(f,Q) = hkD_{21} f(x, y).$$

Hence, there is a point (x, y) in the interior of Q such that

$$\Delta(f,Q) = hkD_{21} f(x, y).$$

Hence the theorem.

Theorem 8.10.3

Suppose f is defined in an open set $E \subset R^2$, suppose that D_1f, $D_{21}f$ and D_2f exist at every point of E and $D_{21}f$ is continuous at some point $(a,b) \in E$. Then, $D_{12}f$ exists at (a, b) and

$$(D_{12}f)(a, \text{b}) = (D_{21}f)(a, b).$$

Proof

Let $A = (D_{21}f)(a, b)$.

Choose $\varepsilon > 0$.

Let $Q \subset E$ be a closed rectangle with sides parallel to the coordinates axes, having (a, b) and $(a + h, b + k)$ as opposite vertices ($h \neq 0, k \neq 0$).

Then, by theorem (8.10.2),

$$\Delta(f,Q) = f(a+h,b+k) - f(a+h,b) - f(a,b+k) + f(a,b).$$

$$= hk(D_{21}f)(x, y), \text{ for some point } (x, y) \in Q. \qquad \text{..........} (8.10.3.1)$$

If h and k are sufficiently small, then we have

$$\left| A - \frac{\Delta(f,Q)}{hk} \right| < \varepsilon.$$

$$\Rightarrow \left| \frac{\Delta(f,Q)}{hk} - A \right| < \varepsilon.$$

$$\Rightarrow \left| \frac{f(a+h,b+k)-f(a+h,b)-f(a,b+k)+f(a,b)}{hk} - A \right| < \varepsilon .$$

$$\Rightarrow \left| \frac{\frac{f(a+h,b+k)-f(a+h,b)}{k} - \frac{f(a,b+k)+f(a,b)}{k}}{h} - A \right| < \varepsilon .$$

Keeping h fixed, let $k \to 0$.

Since $D_2 f$ exists in E, the last inequality implies that

$$\left| \frac{D_2 f(a+h,b)-D_2 f(a,b)}{h} - A \right| < \varepsilon . \qquad \text{..........} \ (8.10.3.3)$$

Since ε is arbitrary, and equation (8.10.3.3) holds for all sufficiently small $h \neq 0$, equation (8.10.3.3) implies that $(D_{21}f)(a, b)$ exists and $(D_{21}f)(a, b) = A$.

Hence $(D_{21}f)(a, b) = (D_{21}f)(a, b)$.

8.11 Problems

1. Every span of a set is a vector space. True or false?
2. Dim. R^3 = ?
3. Prove that $A(A^{-1}x) = x$ for all $x \in X$.
4. Prove that A^{-1} is linear.
5. Let $f = (f_1, f_2)$ be the mapping of R^2 into R^2 given by

   $$f_1(x, y) = e^x \cos y \quad \text{and} \quad f_2(x, y) = e^x \sin y .$$

   Put $a = (0, \Pi/3)$, $b = f(a)$.

   i) Find an explicit formula for g.

   ii) Compute $f'(a)$ and $g'(b)$.

   iii) Verify the formula $g'(y) = \{f'(g(y))\}^{-1}$ when $y = b$.

6. Put $f(0) = 0$ and $f(x, y) = \dfrac{xy(x^2 - y^2)}{x^2 + y^2}$, if $(x, y) \neq (0,0)$.

   Prove that (i) f, $D_1 f$, $D_2 f$ are continuous in R^2.

   (ii) $D_{12}f$ and $D_{21}f$ exist at every point of R^2, and are continuous except at (0, 0)

   (iii) $(D_{12}f)(0, 0) = 1$ and $(D_{21}f)(0, 0) = -1$.

7. If $f(x, y) = \dfrac{xy}{x^2 + y^2}$, if $(x, y) \neq (0,0) = 0$ and if $(x, y) = (10,0)$. Prove that both the partial derivatives exist at (0, 0) but the function is not continuous at (0, 0).

8. Expand $x^3 y + 3y - z$ in powers of $(x - 1)$ and $(y + 2)$.

Unit	
9	**FUNCTIONS OF SEVERAL VARIABLES - INTEGRATION**

9.1 Introduction

In calculus, an iterated integral is finding the integral value of a function with more than one variable, in such a way that each of the integrals considers some of the variables as constants. This chapter begins with the definition of *K*-cell and in the definition of $\int_{I^k} f(x)dx$, the integral is independent of the order in which the *k*-integrations are performed. Also, some of the important theorems like primitive mapping theorem, change of variables theorem and partition of unity are discussed in this chapter.

9.2 Iterated Integrals

Definition 9.2.1

Consider R^k. Consider the subset of R^k of the form,

$$[a_1,b_1]\times[a_2,b_2]\times\ldots\ldots\times[a_k,b_k]$$

where $[a_i, b_i]$ is a closed interval in R, $1\leq i\leq k$. This is denoted by I^k. It is called a *k-cell in* R^k.

If $x=(x_1,x_2,\ldots\ldots,x_k)\in I^k$, then $a_i\leq x_i\leq b_i$, for $i = 1, 2, \ldots\ldots, k$.

Therefore,

$$I^k=\{x\in R^k;a_i\leq x_i\leq b_i,i=1,2,\ldots.,k\}.$$

Each $[a_i, b_i]$ is a compact subset of R. Since cartesian product of compact sets is compact, I^k is a compact subset of R^k.

Let $f\in\mathscr{C}(I^k)$. Therefore, f is a continued real valued function defined on I^k. Since I^k is compact, f is uniformly continuous on I^k.When $f\in\mathscr{C}(I^k)$, we are going to define the integral of f over I^k.

Put $f_k = f$.

Define, $f_{k-1}(x_1, x_2, ..., x_{k-1}) = \int_{a_k}^{b_k} f_k(x_1, x_2, ..., x_k)dx_k$.

Since $f_k = f$ is uniformly continuous, the above definition tells us that f_{k-1} is a real continuous function defined on I^{k-1}.

That is, $f_{k-1} \in \mathscr{C}(I^k)$.

Next define, $f_{k-2}(x_1, x_2, ..., x_{k-2}) = \int_{a_{k-1}}^{b_{k-1}} f_{k-1}(x_1, x_2, ..., x_{k-1})dx_{k-1}$.

Proceeding like this, we get, at each stage a function $f_j \in \mathscr{C}(I^j)$, such that f_{j-1} is the integral of f_j, with respect to dx_j over $[a_j, b_j]$.

Finally, after k-steps, we get a real number f_0, where $f_0 = \int_{a_1}^{b_1} f_1(x_1)dx_1$. This real number f_0 is called the value of the integral f over I^k. We write, $\int_{I^k} f = f_0$.

Therefore, $L(f) = \int_{I^k} f = \int_{a_1}^{b_1} f_1 \int_{a_2}^{b_2} f_2 ...... \int_{a_k}^{b_k} f(x_1, ..., x_k)dx_k dx_{k-1} ...... dx_1$.

Examples 9.2.2

1. $I^k = [-1,1] \times [0,1] \times [3,4] \times [-3,6]$. This is a 4-cell in R^4. $x = (x_1, x_2, x_3, x_4) \in I^k$.
2. $I^2 = [0,1] \times [0,1]$. This is a 2-cell in R^2. This is simply the unit square in R^2.
3. $I^3 = [0,1]^3$. This is the unit cube in R^3.

Problem 9.2.3

Given that $I^3 = [0,1] \times [-1,1] \times [1,2]$. If $f(x_1, x_2, x_3) = 2x_1^2 + 5x_2 - x_3$, evaluate $\int_{I^3} f(x)dx$.

Solution

Take $f_3 = f$.

Define $f_2(x_1,x_2) = \int_1^2 f(x_1,x_2,x_3)dx_3$.

$$= \int_1^2 (2x_1^2 + 5x_2 - x_3)dx_3.$$

$$= 2x_1^2 + 5x_2 - \frac{3}{2}.$$

We have defined $f_2 \in I^2$.

$$f_2(x_1,x_2) = 2x_1^2 + 5x_2 - \frac{3}{2}.$$

Define $f_1(x_1) = \int_{-1}^1 f_2(x_1,x_2)dx_2$.

$$= \int_{-1}^1 \left(2x_1^2 + 5x_2 - \frac{3}{2}\right)dx_2.$$

$$= 4x_1^2 - 3.$$

$$f_0 = \int_0^1 f_1(x_1)dx_1 = \int_0^1 (4x_1^2 - 3)dx_1 = -\frac{5}{3}.$$

Thus, $\int_{I^3} f = -\frac{5}{3}$.

Note 9.2.4

Let $I^3 = [a_1,b_1]\times[a_2,b_2]\times[a_3,b_3]$. If *f* is a constant function, then

$\int_{I^3} dx = [b_1 - a_1][b_2 - a_2][b_3 - a_3]$.

Theorem 9.2.5

In the definition of $\int_{I^k} f(x)dx$, the integral is independent of the order in which the k-integrations are performed. The theorem can also be stated as, If $f \in \mathscr{C}(I^k)$, then $L(f) = L'(f)$, with usual notations.

Proof

Let $L(f)$ denote the integral $\int_{I^k} f(x)dx$, when the integrals are performed in the order x_k, ..., x_1 as given in the definition.

Suppose $L'(f)$ is the result obtained when the integrals are carried out in some other order.

We prove that $L(f) = L'(f)$.

Given, $f \in \mathscr{C}(I^k)$ and $x = (x_1, x_2, \ldots\ldots, x_k) \in I^k$, $a_i \le x_i \le b_i$, $i = 1$ to k.

Choose $h_j \in \mathscr{C}[a_j, b_j]$, $1 \le j \le k$.

Define $h(x) = h_1(x_1)h_2(x_2)\ldots\ldots h_k(x_k)$, for all $x \in I^k$.

Clearly, $h \in \mathscr{C}(I^k)$.

$$\text{Now, } L(h) = \int_{I^k} h(x)dx = \int_{a_1}^{b_1}\int_{a_2}^{b_2}\ldots\ldots\int_{a_k}^{b_k} h(x)dx_k\ldots\ldots dx_1.$$

$$= \int_{a_1}^{b_1} h_1(x_1)dx_1 \ldots\ldots\ldots \int_{a_k}^{b_k} h_k(x_k)dx_k.$$

$$= \prod_{i=1}^{k} \int_{a_i}^{b_i} h_i(x_i)dx_i.$$

Even if we perform the above integrations in some other order of the variables, we get the same result.

$$\text{Therefore, } L'(h) = \prod_{i=1}^{k} \int_{a_i}^{b_i} h_i(x_i)dx_i.$$

Thus, $L(h) = L'(h)$, for the function $h \in \mathscr{C}(I^k)$, which is defined above.

Let $\mathscr{A}$ denote the set of all finite sums of such functions $h \in \mathscr{C}(I^k)$, where $h(x) = h_1(x_1)h_2(x_2)\ldots\ldots h_k(x_k)$ and $h_j \in \mathscr{C}([a_j, b_j])$.

If $g \in \mathscr{A}$, we have $L(g) = L'(g)$, because g is of the form $g = h_1 + h_2$ and $L(g) = L(h_1 + h_2) = L'(h_1) + L'(h_2) = L'(h_1 + h_2) = L'(g)$.

Moreover, $\mathscr{A}$ is a real algebra of function on I^k.

Therefore, Stone Wierstrass theorem can be applied to $\mathscr{A}$.

Let $v = \prod_{i=1}^{k}(b_i - a_i)$ and $\varepsilon > 0$ be given.

Then, there exists a function $g \in \mathscr{A}$, such that $\|f - g\| < \varepsilon / v$, where $\|f\| = \sup\{|f(x)|, x \in I^k\} = \max\{|f(x)|, x \in I^k\}$.

$$
\begin{aligned}
|L(f-g)| &= \left|\int_{I^k} (f-g)x\,dx\right| \\
&\le \int_{I^k} |(f-g)x|\,dx \\
&\le \|f-g\| \int_{I^k} dx \\
&< \frac{\varepsilon}{v}\Pi(b_i - a_i) = \varepsilon .
\end{aligned}
$$

Similarly, $L'(f-g) < \varepsilon$.

Now, $L(f-g) - L'(f-g) = L(f) - L(g) - L'(f) - L'(g)$

$= L(f) - L'(f)$ [Since $L(g) = L'(g)$].

Thus,
$$
\begin{aligned}
|L(f) - L'(f)| &= |L(f-g) - L'(f-g)| \\
&\le |L(f-g)| + |L'(f-g)| \\
&< 2\varepsilon .
\end{aligned}
$$

Thus, $L(f) = L'(f)$.

Hence the theorem.

Definition 9.2.6

Let f be a real or complex function defined on R^k. The closure of the set of points $x \in R^k$, at which $f(x) \neq 0$ is called the support of f. Clearly, it is a closed subset of R^k.

Problem 9.2.7

If $f \in \mathscr{C}(R^k)$, how will you define $\int_{R^k} f$?

Solution

Let $f \in \mathscr{C}(R^k)$. Let S be the support of f. Suppose S is compact. S is closed and bounded.

Then, $S \subseteq I^k$, for suitable k-cell in R^k.

Then, we define $\int_{R^k} f = \int_{I^k} f$.

Note 9.2.8

Let $Q^k = \{(x_1, x_2, ..., x_k) \in R^k, x_1 + x_2 + x_3 \leq 1, x_i \geq 0\}$. Then, Q^k is a convex subset of R^k.

$Q^2 \subset R^2$.

$Q^3 \subset R^3$.

$Q^k \subseteq I^k \subseteq R^k$.

Theorem 9.2.9

If $f \in \mathscr{C}(Q^k)$, then how can it be extended to a function on I^k?

Proof

Now, $I^k = [0,1]^k$.

Define $f(x) = 0$, for all $x \in I^k - Q^k$ and $\int_{Q^k} f = \int_{I^k} f$.

We verify that (1) Right hand side of the integral exists.

(2) It is independent of the order of R-integrations.

Choose δ such that $0 < \delta < 1$.

Define $\phi : R \to [0,1]$ by

$$\phi(t) = \begin{cases} 1, & t \leq 1 - \delta \\ \dfrac{1-t}{\delta}, & 1 - \delta < t \leq 1 \\ 0, & 1 < t \end{cases}$$

Then, define F on I^k by $F(x) = \phi(x_1 + x_2 + ... + x_k) f(x)$.

Claim: $F \in \mathscr{C}(I^k)$.

Let $x = (x_1, x_2, ..., x_k) \in I^k$.

If $x \in Q^k$, then $t = x_1 + x_2 + ... + x_k \leq 1$.

Therefore, $\phi(t) = \dfrac{1-t}{\delta}$.

Hence, $f \in \mathscr{C}(Q^k)$.

Therefore, F is continuous on I^k.

If $x \in I^k - Q^k$, then $t = x_1 + x_2 + ... + x_k > 1$.

Therefore, $f(x) = 0$.

Now, $F(x) = 0$.

Therefore, F is a constant and hence continuous.

Thus, in either case, we get F is continuous on I^k.

Thus, $F \in \mathscr{C}(I^k)$.

Let $\overline{y} = (x_1, ..., x_{k-1})$ so that $\overline{x} = (\overline{y}, x_k)$.

For a given $\overline{y}$, let $A_{\overline{y}}$ denote the set of points x_k, such that $F(\overline{y}, x_k) \neq f(\overline{y}, x_k)$.

Then, $A_{\overline{y}}$ is a set of real numbers which may be empty or a segment whose length doesnot exceed δ.

$$
\begin{aligned}
|F_{k-1}(\overline{y}) - f_{k-1}(\overline{y})| &= |F_{k-1}(x_1, ..., x_{k-1}) - f_{k-1}(x_1, ..., x_{k-1})| \\
&= \left| \int_0^1 F(x_1, ..., x_k)dx_k - \int_0^1 f(x_1, ..., x_k)dx_k \right| \\
&\leq \int_0^1 |F(x_1, ..., x_k) - f(x_1, ..., x_k)| dx_k \\
&= |F(x_1, ..., x_k) - f(x_1, ..., x_k)| \int_{A_{\overline{y}}} dx_k \\
&\leq \delta |\phi(x_1 + x_2 + ... + x_k) f(x_1, ..., x_k) - f(x_1, ..., x_k)| \\
&\leq \delta |\phi(x_1 + x_2 + ... + x_k) f(x) - f(x)| \\
&\leq \delta |f(x)| |\phi(x_1 + x_2 + ... + x_k) - 1| \leq \delta \|f\|.
\end{aligned}
$$

Thus, for all $\overline{y} \in I^{k-1}$, we have proved $|F_{k-1}(\overline{y}) - f_{k-1}(\overline{y})| \leq \delta \|f\|$.

.... (9.2.9.1)

When δ → 0, the above relation yields that, $f_{k-1}(\overline{y})$ is a uniform limit of sequence of continuous functions and hence continuous.

Hence, $\int_{I^k} f$ exists.

Therefore, $\int_{I^k} F = \int_{I^k} f$. [By equation (9.2.9.1)]

Since the left hand side of the integral is independent of the order of integrations, the same is true for the right hand side of the integral.

Thus, we have defined f on the whole of I^k and defined the integral $\int_{I^k} f$ over I^k.

9.3 Primitive Mapping

Definition 9.3.1

Let G be a mapping of an open set $E \subset R^n$ into R^n. Let m be a positive integer and g be a real function defined on E, such that $G(x) = \sum_{\substack{i=1 \\ i \neq m}}^{n} x_i e_i + g(x) e_m$, then g is called a *primitive mapping*.

Note 9.3.2

1) $G(x) = (x_1, \ldots\ldots, x_{m-1}, x_m, x_{m+1}, \ldots\ldots, x_n)$

   Thus, a primitive map changes atmost one coordinate of x, have m^{th} coordinate alone gets changed, it is the image of x under another function g. All other coordinates are kept fixed. We can also write $G(x)$ in another way, that is

   $$G(x) = (x_1, \ldots\ldots, x_{m-1}, x_m, x_{m+1}, \ldots\ldots, x_n) + (0, \ldots\ldots, 0, g(x) - m, 0, \ldots\ldots, 0)$$

   $$= x + (g(x) - m) e_m.$$

2) Primitive map is also a linear operator.

Problem 9.3.3

If $G : E \to R$, $(x_1, \ldots., x_n) \to (x_1, \ldots\ldots, x_{m-1}, g(x), x_{m+1}, \ldots\ldots, x_n)$ and $G = (G_1, G_2, \ldots\ldots, G_n)$, then find the matrix of G.

Solution

Let $G_1, G_2, \ldots\ldots, G_n$ be the components of G.

and, $\quad G_i(x)=x_i, \ i\neq m$.

$G_m(x)=g(x), \ i=m$, where each G_i is a real function on E.

Let $a\in E$.

$$\text{Therefore, } G'(a)=\begin{bmatrix} D_1G_1 & D_2G_1 & \ldots\ldots & D_nG_1 \\ \ldots\ldots & \ldots\ldots & \ldots\ldots & \ldots\ldots \\ D_1g(a) & D_2g(a) & \ldots\ldots & D_ng(a) \\ \ldots\ldots & \ldots\ldots & \ldots\ldots & \ldots\ldots \\ D_1G_n & D_2G_n & \ldots\ldots & D_nG_n \end{bmatrix}.$$

$$=\begin{bmatrix} 1 & 0 & 0 & \ldots\ldots & 0 \\ 0 & 1 & 0 & \ldots\ldots & 0 \\ \ldots\ldots & \ldots\ldots & \ldots\ldots & \ldots\ldots & \ldots\ldots \\ D_1g(a) & D_2g(a) & D_3g(a) & \ldots\ldots & D_ng(a) \\ 0 & 0 & 0 & \ldots\ldots & 0 \\ \ldots\ldots & \ldots\ldots & \ldots\ldots & \ldots\ldots & \ldots\ldots \\ 0 & 0 & 0 & \ldots\ldots & 1 \end{bmatrix}.$$

Hence, $J_G(a)=\det.[G'(a)]=D_mg(a)$.

Then, $G'(a)$ is invertible, if and only if $J_G(a)\neq 0$.

that is, if and only if $D_mg(a)\neq 0$.

Thus, if the primitive mapping $G : E \rightarrow R^n$, is defined by $G(x)=\overline{x}+(g(x)-x_m)e_m$, then $G'(a)$ is invertible if and only if $D_mg(a)\neq 0$.

Definition 9.3.4

A linear operator B on R^n, which interchanges two members of the family of standard basis in R^n, and which keeps the remaining basis vectors unchanged is called a *flip*.

Example 9.3.5

Consider R^5. Let $(e_1, e_2, \ldots., e_5)$ be the standard basis for R^5 and let $B\in L(R^5)$.

If $B(e_1)=e_4, B(e_4)=e_1$ and $B(e_i)=e_i$, for $i=2,3,5$ then B is a flip.

Solution

Let $x = (x_1, x_2, x_3, x_4, x_5) \in R^5$.

$$
\begin{aligned}
B(x) &= B(x_1, x_2, x_3, x_4, x_5) \\
&= B(x_1e_1 + x_2e_2 + x_3e_3 + x_4e_4 + x_5e_5). \\
&= x_1B(e_1) + x_2B(e_2) + x_3B(e_3) + x_4B(e_4) + x_5B(e_5). \\
&= x_1e_4 + x_2e_2 + x_3e_3 + x_4e_1 + x_5e_5. \\
&= (x_4, x_2, x_3, x_1, x_5).
\end{aligned}
$$

Thus, B interchanges the first and fourth coordinate of the vector $\overline{x}$.

Therefore, B is a flip.

Note 9.3.6

A flip interchanges any two coordinates of a vector.

If B is a flip, then $B \circ B = I$.

Definition 9.3.7

Let $x \in R^n$. We define the *Projections* $P_0, P_1, \ldots, P_n$ as follows:

$P_0(x) = 0$

$P_1(x) = x_1e_1 = (x_1, 0, \ldots, 0)$

$P_2(x) = x_1e_1 + x_2e_2 = (x_1, x_2, 0, \ldots, 0)$

....

$P_m(x) = x_1e_1 + x_2e_2 + \ldots + x_me_m = (x_1, \ldots, x_m, 0, \ldots, 0)$

....

$P_n(x) = x_1e_1 + x_2e_2 + \ldots + x_ne_n = (x_1, x_2, \ldots, x_n) = x$.

Theorem 9.3.8

This theorem is known as *Primitive Mapping theorem.*

Suppose F is a $\mathscr{C}'$ mapping of an open set $E \subset R^n$ into R^n. $0 \in E$ and $F(0) = 0$. $F'(0)$ is invertible. Then, F is locally a composition of primitive mappings and flips.

In other words, suppose F is a $\mathscr{C}'$ mapping of an open set $E \subset R^n$ into R^n. $0 \in E$ and $F(0) = 0$. $F'(0)$ is invertible. Then, there is a neighborhood of 0 in R^n in which the representation $F(x)=B_1 \circ B_2 \circ \ldots\ldots \circ B_{n-1} \circ G_n \circ \ldots\ldots \circ G_1(x)$ is valid. Further, each G_i is a primitive $\mathscr{C}'$ mapping in some neighborhood of 0, $G_i(0)=0$ and $G_i'(0)$ is invertible and each B_i is either a flip or the identity operator.

Proof

Put $F = F_1$.

Assume $1 \leq m \leq n-1$.

Assume the induction hypothesis for m.

V_m is a neighborhood of 0, $F_m \in \mathscr{C}'(V_m), F_m(0)=0$.

$F_m'(0)$ is invertible, $P_{m-1}(F_m(x))=P_{m-1}(x), \quad x \in V_m$. .... (9.3.8.1)

By hypothesis, $0 \in E$ and E is open.

Therefore, there exists a neighborhood V_1 of 0, such that $0 \in V_1 \subset E$.

Since $F = F_1$, we have $F_1 \in \mathscr{C}'(V_1)$.

$F_1(0) = 0$, $F_1'(0)$ is invertible and $P_0(F_1(x)) = P_0(x) = 0$.

Thus, the induction hypothesis is true for $m = 1$.

Assuming the induction hypothesis for m and we proceed to prove it for the next integer $m + 1$.

By the induction hypothesis and by equation (9.3.8.1), we have

$$P_{m-1}(F_m(x))=P_{m-1}(x), \quad x \in V_m.$$

Let the components of F_m be $(\alpha_1, \alpha_2, \ldots\ldots, \alpha_n)$.

Then, $F_m(x) = (\alpha_1(x), \alpha_2(x), \ldots\ldots, \alpha_n(x))$, each α_i is a real function on V_m.

Therefore, $F_m(x) = (\alpha_1(x), \alpha_2(x), \ldots\ldots, \alpha_{m-1}(x), 0, \ldots\ldots, 0)$

$$+ (0, \ldots\ldots, 0, \alpha_m, \alpha_{m+1}(x), \ldots\ldots, \alpha_n(x)).$$

$$= P_{m-1}(F_m(x)) + \sum_{\alpha=m}^{n} \alpha_i(x) e_i$$

$$= P_{m-1}(x) + \sum_{\alpha=m}^{n} \alpha_i(x) e_i. \text{ [By equation (9.3.8.1)]}$$

$$= (x_1, \ldots\ldots, x_{m-1}, \alpha_m(x), \alpha_{m+1}(x), \ldots\ldots, \alpha_n(x)).$$

Therefore,

$$[F_m'(0)]=\begin{bmatrix} 1 & 0 & \ldots & 0 & \ldots & 0 \\ 0 & 0 & \ldots & 0 & \ldots & 0 \\ \ldots & \ldots & \ldots & \ldots & \ldots & \ldots \\ D_1\alpha_m(0) & D_2\alpha_m(0) & \ldots & D_m\alpha_m(0) & \ldots & D_n\alpha_m(0) \\ \ldots & \ldots & \ldots & \ldots & \ldots & \ldots \\ D_1\alpha_n(0) & D_2\alpha_n(0) & \ldots & D_m\alpha_n(0) & \ldots & D_n\alpha_n(0) \end{bmatrix}$$

By hypothesis, $F_m'(0)$ is invertible.

In particular, $F_m'(0)e_m \neq 0$.

$\Rightarrow D_m\alpha_m(0)e_m + D_m\alpha_{m+1}(0)e_{m+1} + \ldots + D_m\alpha_n(0)e_n \neq 0$.

Therefore, there exists one k with $m \leq k \leq n$, such that $D_m(\alpha_k) \neq 0$.

We define a flip B_m, which interchanges the mth coordinate with the kth coordinate.

If $k = m$, we take B_m as the identity map.

Then, $G_m(x) = x + [\alpha_k(x) - x_m]e_m, \quad x \in V_m$.

G_m is a primitive mapping, $0 \in V_m, G_m(0) = 0$.

Clearly, $G_m \in \mathscr{C}'(V_m)$.

$G_m'(0)$ is invertible, because $D_m\alpha_k(0) \neq 0$.

Now, apply inverse function theorem to G_m.

Therefore, there exists open sets $U_m \subseteq V_m$ such that $0 \in U_m \subseteq V_m$ and another open set V_{m+1} such that $G_m(0) = 0 \in V_{m+1}$.

G_m is 1-1 on U_m and $G_m(U_m) = V_{m+1}$.

G_m has inverse $G_m^{-1} : V_{m+1} \to U_m$ such that $G_m^{-1}(G_m(x)) = x$.

G_m^{-1} is a $\mathscr{C}'$ mapping on V_{m+1}.

Define $F_{m+1}(y) = B_m \circ F_m \circ G_m^{-1}(y)$, for all $y \in V_{m+1}$. .... (9.3.8.2)

1) V_{m+1} is a neighborhood of 0.
2) $F_{m+1}(0) = B_m \circ F_m \circ G_m^{-1}(0) = B_m(F_m(0)) = B_m(0) = 0$.

3) $F'_{m+1}(0)$ is invertible can be got by chain rule.

4) F_{m+1} is a $\mathscr{C}'$ mapping on V_{m+1}, because it is the composition of such mappings.

5) To prove: $P_m(F_{m+1}(y)) = P_m(y), \quad y \in V_{m+1}$.

$P_m(F_{m+1}(y)) = P_m[B_m F_m G_m^{-1}](y)$

$= P_m B_m F_m(y)$, where $x \in V_m$, such that $G_m(x) = y$.

$= P_m B_m(\alpha_1(x), \alpha_2(x), \ldots\ldots, \alpha_n(x))$.

$= P_m(\alpha_1(x), \ldots\ldots, \alpha_{m-1}(x), \alpha_k(x), \alpha_{m+1}(x), \ldots\ldots, \alpha_m(x), \alpha_{k+1}(x), \ldots\ldots, \alpha_n(x))$.

$= (\alpha_1(x), \ldots\ldots, \alpha_{m-1}(x), \alpha_k(x), \ldots\ldots)$.

$= P_{m-1}(F_m(x)) + \alpha_k(x) e_m$.

$= P_{m-1}(x) + \alpha_k(x) e_m$. [By induction hypothesis and by equation (9.3.8.1)]

$= (x_1, \ldots\ldots, x_{m-1}, \alpha_k(x), 0, \ldots\ldots, 0)$

$= P_m(x_1, \ldots\ldots, x_{m-1}, \alpha_k(x), x_{m+1}, \ldots\ldots, x_n)$

$= P_m(G_m(x))$

$= P_m(y)$.

Therefore, $P_m F_{m+1}(y) = P_m(y)$, for all $y \in V_{m+1}$.

Thus, the induction hypothesis is true for $m + 1$ also.

By equation (9.3.8.2), $F_{m+1}(y) = B_m F_m G_m^{-1}(y)$, for all $y \in V_{m+1}$.

$= B_m F_m(x), \qquad x \in V_m$.

That implies, $B_m F_{m+1}(y) = B_m [B_m F_m(x)]$.

Hence, $B_m F_{m+1} G_m(x) = F_m(x)$, $x \in V_m$, [Since $B_m B_m = I$]

Then, $F_m = B_m F_{m+1} G_m$.

Put $m = 1, 2, \ldots\ldots, n-1$ successively.

$F = F_1 = B_1 F_2 G$.

$= B_1 \circ [B_2 \circ F_3 \circ G_2] \circ G_1$

$= B_1 \circ B_2 \circ [B_3 \circ F_4 \circ G_3] \circ G_2 \circ G_1$

....

$= B_1 B_2 \ldots. B_{n-1} F_n G_{n-1} G_{n-2} \ldots. G_1$.

Claim: $F_n = G_n$.

From equation (9.3.8.1), we have $P_{n-1}(F_n(x)) = P_{n-1}(x)$.

That implies, $(\alpha_1(x),......,\alpha_{n-1}(x),0) = (x_1,......,x_{n-1}(x),0)$.

Therefore, we have $\alpha_i(x) = x_i$, for $i = 1,2,.....,n-1$.

Hence, $F_n(x) = (\alpha_1(x),.....,\alpha_{n-1}(x),\alpha_n(x))$

$$= (x_1,.....,x_{n-1},\alpha_{n-1}(x)) = x + [\alpha_n(x) - x_n]e_n = G_n(x).$$

Thus, $F_n = G_n$.

Therefore, $F = B_1, \ldots\ldots, B_{n-1}, G_n, G_{n-1}, \ldots\ldots, G_1$.

Hence the theorem.

9.4 Change of Variables

Theorem 9.4.1

This theorem is known as **Change of Variables theorem.**

Suppose T is a 1-1 $\mathscr{C}'$ mapping of an open set $E \subset R^k$ into R^k, such that $J_T(x) \neq 0$, for all $x \in E$. If f is a continuous function on R^k, whose support is compact and lies in $T[E]$, then $\int\limits_{R^k} f(y)dy = \int\limits_{R^k} f(T(x))|J_T(x)|dx$.

$$\ldots\ldots\ldots\ldots (9.4.1.1)$$

Proof

J_T is the Jacobian of T.

$J_T(x) \neq 0$ gives T is invertible.

Therefore, T^{-1} is defined on $T(E)$.

Also, by Inverse Function theorem, T^{-1} is also continuous on $T(E)$.

Let K be the compact support of f, which lies in $T(E)$.

Since, "continuous image of a compact set is compact", $T^{-1}(K)$ is also a compact subset of E.

Therefore, the integrand on the right hand side of (9.4.1.1) has compact support on E.

Next, we justify that the absolute sign for $J_T(x)$ is necessary on the right hand side of (9.4.1.1).

Consider the case $k = 1$.

Then, T is a $\mathscr{C}'$ mapping of R^1 onto R^1.

Then, $J_T(x) = T'(x)$.

If T is an increasing function, then $T'(x) > 0$.

So, $\int\limits_{R^1} f(y)dy = \int\limits_{R^1} f(T(x))T'(x)dx$.

But, if T is a decreasing function, then $T'(x) < 0$ and if f is positive in the interior of its support, then left hand side will be positive and right hand side will be negative.

Then, in order to overcome this difficulty, we put $|J_T(x)|$ on the right hand side of the integral.

Therefore, equation (9.4.1.1) is valid, if f is a primitive C 'mapping.

Since $L(f) = L'(f)$, for all $f \in \mathscr{C}(R^k)$, we observe that equation (9.4.1.1) is true for a flip or a linear mapping which merely interchanges two ordinates.

Next, assuming that the equation (9.4.1.1) is valid for two transformations P and Q, we will prove that it is true for their compositions.

Let P and Q be two transformations, such that $s(x) = P(Q(X))$.

Then, $\int f(z)dz = \int f(P(Y))|J_P(y)|dy$.

[Since, we assume (9.4.1.1) is true for T = P]

$$= \int f(P(Q(x)))|J_P(Q(x))||J_Q(x)|dx.$$

Put $|J_P(Q(x))||J_Q(x)| = [\det.P'(Q(x))][\det.Q'(x)]$

$$= \det.[P'(Q(x)).Q'(x)] = \det.s'(x) = J_s(x).$$

Therefore, we get, $\int f(z)dz \int f(s(x))|J_s(x)|dx$.

Thus, equation (9.4.1.1) is true for s also.

Let U be a neighborhood of a.

If x – a is a point in a neighborhood of 0, then by primitive mapping theorem,

$$T(x-a) = B_1B_2.....B_{k-1}G_kG.....G_1(x-a).$$

or $T(x) = T(a) + B_1B_2.....B_{k-1}G_kG_{k-1}.....G_1(x-a)$,

where B_i's are flips and G_i's are primitive $\mathscr{C}'$ mappings.

Let V = T(U).

Then, equation (9.4.1.1) is true for any continuous function *f*, whose support lies inside *V*.

Thus, to each $y \in T(E)$, there exist an open set $V_y \subset T(E)$, such that equation (9.4.1.1) will be valid for continuous function *f* only if its support lies in some V_y.

Thus, V_y is an open cover for *T*(*E*).

By hypothesis, *K* is the compact support of *f*.

Given, $K \subset T(E)$.

Therefore, $\{V_y\}$ covers *K*.

Hence, by previous theorem we get

$f = \sum_{i=1}^{s} \psi_i f$, where each ψ_i is continuous, $0 \le \psi_i \le 1$ and each ψ_i has its support in V_y.

Therefore, equation (9.4.1.1) is valid for each $\psi_i f$.

Consequently, equation (9.4.1.1) is valid for the sum $\sum \psi_i f = f$.

Thus, the theorem is true for the given *f*.

Hence the theorem.

9.5 Partition of Unity

Theorem 9.5

This theorem is known as ***theorem of Partition of Unity.***

Let *K* be a compact subset of R^n. Let $\{V_\alpha\}$ be an open cover for *K*. Then, there exist functions $\psi_1, \psi_2, \ldots\ldots, \psi_s \in \mathscr{C}(R^n)$ such that

1. $0 \le \psi_i \le 1$, for $i = 1, 2, \ldots\ldots, s$.
2. Each ψ_i has its support within some V_α.
3. $\psi_1 + \psi_2 + \ldots\ldots + \psi_s(x) = 1, \quad x \in K$.

Proof

Associate with each $x \in K$, the index $\alpha(x)$, so that $x \in V_{\alpha(x)}$.

Since $V_{\alpha(x)}$ is open, there exist open balls $B(x)$ and $W(x)$, with center at x, such that $\overline{B}(x) \subset W(x) \subseteq \overline{W} \subseteq V_{\alpha(x)}$.

Since K is compact, there exist a finite subcover $\{V_{\alpha(x_1)},......,V_{\alpha(x_s)}\}$ of the open cover $\{V_\alpha\}$ which also covers K.

Therefore, we get a finite set of points $\{x_1,.....,x_s\}$ in K such that $K \subset B(x_1) \cup B(x_2) \cup ...... \cup B(x_s)$ and $B(x_i) \subseteq W(x_i)$ for $i = 1, 2, ..., s$.

Then, $\overline{B}(x_i)$ and $R^n - W(x_i)$ are distinct closed subsets of R^n.

Since R^n is a normal space, there exists a function $\phi_i \in \mathscr{C}(R^n)$, $i = 1,2,.....,s$ such that $\phi_i[B(x_i)] = 1$ and $\phi_i[R^n - W(x_i)] = 0$.

In otherwords, $\phi_i(x) = 1$, if $x \in B(x_i)$ and

$\phi_i(x) = 0$, if x lies outside $W(x_i)$, for $i = 1, 2, ..., s$.

Define $\psi_1 = \phi_1$ and $\psi_{i+1} = (1-\phi_1)(1-\phi_2)......(1-\phi_i)\phi_{i+1}$, $i = 1, 2,....., s - 1$.

Since each $\phi_i \in \mathscr{C}(R^n)$, we get each $\psi_i \in \mathscr{C}(R^n)$, for $i = 1, 2,...., s$.

1) Each φ_i satisfies $0 \leq \varphi_i \leq 1$.

   Therefore, $0 \leq \psi_i \leq 1, \quad i = 1,2,.....,s$.

   This proves (1).

2) Let A_i be the support of ψ_i.

Then, $A_i = \overline{\{x \in R^n ; \psi_i(x) \neq 0\}}$

Claim: $A_i \subseteq W(x_i) \subseteq V_{\alpha(x_i)}$.

Suppose $y \notin V_{\alpha(x_i)}$.

Then, $y \notin W(x_i)$.

That implies, y lies outside $W(x_i)$.

Hence, $\psi_i(y) = 0$.

$\Rightarrow y \notin A_i$.

Thus, support of ψ_i lies in $V_{\alpha(x_i)}$.

This proves (2).

3) First we prove the relation for $i = 1$ to s,

$\psi_1 + \psi_2 + \ldots\ldots + \psi_i = 1 - (1-\phi_1)\ldots\ldots(1-\phi_i)$.

For $i = 1$, $\psi_1 = 1-(1-\phi_1) = \phi_1$.

Therefore, the result is true for $i = 1$.

Assume that the result is true for one $i < s$.

That is, $\psi_1 + \psi_2 + \ldots\ldots + \psi_i = 1 - (1-\phi_1)\ldots\ldots(1-\phi_i)$.

We will prove that the result is true for $i + 1$.

$$\begin{aligned}(\psi_1 + \psi_2 + \ldots\ldots + \psi_i) + \psi_{i+1} &= 1-(1-\phi_1)\ldots\ldots(1-\phi_i) + (1-\phi_1)\ldots\ldots(1-\phi_i)\phi_{i+1} \\ &= 1-\{(1-\phi_1)\ldots\ldots(1-\phi_i)\}(1-\phi_{i+1}) \\ &= 1-(1-\phi_1)\ldots\ldots(1-\phi_{i+1})\end{aligned}$$

Thus, by induction, we get

$$\psi_1 + \ldots\ldots + \psi_s = 1-(1-\phi_1)\ldots\ldots(1-\phi_s).$$

Let $x \in K$.

$$\psi_1(x) + \ldots\ldots + \psi_s(x) = 1 - [(1-\phi_1(x)]\ldots\ldots[1-\phi_s(x)].$$

Since $K \subseteq B(x_1) \cup \ldots\ldots \cup B(x_s)$, x belongs to some $B(x_j)$, for some $1 \le j \le s$.

Then, $\phi_j(B(x_j)) = 1$ and $\phi_j(x) = 1$.

$\Rightarrow 1 - \phi_j(x) = 0$.

Therefore, $[1-\phi_1(x)]\ldots\ldots[1-\phi_s(x)] = 0$.

Hence, the right hand side becomes 1.

Thus, $\psi_1(x) + \ldots\ldots + \psi_s(x) = 1$, $x \in K$.

Hence the theorem.

9.6 Differentiation of Integrals

Note 9.6.1

It will be convenient to use the notation $\varphi^t(x) = \varphi(x,t)$.

Thus, φ^t is a function of one variable, for each t.

Theorem 9.6.2

Suppose

i) φ(x, t) is defined for $a \le x \le b,\ c \le t \le d$,

ii) α is an increasing function on [a, b],

iii) $\phi^t \in R(\alpha)$, for all $t \in [c,d]$,

iv) $c < s < d$, and to every $\varepsilon > 0$ corresponds a $\delta > 0$, such that $|(D_2\phi)(x,t) - (D_2\phi)(x,s)| < \varepsilon,\ \forall x \in [a,b]$ and $\forall t \in (s-\delta, s+\delta)$.

Define

$$f(t) = \int_a^b \phi(x,t)\,d\alpha(x), \qquad c \le t \le d.$$

Then, $(D_2\varphi)^s \in R(\alpha)$, $f'(s)$ exists and $f'(s) = \int_a^b (D_2\varphi)(x,s)\,d\alpha(x)$.

Proof

For $0 < |t-s| < \delta$, consider the difference quotients

$$\psi(x,t) = \frac{\varphi(x,t) - \varphi(x,s)}{t-s}.$$

Since $\varphi^t \in R(\alpha)$ for every $t \in [c,d]$,

$$\psi^t \in R(\alpha) \quad \text{for every } t \in [c,d].$$

By Mean Value theorem,

There exists a u between s and t such that

$$\phi(x,t) - \phi(x,s) = D_2\phi(x,u)(t-s).$$

Hence to each (x, t), there exists a u between s and t such that

$$\psi(x,t) = D_2\varphi(x,u) \text{ whenever } 0 < |t-s| < \delta. \qquad \text{............ (9.6.2.1)}$$

Hence assumption (iv) implies that

$$|\psi(x,t) - (D_2\varphi)(x,s)| < \varepsilon \text{ whenever } a \le x \le b \text{ and } 0 < |t-s| < \delta. \qquad \text{............ (9.6.2.2)}$$

Consider

$$\frac{f(t)-f(s)}{t-s}=\frac{1}{t-s}\left[\int_a^b \varphi(x,t)d\alpha(x)-\int_a^b \varphi(x,s)d\alpha(x)\right]$$

$$=\int_a^b \frac{\varphi(x,t)-\varphi(x,s)}{t-s}d\alpha(x)$$

$$=\int_a^b \psi(x,t)d\alpha(x)\,. \qquad \text{............ (9.6.2.3)}$$

From (9.6.2.2), we get that

$\psi^t \to (D_2\varphi)^s$, uniformly on $[a, b]$ as $t \to s$. .... (9.6.2.4)

Also each $\psi^t \in R(\alpha)$.

Hence, $f'(s)=\lim\limits_{t\to s}\frac{f(t)-f(s)}{t-s}$

$$=\lim_{t\to s}\int_a^b \psi(x,t)d\alpha(x)$$

$$=\int_a^b \lim_{t\to s}\psi(x,t)d\alpha(x)$$

$$=\int_a^b (D_2\varphi)(x,s)d\alpha(x) \qquad \text{[from (9.6.2.4) and theorem (4.4.1)]}$$

Hence the theorem.

R REFERENCES

- Anthony W. Knapp, Basic Real Analysis, 2005, Springer, USA.
- Artin, E., The Gamma Function, 1964, Holt, Rinehart and Winston, Inc., New York.
- Asuman G. Aksoy and Mohamed A. Khamsi, A Problem Book in Real Analysis, 2010, Springer, USA.
- Boas, R. P., A Primer of Real Functions, 1960, Carus Mathematical Monograph No. 13, John Wiley & Sons, Inc., New York.
- Buck, R. C. (ed.), Studies in Modern Analysis, 1962, Prentice – Hall, Inc., Englewood Cliffs, N.J.
- Burkill, J. C., The Lebesgue Integral, 1951, Cambridge University Press, New York.
- Carothers,N.L., Real Analysis, 2000, Cambridge University Press, Cambridge.
- David A. Sprecher, Elements of Real Analysis, 2012, Courier Corporation.
- Dieudonne, J., Foundations of Modern Analysis, 1960, Academic Press, Inc., New York.
- Edward James McShane, Truman Arthur Botts, Real Analysis, 2013, Courier Corporation.
- Elias M. Stein and Rami Shakarchi, Real Analysis: Measure theory, Integration and Hilbert Spaces, 2004, Princeton University Press, Princeton.
- Ethan D. Bloch, The Real Numbers and Real Analysis, 2011, Springer, USA.
- Fleming, W. H., Functions of Several Variables, 1965, Addison – Wesley Publishing Company, Inc., Reading, Mass.
- Gerald B. Folland, Real Analysis: Modern Techniques and their Applications, Second Edition, 1999, John Wiley and Sons, Inc.
- Graves, L. M., The theory of Functions of Real Variables, 2nd ed., 1956, McGraw – Hill Book Company, New York.

- Halmos, P. R., Measure Theory, 1950, D. Van Nostrand Company, Inc., Princeton, N. J.
- Hardy, G. H., Pure Mathematics, 9th ed., 1947, Cambridge University Press, New York.
- Herstein, I. N., Topics in Algebra, 1964, Blaisdell Publishing Company, New York.
- Hewitt, E., and Stromberg, K., Real and Abstract Analysis, 1965, Springer Publishing Co., Inc., New York.
- Jean Paul Penot, Analysis: From Concepts to Applications, 2016, Springer, USA.
- Kellogg, O. D., Foundations of Potential Theory, 1940, Frederick Ungar Publishing Co., New York.
- Kenneth A. Ross, Elementary Analysis, The theory of Calculus, II edition, 2013, Springer, USA.
- Knop, K., Theory and Application of Infinite series, 1990, Dover Publications Inc.
- Knopp, K., Theory and Application of Infinite Series, 1928, Blackie & Son, Ltd., Glasgow.
- Kolmogorov, A.N. and Fomin, S.V., Introductory Analysis, 2012, Dover Publications Inc.
- Landau, E. G. H., Foundations of Analysis, 1951, Chelsea Publishing Company, New York.
- Mariano Giaquinta and Giuseppe Modica, Mathematical Analysis: Linear and Metric structures and Continuity, 2007, Springer, USA.
- Mathew A. Pons, Real Analysis for the Undergraduate, with an invitation to Functional Analysis, 2014, Springer, USA.
- Mcshane, E. J., Integration, 1944, Princeton University Press, Princeton, N.J.
- Miklos Laczkovich and Vera T. Sos, Real Analysis: Foundations and Functions of one variable, 2015, Springer, USA.
- Niven, I. M., Irrational Numbers, 1956, Carus Mathematical Monograph No. 11, John Wiley & Sons, Inc., New York.
- Peter A. Loeb, Real Analysis, 2016, Birkhäuser, Springer, USA.
- Rajnikanth Sinha, Real and Complex Analysis, vol. 2, 2018, Springer, USA.

- Robert G. Bartle and Donald R. Sherbert, Introduction to Real Analysis, Fourth edition, 2014, John Wiley & Sons Inc.
- Royden, H. L., Real Analysis, 1963, The Macmillan Company, New York.
- Rudin, W., Real and Complex Analysis, 2nd ed., 1974, McGraw–Hill Book Company.
- Simmons, G. F., Topology and Modern Analysis, 1963, McGraw–Hill Book Company.
- Singer, I. M. and Thorpe, J. A., Lecture Notes on Elementary Topology and Geometry, 1967, Scott, Foresman and Company, Glenview, Ill.
- Smith, K. T., Primer of Modern Analysis, 1971, Bogden and Quigley, Tarrytown–on- Hudson, New York.
- Spivak, M., Calculus on Manifolds, 1965, W. A. Benjamin Inc., NewYork.
- Sudhir R. Ghorpade and Balmohan V. Limaye, A Course in Calculus and Real Analysis, 2006, Springer, USA.
- Teodora – Liliana T. Radulescu, Vicentu D. Radulescu and Titu Andreescu, Problems in Real Analysis: Advanced Calculus on the Real Axis, 2009, Springer, USA.
- Thurston, H. A., The Number System, 1956, Blackie & Son, Ltd., London-Glasgow.
- Walter Rudin, Principles of Mathematical Analysis, Third Edition, 1976, McGraw – Hill, Inc.
- William P. Ziemer, Modern Real Analysis, 2017, Springer, USA.

■